Sustainable Energy and Fuels

Sustainability refers to the concept that all people should be able to meet their basic needs indefinitely, without compromising future generations. Sustainability, in terms of energy, embraces the same principles. One day the world will run out of fossil fuels. We need to realize how important sustainable energy is and its significance when it comes to the future of our planet. Sustainable energy includes any energy source that cannot be depleted and can remain viable forever. It does not need to be renewed or replenished; sustainable energy meets our demand for energy without any risk of failing or running out. This is why sustainable energy is the answer to our energy needs. Furthermore, sustainable energy doesn't harm the environment (or at most, there is a minimal risk), increase climate change, or cost a heavy price. Although there is a cost associated with creating and building ways to capture sustainable energy, the energy sources themselves are typically free. The main objective of this book is to provide an up-to-date review of conduction mechanisms, structure construction, operation, performance evaluation, and applications of various renewable energies and fuels. The current trend in innovation is likely to explore the potential to connect novel materials, design methods, and new techniques, which would allow us to maintain existing resources and develop new methods by employing smart technologies. This book provides a complete insight into recent advancements in nanomaterials, renewable energy design, and applications. The purpose of this book is to provide relevant theoretical frameworks that include materials, modeling, circuit design, and the latest developments in experimental work in the field of renewable energy and fuels.

This book:

- Presents solar energy conversion including photovoltaics and artificial photosynthesis.
- Discusses important topics such as energy management standards, biofuels, biorefining, and capacitive desalination.
- Illustrates the importance of novel materials and process improvements for sustainable energy and fuels.
- Includes research problem statements with specifications and commercially available industry data.
- Covers catalysis for energy technologies, including the sustainable synthesis of fuels and chemicals, molecular, and bioinspired catalysis.

The text is primarily written for senior undergraduate and graduate students, and academic researchers in the fields of electrical engineering, electronics and communication engineering, environmental engineering, and renewable energy.

Sustainable Energy and Fuels

Materials, Processing Methods, and Development

Edited by
Piush Verma, Ahmed Boubakeur,
Leila Mokhnache, and Balwinder Raj

CRC Press
Taylor & Francis Group
Boca Raton London New York

CRC Press is an imprint of the
Taylor & Francis Group, an **informa** business

First edition published 2025
by CRC Press
2385 NW Executive Center Drive, Suite 320, Boca Raton FL 33431

and by CRC Press
4 Park Square, Milton Park, Abingdon, Oxon, OX14 4RN

CRC Press is an imprint of Taylor & Francis Group, LLC

© 2025 selection and editorial matter, Piush Verma, Ahmed Boubakeur, Leila Mokhnache and Balwinder Raj; individual chapters, the contributors

ISBN: 9781032480916 (hbk)
ISBN: 9781032783994 (pbk)
ISBN: 9781003487692 (ebk)

DOI: 10.1201/9781003487692

Typeset in Sabon
by Deanta Global Publishing Services, Chennai, India

Contents

Preface

Sustainability refers to the concept that all people should be able to meet their basic needs indefinitely, without compromising future generations. Sustainability, in terms of energy, embraces the same principles. One day the world will run out of fossil fuels. We need to realize how important sustainable energy is and its significance when it comes to the future of our planet. Sustainable energy includes any energy source that cannot be depleted and can remain viable forever. It does not need to be renewed or replenished; sustainable energy meets our demand for energy without any risk of failing or running out. This is why sustainable energy is the answer to our energy needs. Furthermore, sustainable energy doesn't harm the environment (or at most, there is a minimal risk), increase climate change, or cost a heavy price. Although there is a cost associated with creating and building ways to capture sustainable energy, the energy sources themselves are typically free. A brief summary of the chapters is presented below.

Chapter 1 Since the latter half of the 20th century, the growth of the world's population and industrial development has gained momentum and has had a clear impact on the expansion of the global economy as well as gross world product (GWP). This has resulted in steady growth in energy demand and the influence of all these variables on the environment. Concerns have grown about the potential ramifications of the relentless association of expansion with development. Sustainability itself, defined as "the pursuit of robust and enduring social structures and activities for both humans and the environment that sustains their life", is relatively new. It was the efforts of the Club of Rome, established in 1968, as well as "The Limits of Growth" report, published in 1972, that first presented the concept that economic and environmental issues should not be treated individually, providing a grave warning about endless growth of material and unrestricted utilization in a world of limited resources. The concept of evaluating the amount of energy required to manufacture a product or service while taking into account all significant factors (economic investment, primary energy, environmental loading, labor, etc.) resulted in the creation of a methodology known as "emergy", which is primarily derived from the seminal work of Howard T. Odum.

Chapter 2: Earth is on the verge of becoming an energy-deprived planet. Researchers are desperately looking for alternative resources. It would be best to start with what we have and see what we can achieve. In this context, what is available is an abundance of solar energy, a free source of energy. The problem is balancing the increasing demand for energy with the most available source without affecting natural cycles. The efficiency of solar cells has increased from 3.8% to 30% and beyond. In this chapter, a perovskite solar cell (PSC) is simulated using open-access SCAPS software and analyzed for efficient results. The device design is analyzed by varying some of the morphological parameters for respective layers in the design structure in order to achieve better results. The parameters chosen boost the design performance up to 36.44% power conversion efficiency (PCE).

Chapter 3: Rapid urbanization presents significant environmental challenges for cities. Smart cities offer a solution to these issues by enhancing urban infrastructure and mitigating related problems. India's Smart City Mission aims to create 100 smart cities to address these concerns. This study fills a research gap by examining smart city indicators within the "smart environment" component from the perspectives of city officials and citizens. The study utilizes online citizen survey data collected from 64 Indian cities via the MyGov.in platform. The collected data encompass development priorities identified by municipal authorities for smart cities and the most critical issues as perceived by citizens. We identify four key dimensions of a smart environment: wastewater management, pollution control and monitoring, solid waste management, and environmental protection and sustainability. The indicators within these dimensions are categorized through relevant keywords associated with a smart environment. The results indicate that "solid waste management, sanitation and sewerage" are the most prioritized indicators, reflecting a consensus between officials and citizens. However, indicators such as "biodiversity", "climate change" and "construction waste management" receive limited attention across cities, highlighting context-specific challenges in Indian smart cities. Our study reinforces aligning city development goals with public aspirations for sustainable inclusive urban development in India's smart cities.

Chapter 4: Solar panels made of semiconductors have entirely altered how renewable energy is produced. The creation of novel materials, enhanced efficiency, and new uses of semiconductor solar cells are all highlighted in this chapter's thorough summary. The history of semiconductor solar panels is traced in the first section of the study, emphasizing how they developed from initial experimentation configurations to the extremely effective and economically feasible systems that are in use today. The basis of the photovoltaic sector, crystalline silicon solar cells, is thoroughly addressed, showing how they have changed from their original ideas into their present powerful arrangements. The chapter explores the use of novel semiconductor compounds, including multi-junction solar cells, developing perovskite materials, and thin-film. These substances are investigated for their special qualities, enhancements to efficiency, and appropriateness for a range of uses, such as grid integration, mobile electronics, and space travel. This chapter's main focus is improvements in efficiency, a crucial area for solar cell advancement. It includes cutting-edge methods and inventions such as tandem cell arrangements, anti-reflective paints, and light-trapping architectures targeted at improving the conversion of energy efficiency. The chapter also explores the difficulties and opportunities of expanding these highly efficient techniques for use in business.

Chapter 5: AlGaAs are used in the window layer of the top cell to optimize the structure of the device. Moreover, efficiency has been improved by altering the top window layer's thickness. III-V compound semiconductor-based multi-junction solar cells reach the highest efficiency for the conversion of solar energy today. First, the design of the solar cell is presented by GaAs as the tunnel diode/junction. Then the window layer is optimized using AlGaAs as the window layer after comparison with the reported work in the literature. Then the thickness of the window layer is optimized taking different values. AlGaInP is the back-surface field (BSF) layer of the top subcell, and a single BSF layer is taken as in the previous two BSF layers. The mole fraction of the window layer and the bottom BSF layer's thickness and doping are adjusted. We then obtain the following results for the electrical performance parameters: Jsc=28.65 mA/cm2, Voc=1.49 V, and FF=86.94, and the conversion efficiency is 37.27%. It is observed from the results that AlAs/GaAs dual-function (DJ) solar cell is the best combination of III-V compound multi-junction solar cells.

Chapter 6: In recent decades, photovoltaic technology has significantly advanced, leading to cost- and energy-efficient solutions. One of the promising approaches that has led to significant breakthroughs in this field is tandem-based solar cells. Tandem means an arrangement pattern of something, one behind another or stacked over one another. In order to harvest most of the bandgap falling onto the surface of solar cells, researchers have preferred this methodology, and results have been promising. Tandem cells strive to enhance the conversion efficiency of solar cell stacks one over another, varying the bandgap for complete or partially complete absorption bands of energy incidents on it. Designing a solar cell involves complex engineering and scientific considerations.

Chapter 7: The study provides insights into the specific challenges and limitations of wind energy production in the unique and harsh environment of the desert. One of the main challenges of wind energy production in desert climates is the impact of sand particles on the performance of wind turbines. The abrasive and erosive properties of sand can cause significant wear and tear on the blades, reducing their efficiency and lifespan. The review summarizes recent studies that have investigated the effects of sand on wind turbine aerodynamics, including the characterization of sand particle size and distribution, and the development of accurate numerical methods allowing the effect of sand on the aerodynamic performance of wind turbines to be predicted. Another important factor that would affect wind turbine performance in desert climates is excessive ambient air temperature. High temperatures can cause a reduction in air density, which in turn affects the lift and drag forces acting on the blades. The impact of ambient air temperature on wind turbine aerodynamics as well as methods for mitigating this effect are presented and discussed, such as the use of cooling systems and the optimization of blade shape and surface area. In addition to summarizing the current state of research, the need for further studies to better understand and optimize wind turbine aerodynamics in desert climates has been emphasized.

Chapter 8: A device that transforms the chemical energy of a fuel (often hydrogen) and an oxidant (often oxygen) into electrical energy through two redox reactions is called a fuel cell. Fuel cells differ from most batteries because they require a continuous supply of fuel and oxygen (often from the air) to maintain the chemical reaction. Batteries, on the other hand, usually have chemical energy stored in the materials inside them. Fuel cells can generate electricity as long as fuel and oxygen are present. The first fuel cell was created by Sir William Grove in 1838. Francis Thomas Bacon developed the hydrogen-oxygen fuel cell in 1932, which was the first fuel cell to be commercially used. This fuel cell is sometimes named after its inventor, a "Bacon fuel cell". Since the mid-1960s, NASA has used alkaline fuel cells to power satellites and spacecraft. Fuel cells have also been applied for various purposes, such as primary and backup power sources for commercial, industrial, and residential buildings and remote or harsh locations. Fuel-cell vehicles, such as forklifts, cars, buses, trains, boats, motorcycles, and submarines, also employ them.

Chapter 9: This chapter presents static random access memory (SRAM) cell design for low power applications. It is well known that the mathematical modeling and manual computations of transistor design parameters are sometimes impractical and remain a challenge for researchers. In submicron technologies, the basic component of an SRAM circuit such as MOSFET is modeled by various complex nonlinear equations. The modeling equations include parameters such as channel length (L), channel width (W), node voltages, and branch currents. However, the design and analysis of such complex nonlinear equations depend on the expertise of the designer. In this chapter, an artificial neural network is used to design and implement an SRAM circuit design. The channel length and width that are most suitable for circuit characteristics are calculated using neural network training. The LTspice tool is used to simulate the various circuit diagrams. The neural network model is developed and trained using MATLAB R2021a software platform. The effectiveness of the proposed neural network model is tested using the various circuits.

Acknowledgements

We would like to thank and acknowledge the efforts and support of all contributors in completing this book. This book is based on research work carried out by a number of authors. Many people have contributed greatly to this book, *Sustainable Energy and Fuels: Materials, Processing Methods, and Development*. We are grateful to a number of friends and colleagues for encouraging us to start work on this book. We, as the editors, would like to acknowledge all of them for their valuable help and generous ideas in improving the quality of the book. With our feelings of gratitude, we would like to introduce them in turn. The first mention is for the authors and reviewers of each chapter of this book. Without their outstanding expertise, constructive reviews, and devoted effort, this book would not been as comprehensive as it is. The second mention is for the CRC Press, Taylor & Francis staff for their constant encouragement, continuous assistance, and untiring support. Without their technical support, this book would not have been completed. The third mention is for the editors' families for being the source of continuous love, unconditional support, and prayers, not only for this work but throughout our lives. Last but far from least, we express our heartfelt thanks to the Almighty for bestowing over us the courage to face the complexities of life and complete this work.

About the editors

Piush Verma is working as Registrar at Guru Kashi University (GKU), Bathinda. Before joining GKU, Bathinda, he worked as a professor in electrical engineering at National Institute of Technical Teachers Training and Research Chandigarh (NITTTR) India, from 2019 to 2023. Before joining NITTTR, Dr Verma was Vice Chancellor of Bahra University Shimla Hills. Dr. Verma was Campus Director at Rayat-Bahra Group of Institutions, Patiala Campus, for more than ten years. He has served Thapar Group for more than 17 years, which includes Thapar University, Thapar Research & Development Centre, and Patiala and M/s Crompton Greaves Hydro Division Ltd. Dr. Verma completed his PhD and Master of Engineering with honors from Thapar University, Patiala. He has nearly 36 years of teaching, research, industry, and administrative experience. During these periods, he published many research papers and filed one patent. He has supervised a number of ME theses and PhDs in the area of electrical, electronics, and communication engineering. He is a member of IEEE, USA, a Fellow of the Institution of Engineers, India, and a Senior Member of the International Association of Computer Science and Information Technology (IACSIT), Singapore. He was honored with the Best Administration Award from the MHRD Ministry (AIMTC) and the Institution of Engineers (India). He has visited more than 20 countries, including the USA, UK, France, Indonesia, Dubai, Germany, Singapore, Malaysia, Thailand, Hong Kong, and Canada for collaborative research projects and invited lectures. He is a member of the Board of Studies for Electrical Engineering, Punjab Technical University (PTU), Jalandhar. He is a member of the Board of Studies GNE (Deemed University) Ludhiana, and a member of the Board of Research Studies at the National Institute of Technology (NIT), Hamirpur. He is the recipient of the Best Paper Award from the Institution of Engineers. Dr. Verma is a reviewer for *IEEE Transactions*, USA, and *Electrical Power Components and Systems* of Taylor & Francis, UK.

Ahmed Boubakeur was born on April 1, 1952, in Biskra, Algeria. In June 1975, he obtained an engineering degree in electrical engineering from the École Nationale Polytechnique (ENP) in Algiers. He obtained a certificate of specialization in high-voltage engineering, and in May 1979, he defended his doctoral thesis in technical sciences (PhD) at the Institute of High Voltages of the Polytechnic School of Warsaw, Poland (currently Technical University of Warsaw (TUW)). He has a certificate in intellectual property from the WIPO Worldwide Academy (2002) and a certificate in eco-management from ENP/EPF Lausanne (2003). He is currently a full professor at ENP, where he has taught and directed research works in the field of high-voltage engineering since 1979. In the same field, he has provided graduation and post-graduation (master's level) courses in some Algerian universities (Béjaïa, Batna, Jijel) and has been a visiting professor at the Polytechnic School of the University of Nantes (France) and at the Warsaw University of Technology (WUT). He has given conferences at the Wroclaw University of Sciences and Technology (Poland) and at Cardiff University (UK). He has supervised 24 master's theses since 1984 (ENP, U-Tizi-Ouzou, U-Béjaïa, U-Batna) and 14 doctorate theses (PhD) since 2000 (ENP and U-Béjaïa). He is currently supervising several theses, some of which are in the new system (LMD).

Since 1991, he has taught intellectual property and standardization at ENP. He taught the graduated intellectual property course at the Department of Industrial Engineering of the University of Batna, as well as at the ENSET d'Oran (ENP-Oran) as part of his post-graduate training in innovation. He has also taught legal metrology at ENP since 2011. He organized and participated in the supervision of an ENP/EPF Lausanne eco-management training course in 2002 and 2003, contributing to the introduction of environmental management tools in the training of ENP engineers.

From 1980 to August 2021, he published 56 publications in internationally renowned journals, 120 papers at international congresses (IEEE conferences, ISH, CLDP, etc.) and 72 papers at national conferences (CNHT, etc.), all of which have been published in edited proceedings. He is a member of scientific committees and organizing committees of several national and international conferences (CNHT, CNCEM, ICEE, COMADEM, CISTEM, etc.). His research axes concern various topics of insulation technology (partial discharges, breakdown, aging of insulating materials, diagnosis, etc.) and insulation coordination (pollution of insulators, earthing, lightning and lightning protection, etc.).

He is a member of the Association of Graduates from ENP (AD-ENP) and a member of the Association of High Voltage Electric Networks ARELEC (Algeria). He is a CIGRE member (through the ARELEC as Algerian NC), Senior Member of IEEE, and a member of the DEI-Society, and for a few years, he was a member of the Education Society and of the NPS Society. He is a member of the IEEE Algeria Section and of the DEI IEEE Algeria Section Chapter. He is a reviewer for various journals including those of IET (MST and GTD) and IEEE (transaction on DEI). He has been a member of the Editorial Board of the *International Journal of Electrical and*

Power Engineering (created in 2007), and from July 2014 to August 2019, he was a member of the Editorial Board and Associate Editor of the journal *IET Science Measurement and Technology*. Since January 2021, he has been Associate Editor and member of the editorial board of the *ENP Engineering Science* journal. He has served at ENP as Head of Department of Fundamental Sciences (1983–1985), Head of Department of Electrical Engineering (1985–1986), Director of (Undergraduate) Studies (1986–1989 and 1995–2003), and finally, Deputy Director in charge of continuing education and external relations (2005–2019). Since 1995, he has contributed to the rapprochement between ENP and industry and has set up some models of university-industry relationships in Algeria which have been the subject of international communications (UNESCO congresses, QRM/I.MEch.E in Oxford, etc.) and which continue to serve to simplify the realization of scientific and educational works in collaboration with industrial enterprises. He participated in the organization of the ENP Young Innovator competition and helped to set up an innovation center at ENP (opened in June 2007). He has collaborated in European Tempus projects on the quality of training, innovation, and the employability of engineers. The latter led to the establishment of the "House of Industry" at ENP. He participated in various commissions of the Ministry of Higher Education and Scientific Research: President of the National Pedagogical Committee of Electrical Engineering from 1986 to 1988, President of the Commission on Training-Employment Relations from 1995 to 1997, member of the Commissions on Training Programs, and member of the National Equivalences Commission and other think tanks on higher education systems. He has also participated in working groups at the ministries responsible for SMEs, industry, trade, and ICT. He was a member of the Scientific Committees of USTHB (three years) and EMP Engineers School (six years).

Leila Mekhnache is a professor in the Department of Electrical Engineering at the University of Batna 2, Algeria, and Head of the National Higher School for Renewable Energies, Environment & Sustainable Development (RE^2=SD), opened in June 2020. She received her master's and doctorate from the École Nationale Polytechnique of Algiers in 2004, which was based on the application of neural networks in the diagnosis and aging prediction of high-voltage insulations, and it was the first application of AI in the field of high voltage. Her research work is in the fields of diagnosis, prediction, and maintenance of high-voltage systems using artificial intelligence. She also worked on numerical methods (FEM) in high-voltage insulation and electromagnetics. She has participated in many international conferences, and she has published more than 100 papers and publications. She is a member of many technical and scientific boards for international journals and conferences. She is now leading a project for a renewable power plant for 2 MWc in Batna 2 with many applications (smart grid, smart home, and applications in agriculture, traffic, and water desalination).

Balwinder Raj is currently working as an associate professor in the ECE Department, at NIT, Jalandhar, India. He has more than 15 years of teaching/research experience. He completed a B.Tech in electronics engineering (PTU Jalandhar), M.Tech in microelectronics (PU Chandigarh), and a PhD in VLSI design (IIT Roorkee), India, in 2004, 2006, and 2010, respectively. For other research work, the European Commission awarded him a Mobility of Life research fellowship for postdoc research work at the University of Rome, Tor Vergata, Italy, in 2010–2011. He also worked as a visiting researcher at KTH University, Sweden, from October to November 2013, and at Alto University, Finland, in July 2017. He has visited countries such as Japan, Australia, Malaysia, Italy, Sweden, Finland, and Thailand for research collaborative projects, invited lectures, and conferences. Dr. Raj has authored/co-authored five books, 12 book chapters, and more than 100 research papers in peer-reviewed international/national journals and conferences. He has guided eight PhDs, and currently four PhD scholars are working with him. He guided more than 40 M.Tech students. Dr. Raj completed five research projects from DST Delhi, SERB Delhi, and CIMO Finland. Currently, he is handling two research projects from AICTE and ISRO. His areas of interest in research are classical/non-classical nanoscale semiconductor device modeling, memory design for AI applications, low-power VLSI design, digital/analog VLSI design, and FPGA implementation.

List of contributors

Akam
NITTTR Chandigarh
Chandigarh, India

Abdelhamid Bouhelal
Laboratory of Green and
Mechanical Development (LGMD)
Algeria

Srishtee Chaudhary
NITTTR Chandigarh
Chandigarh, India

Meenakshi Devi
University of Oxford Brooks, UK

Dharmendra Gill
Jal Shakti Vibhag, Government of
Himachal Pradesh, Mandi
Mandi, Himachal Pradesh, India

Shahraan Hussain
NITTTR Chandigarh
Chandigarh, India

Arun Kumar
Dept of ECE, GITAM University,
Bangalore
Bangalore, India

Hemant Kumar
NITTTR Chandigarh
Chandigarh, India

Mohammed Amokrane Mahdi
École Nationale Polytechnique, B.P.
182, El-Harrach
El-Harrach, Algeria

Rajesh Mehra
NITTTR Chandigarh
Chandigarh, India

Pawanram
Dr B R Ambedkar National
Institute of Technology Jalandhar
Jalandhar, Punjab, India

Balwant Raj
UIET Panjab University
Chandigarh, S S Giri PU Campus
Hoshiarpur
Hoshiarpur, Punjab, India

Balwinder Raj
Dr B R Ambedkar National
Institute of Technology Jalandhar
Jalandhar, Punjab, India

Raj Kumar Saini
Shoolini University
Solan, India

Charul Sharma
NITTTR Chandigarh
Chandigarh, India

Divya Sharma
AKGEC, Ghaziabad, India

Sanjay Kumar Sharma
NITTTR Chandigarh
Chandigarh, India

Mandeep Singh
Dr B R Ambedkar National
Institute of Technology Jalandhar
Jalandhar, Punjab, India

Arezki Smaili
Laboratory of Green and
Mechanical Development (LGMD)
Algeria

Amita Verma
Shoolini University
Solan, India

Piush Verma
Rayat Bhara University Mohali
Kharar, Mohali, India

Chapter 1

Emerging organic and inorganic solar cell materials for energy

Hemant Kumar, Piush Verma, and Balwinder Raj

1.1 INTRODUCTION

Since the latter half of the 20th century, the growth of the world's population and industrial development has gained momentum and has had a clear impact on the expansion of the global economy as well as gross world product (GWP). This has resulted in steady growth in energy demand and the influence of all these variables on the environment. Concerns have grown about the potential ramifications of the relentless association of expansion with development. Sustainability itself, defined as "the pursuit of robust and enduring social structures and activities for both humans and the environment that sustains their life" [1], is relatively new. It was the efforts of the Club of Rome, established in 1968, as well as "The Limits of Growth" report, published in 1972, that first presented the concept that economic and environmental issues should not be treated individually, providing a grave warning about endless growth of material and unrestricted utilization in a world of limited resources. The concept of evaluating the amount of energy required to manufacture a product or service while taking into account all significant factors (economic investment, primary energy, environmental loading, labor, etc.) resulted in the creation of a methodology known as "emergy", which is primarily derived from the seminal work of Howard T. Odum [2].

Figure 1.1 depicts global energy consumption, GWP, and population growth rates from 2010 to 2019. Even though the growth of the human population has decelerated recently (in 2019, it was 1.07%, compared with 2010 when it was 1.23%), the human population has reached over 7.7 billion. Annual GDP varied but remained over 3%, reaching 3.6% in 2017 and 2018 (values calculated by discrepancies in living expenses throughout nations) [3].

Consumption of energy rose at a slower rate compared to GDP in 2017 and 2018, owing to improved efficiency of industrial and technological processes. According to the International Energy Agency's (IEA) Global Energy & CO_2 Status Report 2017, global energy consumption in 2017 was more than 14,000 Mtoe (million tons of oil equivalent), as compared to 2000's

DOI: 10.1201/9781003487692-1

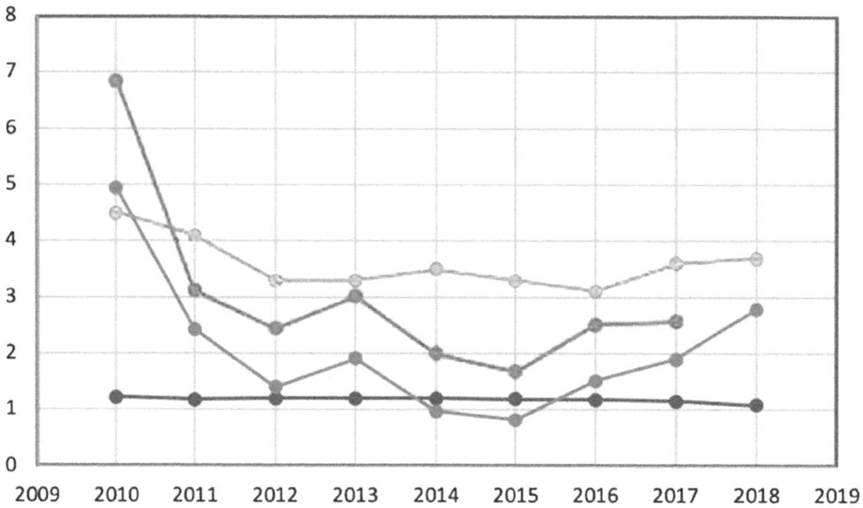

Figure 1.1 Gross world product (level 1), worldwide electricity consumption growth rates in the current decade (level 2), worldwide energy consumption (level 3), and population (level 4).

10,035 Mtoe [4]. The worldwide trend of electricity consumption, in particular, slowed after a continual significant rise from 2000 to 2010, with a minimal rate of growth in 2015 coinciding with minimal GDP growth [5], and then began to climb again, with a rise of 2.6% in 2017. In accordance with the IEA's World Energy Outlook 2018 [6], global energy generation is expected to increase by 60% between 2017 and 2040, covering a quarter of the primary consumption. Similarly, assessments of the natural aggregation of CO_2 in our environment revealed that the yearly rise rate over the last five decades was approximately 100 times greater than prior natural increases caused by temperature oscillations in geological ages. Figure 1.2 depicts the historical changes in atmospheric CO_2 concentrations as recovered from ice cores [7]. "World energy-related CO_2 emissions increased by 1.4% in 2017, hitting a record level of 32.5 Gt (gigatonnes), resuming growth after three years of stagnation [4]; the average CO_2 concentration in the atmosphere also in 2017 reached a new record: 405 ppm". Figure 1.3 shows CO_2 levels in our environment recorded at the Observatory of Mauna Loa in Hawaii from 2005 to 2020.

Despite some disagreement, there is, however, scientific agreement on two points: first, that greenhouse gases are the primary driver of climate change, and second, that the emissions of carbon dioxide (CO_2) resulting from the combustion of fossil fuels are the most significant contributors to these greenhouse gases. Consequently, most government policies are designed to ensure universal access to energy, reduce pollution, and achieve

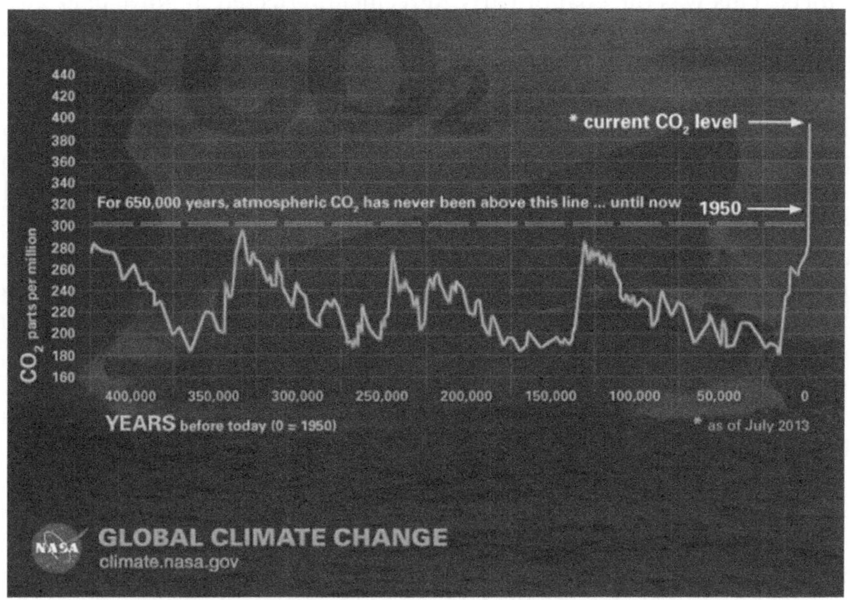

Figure 1.2 Carbon dioxide concentrations in the atmosphere over the previous three glacial cycles, as inferred from ice cores, as well as present levels.

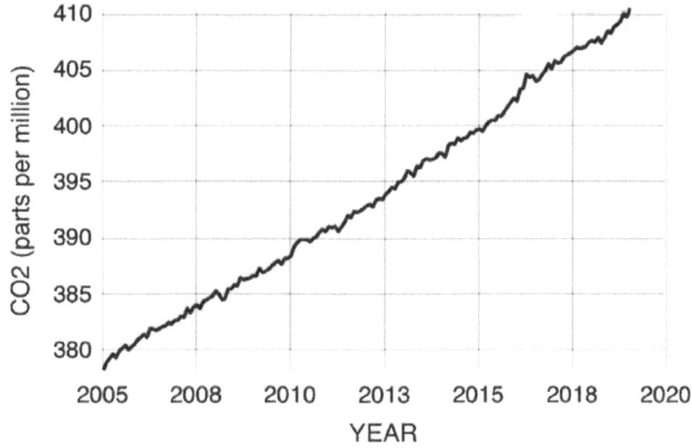

Figure 1.3 Carbon dioxide content in the atmosphere up to 2020, measured by averaged infrared absorption observations every month at the Observatory of Mauna Loa, Hawaii.

long-term climate objectives. One central objective of these policies is to prevent the global temperature from rising more than 2 degrees Celsius above preindustrial levels during this century, with an even more ambitious aim of limiting the increase to 1.5 degrees Celsius by making diligent efforts. This was the goal of Paris Agreement, effected in November 2016, which has now been ratified by 185 of the 197 United Nations Framework Convention on Climate Change (UNFCCC) parties [8].

The objective of lowering the emissions of CO_2 while maintaining and delivering the required energy services to back global economic growth can only be reached by significantly decreasing the effects of power-generating sources that produce the most carbon dioxide, like fossil fuels. According to the IEA, electric mobility, electricity access, electric heating, and industrial electric motor systems might contribute to a 90% increase in energy consumption from now to 2040 [6]. Renewable energy sources can play a significant role here. As seen in Figure 1.4, the amount of generated electricity by renewable sources attained a historical high of 25% in 2017; additionally, the IEA predicts that the percentage of generation through renewable sources could increase from 25% currently to about 40% by 2040 [4].

Presently, hydropower, photovoltaic (PV) solar cells, and wind are major low-emission electricity resources. It is anticipated by the IEA that rapid expansion of PV cells will lead to it surpassing wind-installed capacity before 2025, hydropower-installed capacity by around 2030, and coal-installed capacity before 2040. Additional issues must be addressed, such as the increased solar-generated electricity distribution, which may need new standards for the reliable running of power networks. Furthermore, while electrification certainly decreases pollution, supplementary efforts to regulate its generation and operation, and the decommissioning of power systems may be required from an environmental standpoint; otherwise, CO_2 emissions and pollution may simply be transferred from consumer sectors to the generating, decommissioning, and recycling sectors.

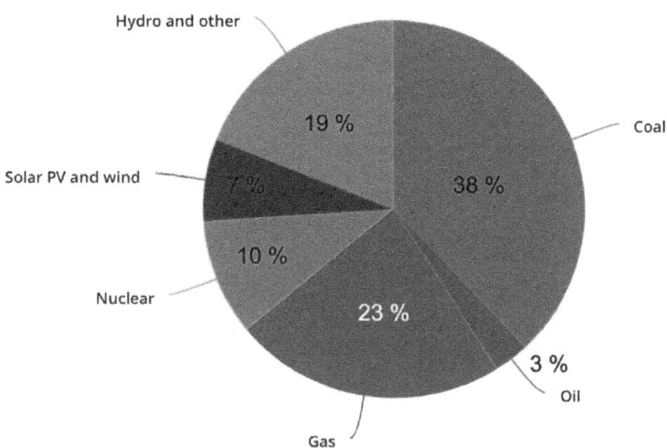

Figure 1.4 Electricity generation from various energy sources totaled 25,570 TWh in 2017. PV stands for photovoltaics.

This chapter provides a short review of the current state of research on the essential building blocks of photovoltaic solar cells, without delving into issues on concentrated photovoltaics (CPV) and solar panels.

1.2 SOLAR CELLS: HISTORICAL NOTES

Our earth's upper atmosphere receives around 174 PW of solar radiation, of which the yearly average radiation is approximately 1.36 W/m^2 because about 30% of the total is bounced back into space, and the peak of typical surface irradiance at sea level is approximately 1 kW/m^2 on a clear day. This means that, following the ASTM G-173 standard, the direct tilted spectra are 1000.4 W/m^2 and the total integrated hemispherical irradiance is 900.1 W/m^2, which quantifies the intensity of solar radiation across the 280 to 4000 nm band. The latter figure corresponds to what is known as the atmospheric condition AM1.5, which assumes an ultimate air mass of 1.5 and is appropriate for representing the average annual data for temperate regions.

Solar power is a colossal power source that is readily and abundantly accessible, and it has been utilized for millions of years in multiple ways, both directly and indirectly. In the 18th century, scientists such as Samuel Pierpont Langley, Horace de Saussure, and John Herschel discovered ways to make water boil by increasing the temperature by making a correctly built glass-covered box that allowed water to reach its boiling point. These hot pots can be regarded as prototypes for modern solar collectors [63–69].

Furthermore, the breakthrough came with the discovery of the photovoltaic effect (1839). This natural phenomenon was discovered by French physicist Edmond Becquerel, who experimented with electrolyte liquid electrolyte-immersed metallic electrodes. The first "real" solid-state photovoltaic cell was constructed by Charles Fritts in 1883, who used junctions formed by coating a copper plate with selenium and then covering it with an ultrathin, nearly transparent layer of gold. Albert Einstein later wrote a paper in 1905 entitled "On a Heuristic Viewpoint Concerning the Transformation and Production of Light", which won the Nobel Prize in Physics in 1921 [9], clarifying all the theoretical concepts of the fundamental photovoltaic effect. A prototype was further developed in 1916 by Robert Millikan [10].

Ultimately, the 1950s saw the birth of photovoltaic solar power. Calvin Fuller, Daryl Chapin, and Gerald Pearson showed the very first practical photovoltaic cell, silicon-based, in early 1954 at Bell Labs, with an efficiency of 4% (eventually increasing to 11%). Considering the low efficiency and high cost, the initial period of solar cell research concentrated on space applications; Vanguard 1, deployed in 1958 as part of the program led by the US Naval Research Laboratory, was the first spacecraft to be deployed using solar-powered radios (less than 1 W). A few years later, in 1962, Telstar 1, the first commercial communications satellite, was conceived/

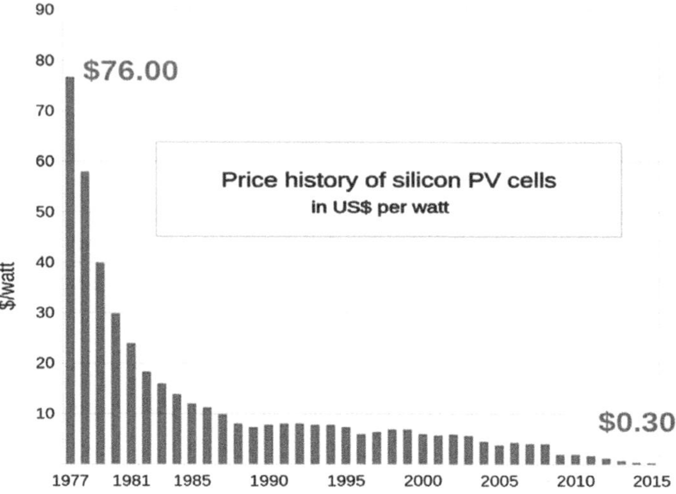

Figure 1.5 From 1977 to 2015, the price of solar cells made from silicon in US $/W was charted.

manufactured by a Bell Labs-based team and deployed by NASA, fueled by 3600 PV cells totaling 14 W. By 2010, with the efforts of E. Berman at Exxon, the advancements in methods and materials drove the price of a silicon PV cell down to a fraction of a US dollar per generated electrical watt. This reduced the cost and practicality of using PV devices in big PV panels for both commercial applications as well as residential [11].

Figure 1.5 depicts the reduction in costs of silicon photovoltaic cells from 1977 to 2015. Throughout the previous decade, PV development and research have progressed with a sole focus on new materials and sophisticated PV design [70–76]. Scientists have been exploring materials capable of functioning at comparable or even greater levels than silicon (in its different forms, such as amorphous silicon, polycrystalline, and monocrystalline) but which are more readily available and less expensive to create. Organics, perovskites, and carbon nanotubes are among the novel materials; their essential concepts are briefly reviewed in the following section.

1.3 EMERGING SOLAR CELL MATERIALS

1.3.1 ST-OSCs (semitransparent organic solar cells)

An OSC's active layer is composed of molecules based on organic donors and acceptors. Some donor materials, such as [polythieno[3,4-b]-thiophene/benzodithiophene (PTB7), 4-phenylenevinylene (MEH-PPV), poly(3-hexylthiophene) (P3HT)], poly 2-methoxy-5-(2-ethylhexyloxy)-1 and the newly discovered poly (PffBT4T-C_9C_{13}) [(3,3′ ″-di(2-nonyltridecyl-2,2′;5′, 2″;5″,2′ ″-quaterthiophen-5,5′ ″-diyl)-alt-(5,6-difluoro-2,1,3-benzothia diazol-4,7-diyl)], have been effectively employed in organic solar cells, with corresponding OSC efficiencies increasing from 3% to over 12% [12]. Acceptors previously utilized in high-efficiency organic solar cells are primarily derivatives of fullerene, for example [6,6]-phenyl C61/71 butyric acid methyl ester ($PC_{61}BM/PC_{71}BM$) and indene-C_{60} bisadduct (ICBA), whereas acceptors that are non-fullerene-based have recently gained attention because they have also demonstrated 12% conversion efficiencies [13]. Figure 1.6a displays the construction of a typical organic solar cell of a single junction. An intermingled bicontinuous layer of electron acceptor and donor substance forms a bulk heterojunction (BHJ) in the device. This layer is placed between an anode and a cathode, each having its own transport layer for selective charge. The operating mechanism of an organic solar cell is mostly attributable to exciton dissolution at the organic heterojunction, as seen in Figure 1.6b, and has been extensively studied in several works [14]. The device may be made using either an inverted structure with ITO as a cathode as well as a high-work-function metal (e.g., Ag or Au) as an anode, or a normal composition of

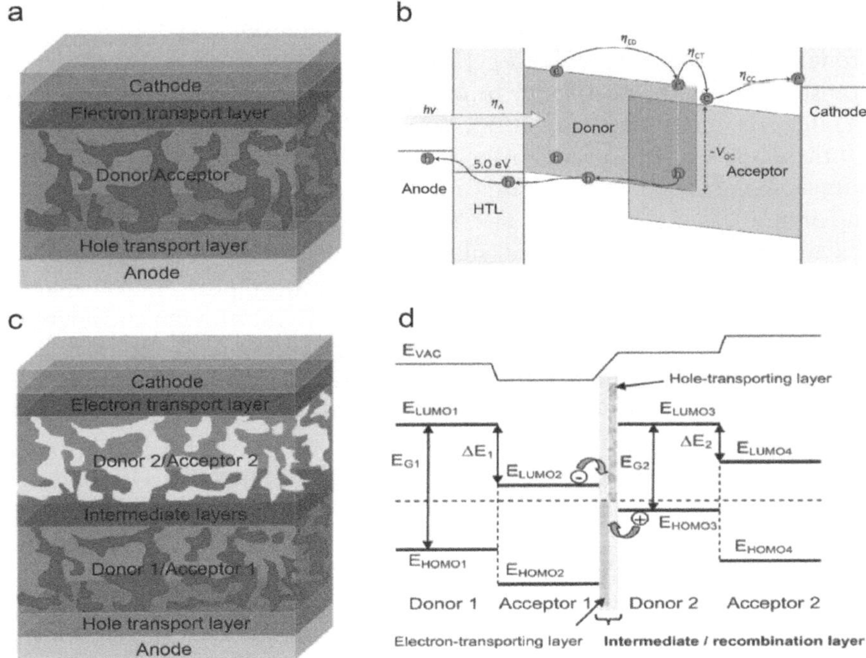

Figure 1.6 General device architectures and energy diagrams for single junction (a,b) and tandem (c,d) bulk heterojunction OSCs.

indium tin oxide (ITO) as an anode as well as a cathode as a low work function metal (e.g., Al) [14]. Yan's group recently achieved the highest verified power conversion efficiency (PCE) of 11.5% on a single junction OSC. The thermalization losses of holes and electrons created by photons having energies larger than the donors' bandgaps are one significant factor for the loss of V_{OC} [15]. A tandem device configuration (Figure 1.6c) comprised of more than one independent photoactive layer with complementing absorption is projected to yield higher PCEs [16]. Figure 1.6d depicts the equivalent energy diagram. Whenever transparent upper electrodes are used, both tandem and single junction OSCs may be rendered semitransparent. As shown in Table 1.1, several varieties of successfully created semitransparent organic solar cells may be loosely classed based on transparent upper electrode materials, as transparent electrodes are indeed an important portion of the cell structure that distinguishes ST-OSCs from standard OSCs [77–84].

1.3.2 Transparent top electrodes based on silver nanowire (AgNW)

"Because of their high electrical conductivity as well as sheet resistance <15 sq^{-1} with 90% transparency, great flexibility, low temperature, and solution processability, AgNWs are regarded as a promising option for electrode substances to create high-performance ST-OSCs" [32]. AgNW-based transparent top electrodes may be incorporated into ST-OSCs using doctor blading, drop casting, lamination, spray coating, and inkjet printing [33]. Of these approaches, spray coating is the most commonly utilized since it is simple, scalable, and, more significantly, has a low influence on the underlying organic layer [82–85].

Guo et al. (2015) reported significant success in constructing mass ST-OSC units on plastic substrates and glass, employing transparent AgNWs as the top electrode material (Figure 1.7 a, b). One considerable advancement recorded in their research is the use of a high-precision laser designing system, which might efficiently enhance electrical terminals between discrete cells while promoting dead area reduction in the module, enabling manufacturing of mass area production of ST-OSC modules (64 cm^2) having FF of 63% and geometric FF higher than 95%. This makes it a possible contender for the eventual commercialization of ST-OSCs [34]. Min et al. recently described the first solution-processed small-molecule-based ST-OSCs with AgNW top electrodes. Figure 1.7c depicts the device construction, which includes a PDINO (ZnO/perylene diimide derivative) bilayer as a cathode buffer, four small molecules with varying absorptions (colors) as electron donors (Figure 1.7d), and PC$_{71}$BM as an electron acceptor. The ST-OSC with the 2D-conjugated molecules based on the support of thienyl-conjugated secondary chains (BDTT-S-TR) provided

Table 1.1   Several ST-OSCs: a summary

Top electrode	Device structure	PCE (%)	AVT (%)	Ref.
Thin-film Ag	ITO / ZnO / C_{60} - SAM / PCPDTFBT:$PC_{71}BM$ / PEDOT: PSS / Ag	5	47.3	[17]
	ITO / ZnO / C_{60} - SAM / PBDTTT-C-T:$PC_{71}BM$ / MoO_3 / Ag	6	25	[18]
	ITO / ZnO / PDBTT - DPP:$PC_{71}BM$ / MoO_3 / Ag / Si NPs / Alq_3	6.22	32	[19]
Thin-film Au	AZO / ZnO / P3HT:$PC_{61}BM$ / MoO_3 / Au / MoO_3	2	–	[20]
	ITO / Cs_2CO_3 / P3HT:$PC_{61}BM$ / V_2O_5 / Au	0.85	–	[21]
TCO	ITO / PEDOT:PSS / PCDTBT:$PC_{71}BM$ / TiO_x / AZO	4	34	[22]
	ITO / Cs_2CO_3 / P3HT:$PC_{61}BM$ / PEDOT:PSS:D-sorbitol / ITO	3	70 (650–800 nm)	[23]
AgNWs	ITO / PEDOT:PSS / BDTT - S - TR:$PC_{71}BM$ / ZnO / PDINO / AgNWs	3.62	≈30	[24]
	Graphene / PEDOT:PSS / PSEHTT:ICBA / ZnO / PEDOT: PSS / PBDTT - DPP:$PC_{71}BM$ / TiO_2 / AgNWs	8.02	45	[25]
CNTs	Glass / PEDOT:PSS / p - BF - DBP / ZnPc:C_{60} / C_{60} /n - C_{60} / CNTs	1.1	22	[26]
	ITO / p - BF - DBP / ZnPc:C_{60} / C_{60} / n - C_{60} / CNTs	1.3	24	[26]
	Glass / $AgNW_s$ / TiO_x / ZnO / P3HT:$PC_{61}BM$ / PEDOT:PSS / GO / PH1000	2.3	≈60 (650–800 nm)	[27]
PEDOT:PSS	Glass / PH1000 / ZnO / P3HT:PC61BM / PEDOT:PSS / PH1000	1.8	≈55 (650–800 nm)	[28]
	Glass / graphene / PEDOT:PSS / ZnO / PTB7:$PC_{71}BM$ / PEDOT:PSS / Graphene	3.4	40	[29]
Graphene	ITO / ZnO / P3HT:$PC_{61}BM$ / GO / graphene	1.8–2.5	–	[30]
	ITO / ZnO / P3HT:$PC_{61}BM$ / PEDOT:PSS / graphene	3	–	[31]

the maximum PCE of 3.62% under optimum circumstances [93–98]. Furthermore, the transparency perception by human eyes of these photovoltaic cells was investigated, and all of the devices displayed color coordinates that were close to white light (Figure 1.7e) [24].

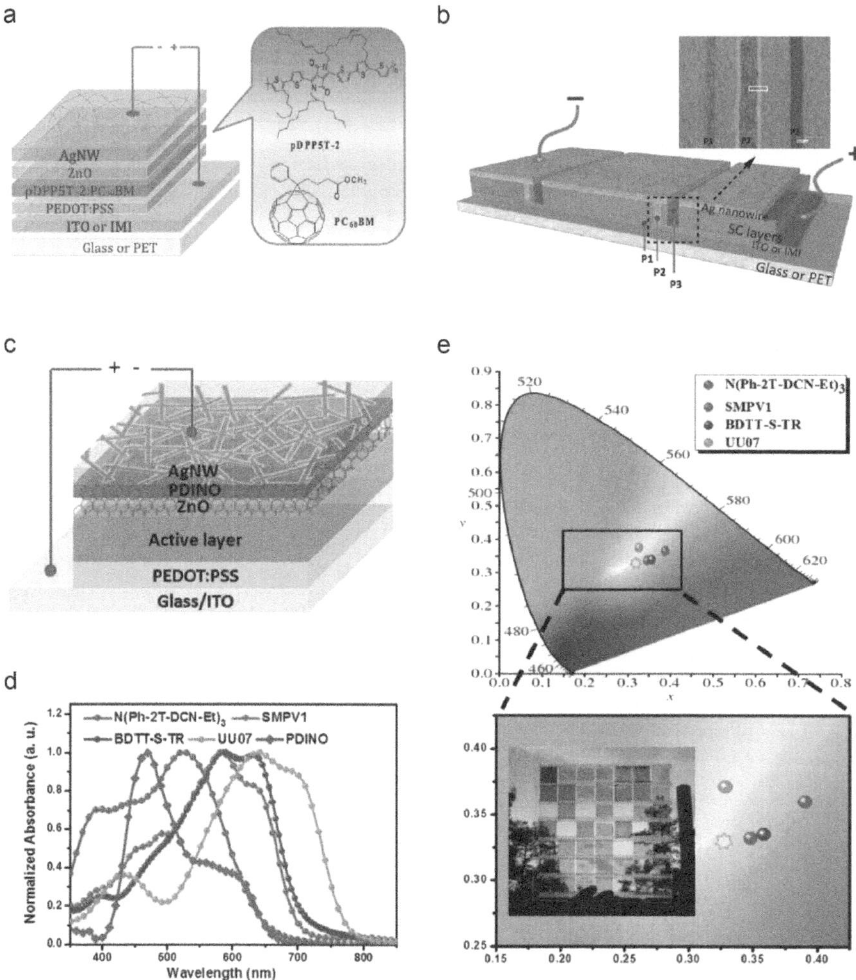

Figure 1.7 (a) Device architecture and (b) semitransparent-based unit schematics for an ST-OSC using pDPP5T-2:PC60BM [34]. c) Device construction schematic, d) absorption spectrum from various donors, and e) ST-OSC color coordinates determined by various donors in CIExy 1931 the chromaticity diagram, with an image shot via a ST-OSCs inserted module [24].

1.3.3 Top electrode based on graphene

In recent years, graphene was discovered to be the thinnest 2D carbon material and has been rigorously investigated for optoelectronic device transparent conductive electrodes because of its high conductivity and transparency, chemical stability, and excellent mechanical flexibility [32]. Lee et al. (2011) proposed an ST-OSC based on a top-layered graphene electrode that has an inverted structure of ITO/ZnO/P3HT:PC$_{61}$BM/graphene oxide (GO)/graphene, wherein the GO acts as the HTL. They discovered that as the number of layers made of graphene increased, the sheet resistance for the graphene film fell dramatically, but the film transmittance declined by roughly 2–3%. When lit from the ITO side, the finished device achieved the greatest PCE of 2.5% by using a ten-layer graphene electrode, whereas the best performance of the device (2.04%) was achieved when the total layer count of graphene was eight [30]. Because of the changing of the Fermi level within the graphene layer, the research discovered that HAuCl$_4$ and PEDOT:PSS doping may considerably increase the electrical conductivity of graphene film. For example, in the lab, a PEDOT:PSS-covered single-layered graphene layer showed an 80% reduction in sheet resistance (from 580 to 96 cm^2). This discovery enabled researchers to create ST-OSCs having single-layered graphene top electrodes (Figure 1.8a,b). Due to the

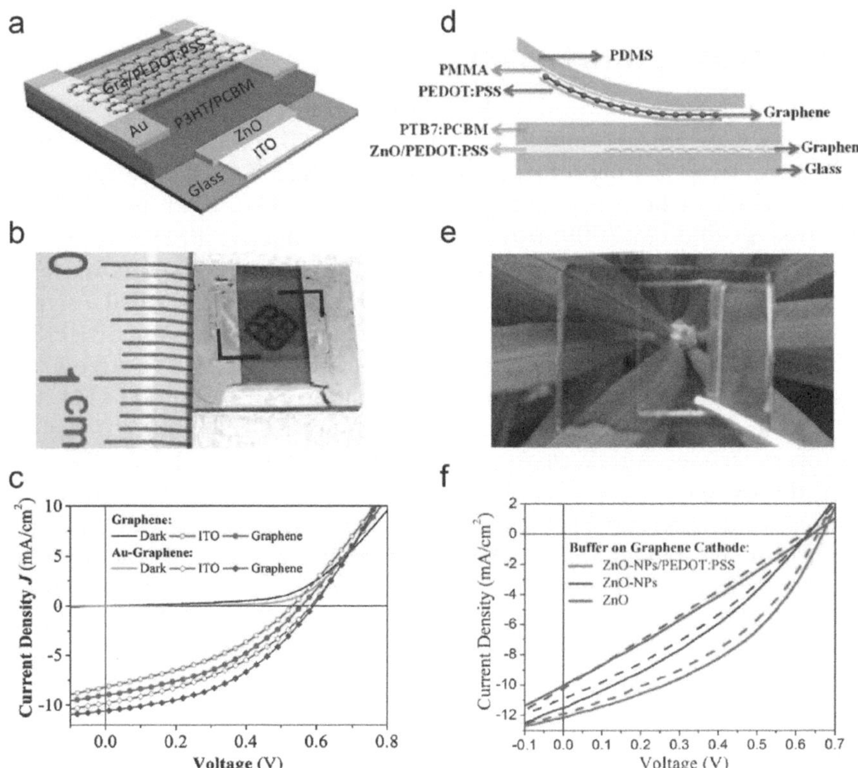

Figure 1.8 a) Schematic diagram, b) image, and c) J-V curves of a transparent graphene-based top electrode based on ST-OSC [31]. American Chemical Society. All rights reserved. d) Diagram, e) picture, and f) J-V curves from an all-graphene-based ST-OSC [29].

exceptional transmittance (AVT > 90%) of a graphene electrode, devices (based on P3HT and PC$_{61}$BM) demonstrated PCEs of 3% when lit via the graphene end, which was greater than those produced from the ITO end (Figure 1.8c). Because the multi-step layer transfer technique for creating graphene electrodes is no longer required, for the large-scale manufacture of graphene-based semitransparent devices this method would be desirable [31]. Furthermore, the researchers discovered that devices with top electrodes as graphene had outstanding device stability. Their team produced a completely transparent OSC employing graphene as both the top and bottom electrodes (Figure 1.8d, e).

With irradiation from either side, PCEs of up to 3.4% and transparency of 40% were attained utilizing PTB7 as an electron donor, and PC$_{71}$BM as an electron acceptor (Figure 1.8f). It should be noted that the devices were mostly constructed of carbon-based materials. The acquired PCE is, to the best of our knowledge, the greatest value ever reported for all based on carbon solar cells [29].

1.3.4 ST-OSCs with no ITO

Aside from being employed as ST-OSCs' top electrodes, the transparent electrodes have the potential of being employed as lower electrodes as well, permitting the manufacture of ITO-free ST-OSCs with appropriate combinations of both top and bottom [99–105]. Hau et al. presented an ITO-free ST-OSC in 2009, using extremely conductive PEDOT:PSS films as top and bottom electrodes with a P3HT:PC$_{61}$BM mix as an active layer. However, just 0.47% PCE was achieved [35]. Zhou et al. attained a significantly greater PCE of 1.8% with a comparable device design in 2010 [28]. Kim et al. studied several electrode configurations for ST/ITO-free OSCs based on a ZnPc:C$_{60}$ mix with silver (opaque), PEDOT:PSS, ITO, and AgNWs as bottom electrodes, and a thin film of f-CNT and Ag (t-Ag) for the top electrodes. AgNWs/t-Ag, PEDOT:PSS/f-CNT, and PEDOT:PSS/t-Ag were discovered as promising ITO-free ST-OSC pairings [26]. In 2013, Guo et al. detailed an ITO-less ST-OSC employing AgNWs as both top and bottom electrodes and a ternary blend that included P3HT, poly[[(4,40 -bis(2-ethylhexyl)dithieno[3,2-b:2′,3′-d]-silole)-2,6-diyl-alt-(4,7-bis(2-thie nyl)-2,1,3-benzothiadiazole)-5,5′-diyl] (Si-PCPDTBT), and PC$_{61}$BM as the active layer. The resultant device had a PCE of 2.2% at an AVT of 33%, which was equivalent to an ITO bottom electrode reference device (Figure 1.9a, b) [36]. In a subsequent trial, they produced a 2.9% PCE with a significantly greater AVT of 41% using the BHJ mix of pDDP5T-2:PC$_{61}$BM (Figure 1.9c,d). Similarly, Yim et al. described a P3HT:PC$_{61}$BM-based ST-OSC with a transparent AgNW film as the PEDOT:PSS, and the bottom electrode as the top electrode. The finished device demonstrated outstanding transparency and respectable PCEs of 2.3% and 2.2% when illuminated from the PEDOT:PSS sides and AgNWs, respectively (Figure

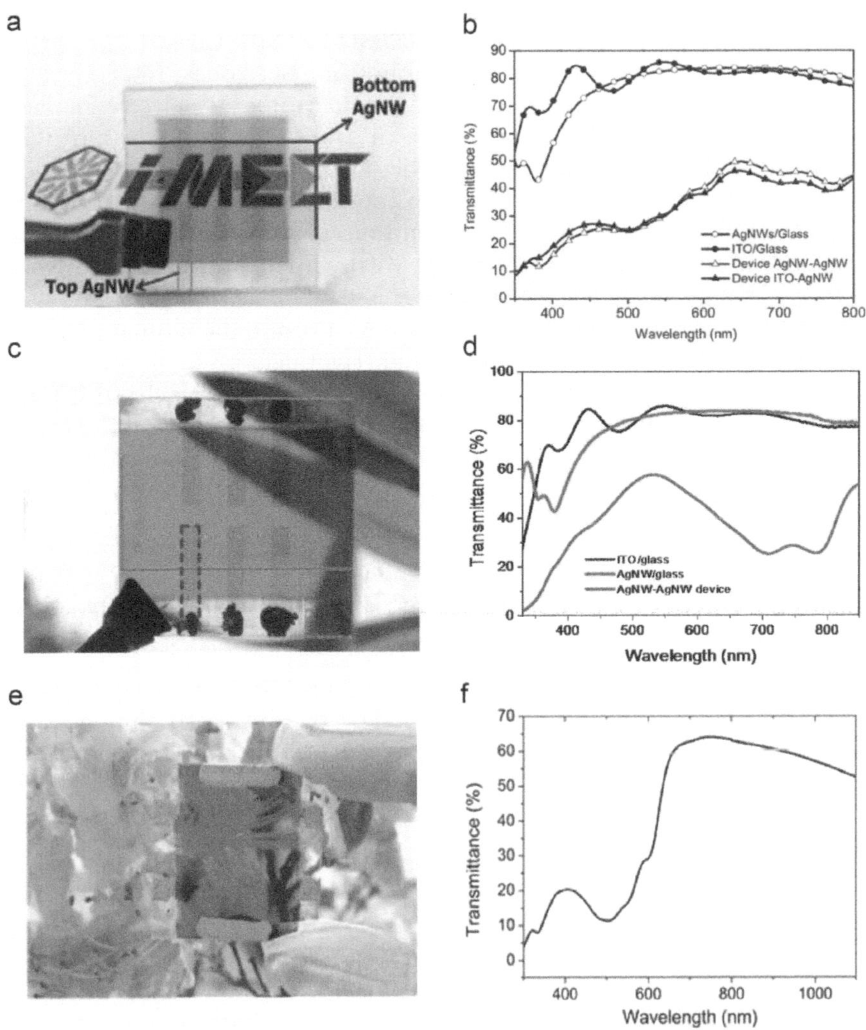

Figure 1.9 a) Image and b) transmittance of ST-OSCs utilizing the AgNWs/AZO/P3HT: Si-PCPDTBT:PC61BM/PEDOT:PSS/AgNWs configuration [30]. c) Photo and d) transmission characteristics of ST-OSCs with the AgNWs/PEDOT:PSS/ PDD5T-2:PC61BM/ZnO/AgNWs setup. e) Photograph and f) transmission properties of ST-OSCs in the AgNWs/TiOx/ZnO/P3HT:PC61BM/PEDOT:PS S:GO/PEDOT:PSS solar cell device structure.

1.9e,f) [27]. Yusoff et al. recently demonstrated an effective ITO-free tandem ST-OSC based on graphene-mesh/PEDOT:PSS/PSEHTT:IC$_{60}$BA/ZnO/ PEDOT:PSS/PBDTT-DPP:PC$_{71}$BM/TiO$_2$/AgNWs device structure. They achieved a phenomenal AVT of 45% after optimizing the manufacturing conditions for both front and back subcells, as well as PCEs of 6.47%

and 8.02% on the AgNW and graphene sides, respectively [25]. Wilken et al. recently announced an ITO-free P3HT:PC$_{61}$BM-based OSC using MoO$_3$/Au/MoO$_3$ and AZO as top and bottom electrodes, respectively. A ZnO overlayer was placed on top of the AZO to improve electron collecting efficiency. Two AZO electrode thicknesses (1270 and 625 nm) were investigated. It was discovered that devices with narrower AZO electrodes had somewhat greater J$_{SC}$ compared to those with thicker ones, probably because of their greater average transmittance; however, no statistical data was offered to support this discovery. Meanwhile, because of their lower sheet resistance, thicker AZO electrodes had significantly lower R$_S$ (series resistance) compared to the thinner ones. As a result, the ultimate PCEs of these photovoltaic cells were quite similar. The thickness of active layers in devices with MoO$_3$ (12 nm)/Au (12 nm)/MoO$_3$(50 nm) top electrodes and 1270 nm AZO bottom electrodes was tuned. Under bottom illumination, an optimum PCE of 2% was attained, with a maximum transmittance of the device of 60% at wavelengths over 650 nm, equivalent to a thickness of the active layer of 140 nm [20].

1.3.5 ST-PSCs (semitransparent perovskite solar cells)

Figure 1.10a depicts the fundamental crystal arrangement of organometal halide perovskite substances. The unit cell of an ABX$_3$ perovskite material is formed from five atoms within a cubic structure, with cation A having 12 closest neighbors and cation B having six closest- neighbor anions X. The lattice constants a, b, and c must be the same to construct an ideal cubic structure, which gives

$$\frac{r_A + r_x - r}{\sqrt{2}} \to B \to X \tag{1.1}$$

where both r_A and r_B represent the ionic radius for both A and B cations, and r_X is the anion's ionic radius. Goldschmidt's tolerance factor (t) is also defined as

$$t = \frac{r_A + r_X}{\sqrt{2}(r_B + r_X)} \tag{1.2}$$

The optimal cubic arrangements can be created when t is 0.9–1. Otherwise, deformed forms such as rhombohedral, tetragonal, and orthorhombic are likely possible. The bandgaps of perovskite components vary depending on their composition, as seen in Figure 1.10b. Currently, the most often utilized perovskite materials for PSCs are formamidinium (HC(NH$_2$)$_2$$^+$,FA) and methylammonium (CH$_3$NH$_3$$^+$,MA) -based lead halides (FAPbX$_3$ and

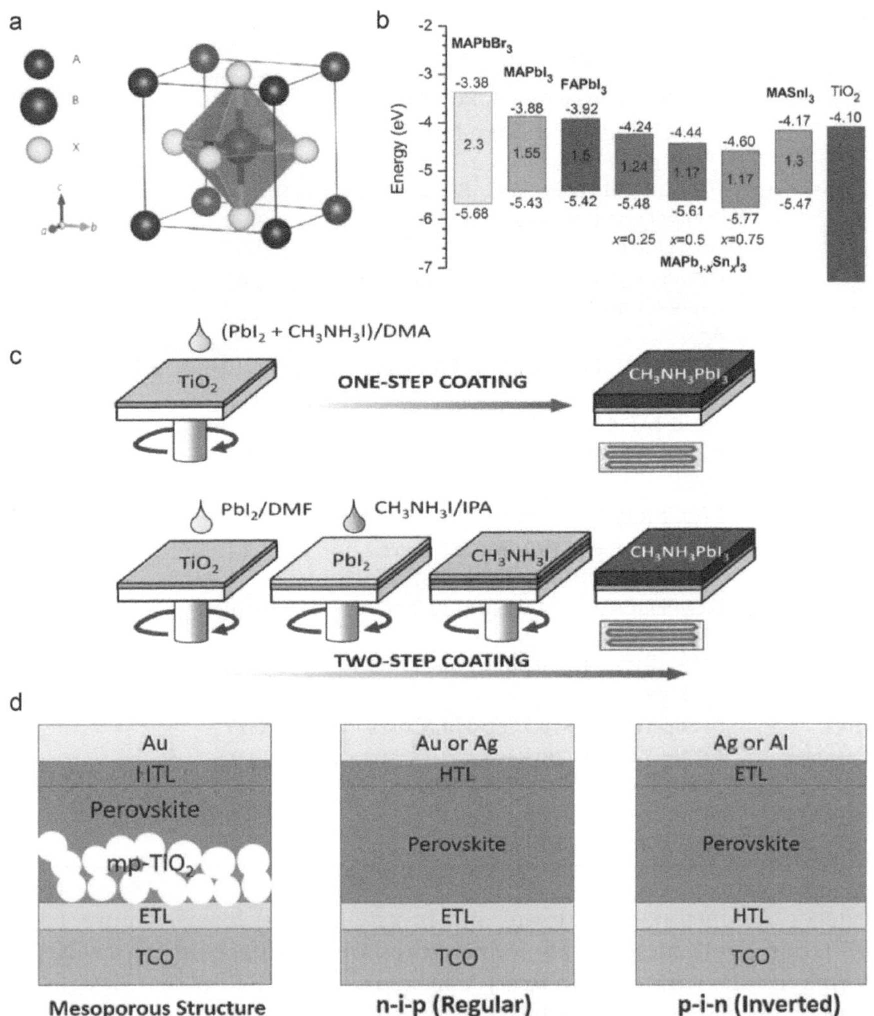

Figure 1.10 a) ABX3 perovskite crystal structure, where A is normally HC(NH2)2 $^+$ (FA), CH3NH3 $^+$ (MA), or a mixture thereof, B is Pb or Sn, and X is typically I, Cl, Br, or a mixture thereof [53]. Nature Publishing Group 2014, all rights reserved. b) The energy level diagrams of perovskites of various compositions [42]. c) Schematic representation of one-step and two-step perovskite film deposition processes [54]. d) Schematic representation of three common PSC architectures [37].

MAPbX$_3$, where X = I$^-$, Cl$^-$, Br$^-$, or a combination thereof). Perovskite films are typically created by the reaction of PbX$_2$ with MAX or FAX using one- or two-step deposition techniques (Figure 1.10c). PSCs may be made using either planar or mesoporous structures; the latter type may be further

Table 1.2   Recent ST-PSCs: a summary

Top electrode	Device structure	PCE (%)	AVT (%)	Ref.
Thin-film Au	ITO/PEDOT:PSS/MAPbI$_3$/PC$_{61}$BM/Au/LiF	7.3 (6.4)	22 (29)	[38]
	FTO/c-TiO$_2$/MAPbI$_3$/Spiro-MeOTAD/ MoO$_3$/Au/MoO$_3$	13.6 (5.3)	7 (31)	[39]
Thin-film Ag	ITO/CuSCN/MAPbI$_3$/PC$_{61}$BM/Bis-C$_{60}$/Ag	10.7 (7.5)	13 (37.5)	[40]
	ITO/PEDOT:PSS/MAPbI$_3$/PC$_{61}$BM/ MUTAB/Ag	11.8	20.8	[41]
	ITO/PEDOT:PSS/MAPbI$_3$/PC$_{61}$BM/PTCBI/ Ag/WO$_3$/PTCBI/ Ag	3.86	–	[42]
TCOs	FTO/c-TiO$_2$/MAPbI$_3$/PTAA/ITO/glass	15.8(12.6)	6.3 (17.3)	[43]
	FTO/c-TiO$_2$/mp-TiO$_2$/MAPbI$_3$/Spiro- MeOTAD/MoO$_x$/ITO	6.2	–	[44]
	ITO/PEDOT:PSS/MAPbI$_3$/PC$_{61}$BM/AZO/ ITO	12.3	–	[45]
	FTO/ZnO/PC$_{61}$BM/MAPbI$_3$/MoO$_3$/IO:H	14.1	–	[46]
	FTO/c-TiO$_2$/mp-TiO$_2$/MAPbI$_3$/Spiro- MeOTAD/MoO$_3$/AZO/Ni-Al grid/MgF$_2$	12.1	–	[47]
AgNWs	ITO/PEDOT:PSS/MAPbI$_3$/ALD-ZnO/ AgNWs/ALD-Al$_2$O$_3$	10.55	22.5	[48]
	FTO/c-TiO$_2$/MAPbI$_3$/Spiro-MeOTAD/Au/ AgNWs	11.07	–	[49]
CNT	FTO/c-TiO$_2$/mp-TiO$_2$/MAPbI$_3$/CNTs	6.29	–	[50]
Graphene	FTO/c-TiO$_2$/MAPbI$_{3-x}$Cl$_x$/Spiro- MeOTAD/PEDOT:PSS/Graphene	12.02	–	[51]
PEDOT:PSS	FTO/c-TiO$_2$/mp-TiO$_2$/MAPbI$_3$/Spiro- MeOTAD/PEDOT:PSS	10.1/2.9	7.3	[52]

categorized into normal (n-i-p) and inverted (p-i-n) forms (Figure 1.10d) [37]. In general, mesoporous architectures are advantageous for device performance and can result in less J-V hysteresis, although planar designs are easier and can be used to build devices at low temperatures. The transparency of the top electrodes determines the performance of ST-PSCs, like that of ST-OSCs and ST-DSCs. Table 1.2 summarizes some of the recent advancements in ST-PSCs.

1.3.6 PSCs based on TCO top electrodes

TCOs, such as ITO [43], AZO (aluminum-doped zinc oxide), [47] IZO (indium zinc oxide), [55], and hydrogenated indium oxide (IO:H) [46], were examined as ST-OSC transparent electrodes. Because these electrodes are often made by sputtering, inorganic buffer layers (such as MoO$_3$ or MoO$_x$) are necessary to protect the underneath perovskite layer from damage [56]. It is noted that the majority of the research is oriented toward the production of NIR-transparent semitransparent-PSCs for applicability in tandem solar cells. Heo et al. (2015) reported the construction of planar ST-PSCs with the device configuration FTO/TiO$_2$/MAPbI$_3$/P3HT (or PTAA)/PEDOT:PSS/ITO, wherein the ITO electrode was inserted by a lamination approach using PEDOT:PSS as an interfacial layer [106–112]. While P3HT was employed as the HTM, these devices had a PCE of 12.8% on average. However, due to P3HT absorption, the spectral transparency of these devices was low. In comparison, devices using PTAA as the HTM demonstrated improved device transparency (AVT = 6.3–17.3%) and PCEs of 15.8–12.55%. Furthermore, devices without encapsulation demonstrated outstanding stability in humid air (>20 days) [43]. Löper et al. described the use of sputtered ITO-based top electrodes in ST-PSCs having a standard device construction. MoO$_x$ was covered by thermal evaporation before ITO deposition to protect the perovskite active layer and also served as a buffer layer (hole-collecting) in the devices. The ST-PSCs, on the other hand, had a PCE of merely 6.2% [44]. Bush et al. recently created inverted devices with a significantly higher PCE of 12.3% employing a solution-processed AZO layer beneath the ITO electrodes and the identical sputtered ITO top electrodes. The top ITO electrodes also significantly increased the devices' ambient and thermal stability [45]. IO:H is also a good TCO material that may be utilized to make transparent electrodes in semitransparent photovoltaic cells. Fu et al. created ST-PSCs featuring a planar arrangement of FTO/ZnO/PC$_{61}$BM/ MAPbI$_3$/MoO$_3$/IO:H, which demonstrated PCEs of 14.1% and 9.5% under FTO and IO:H illumination, respectively [46]. Furthermore, IO:H/ITO bilayers were employed in ST-PSCs with enhanced PCEs (16.3%), most likely due to higher electrode conductivity [57]. Wener et al. (2015) demonstrated the construction of ST-PSCs using sputtered IZO electrodes on a typical mesoporous structure. When the IZO was applied directly on Spiro-MeOTAD, the PCE was 9.7%, whereas when a thin MoOx-based buffer layer was added beneath the IZO electrodes, the PCE increased to 10.36% [55]. Kranz et al. have recently shown a promising PCE of 12.1% in mesoporous ST-PSCs having sputtered AZO transparent electrodes. As previously stated, AZO can be employed as a buffer layer (electron-selective) prior to the deposition of other TCO electrodes [47].

1.3.7 Tandem solar cells using ST-PSCs

Because of the wider bandgaps of perovskite materials, semitransparent perovskite solar cells may be utilized for building tandem-based photovoltaics that have narrow bandgap CIGS solar cells and crystalline silicon (c-Si), providing an easy and cheaper solution for improving the performances of solar cells [58]. PCEs of over 30% have been anticipated for perovskite/c-Si tandem photovoltaic cells, which is significantly better than the existing record efficiency of c-Si solar cells (25.6%) [44]. Excellent NIR transparency when mounting ST- PSCs on the top of the inorganic photovoltaic cells is required in the manufacture of effective tandem solar cells. The tandem device can be built in a four-terminal manner by the mechanical mounting of the top (ST-PSCs) with the bottom (c-Si/CIGS) cells, or in a two- terminal monolithic manner [57, 59]. A four-terminal device is simple to build and test in a laboratory. The highest recorded conversion efficiency for four-terminal perovskite/c-Si tandem photovoltaic cells is 25.2% and 20.5%, respectively [46, 57, 58]. A two-terminal mono-lith device necessitates a more complex fabrication process, but it is better suited for commercial products and applications. Werner et al. (2015) pre-sented a monolithic perovskite/c-Si tandem solar cell with the configuration Ag/ITO/c-Si/IZO/PCBM/PEIE/MAPbI$_3$/Spiro-MeOTAD/MoO$_x$/IO:H/ITO, where IZO was the intermediate layer and IO:H/ITO was a trans-parent electrode. PCEs of 21.2% and 19.2% were found in devices hav-ing active regions of 0.17 and 1.22 cm^2, respectively [60]. They increased the PCE to 20.5% for a bigger device with an area of 1.43 cm^2 by opti-mizing the manufacture of perovskite top cells. Todorov et al. described a 10.98% PCE monolith tandem photovoltaic cell based on a CIGS cell with the structure glass/Si$_3$N$_4$/Mo/CIGS/CdS/ITO/PEDOT:PSS/MAPbI$_3$/PCBM/BCP/Ca. Although the device's efficiency is rather low, its efficiency is predicted to increase by optimizing perovskite top cells' transparency and utilizing a higher-performing CIGS cell [59].

1.3.8 Self-healing lead-free perovskites

For solar applications, FASnI$_3$ (formamidinium tin triiodide) is the primary contender in the lead-free materials category, providing reasonably low exci-ton bind energy and band gap, and great carrier mobility. Combined with large-volume amines (LVAs), 3-phenyl-2-propane-1-amine (PPA) additives partially substitute FA from FASnI$_3$ and generate a precursor solution with the formula PPA$_x$FA$_{1-x}$SnI$_3$, which can be used to synthesize 3D perovskite grains [61]. After heating at 100 °C for 100 minutes, the PCE of this 15% perovskite-based device decreased to 60% of its initial value and increased to 80% after 24 hours in an N$_2$ environment (Figure 1.11) The self-healing cycle may be replicated up to three times, and this was also observed in the air-N$_2$ environment [113–130].

The phenomenon of self-healing wasn't detected in devices with 0% PPA [131–145]. This was explained using the steric hindrance idea of LVAs. To validate these devices, two more LVAs, phenylethylammonium (PEA) and butylammonium (BA) [146–157], were used as additives, and the self-healing effect was demonstrated in both heat-N$_2$ and air-N$_2$ cycles. PPA>PEA>BA was the sequence of self-healing of LVAs, which corresponds to the molecular volume order of PPA>PEA>BA. Negative photoconduc-tivity occurs when Cs$_3$Bi$_2$Cl$_9$ perovskite single crystals are illuminated and display conductivity that is lower than that in the dark (NPC) [158–167]. After 15 minutes of light, these crystals indicate a drop in current, and

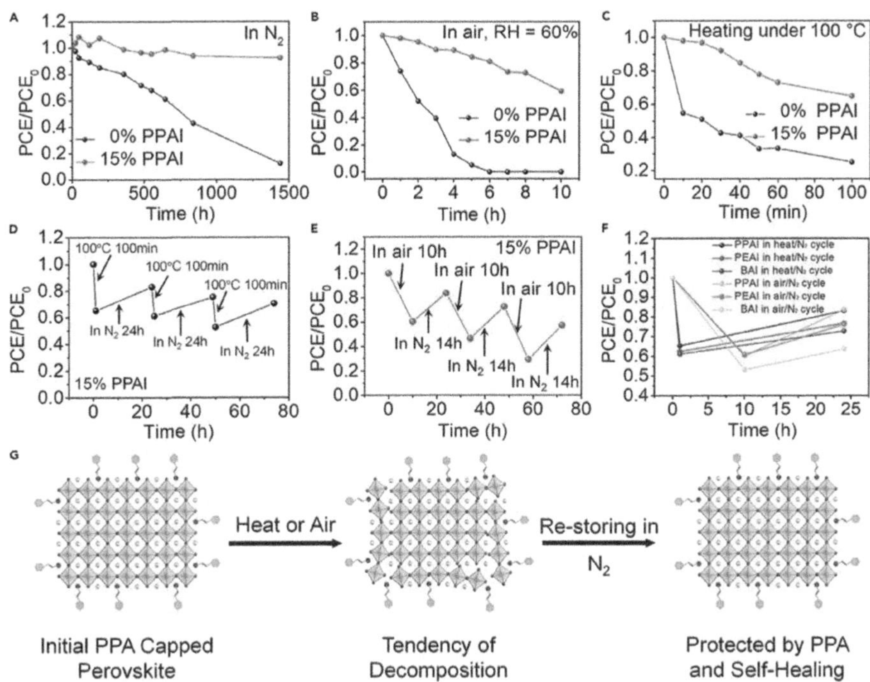

Figure 1.11 FASnI3 PPA cap reliability and self-repair capabilities demonstration. Figures show the normalized PCE of FASnI3 PSCs with and without PPAI additions under the conditions of (A) nitrogen storage, (B) air storage, (C), and (D) heating at 100 °C (A-E). The ability of the 15% PPAI device to self-heal during heating-nitrogen cycles (D) and air-nitrogen cycles (E). (F) PCE device variations are affected by PEA, PPA, and BA throughout several testing cycles. (G) Schematic of a putative self-healing mechanism in PPA-capped FASnI3 [75].

when left in dark conditions, the current progressively recovers to its previous amount. By trapping charge carriers that were progressively released in the dark, the creation of a metastable state produces NPC and dark self-recovery [62].

1.4 CONCLUSION AND OUTLOOK

In conclusion, PSCs and OSCs are examples of new PV technologies that have piqued the interest of researchers in the last 20 years. When compared to traditional inorganic silicon along with thin-film solar panels, one distinct benefit of developing solar cells is their potential for semitransparency, which enables numerous unique uses. In order to fabricate ST photovoltaic cells, two transparent electrodes on each side are required, as opposed to standard opaque sun cells. Although thin-film metal electrodes (ST-PSCs, Pt for

ST-DSCs, and Au or Ag for ST-OSCs) were first employed in device manufacturing, they are not good options for high-performance devices. The trade-off between metal electrode transmittance and conductivity and their high reflectance is harmful to device performance. Although the manufacturing procedures of these materials have yet to be refined, encouraging outcomes have been observed in ST devices using different transparent electrodes composed of AgNWs, graphene, CNTs, TCOs, and conductive polymers. In reality, the electrical conductivity of graphene, carbon nanotubes, and polymers with conductive properties is still insufficient for such applications. The majority of transparent electrodes can be easily built into ST-OSCs and ST-PSCs by laminating and coating processes, paving the path for large-area device mass manufacturing. Along with transparency and conductivity, adequate catalytic activity on the surfaces of ST-DSC CEs is required. As a result, the CEs may be changed with increased catalytic activity nanomaterials without losing transmittance and conductivity. Furthermore, PANI, PPy, and a few metal sulfides as well as metal selenide alloys have been investigated as suitable transparent CEs for ST-DSCs. ST solar cells must have crucial qualities such as PCE, transparency, and color. Furthermore, human perceptions of color and transparency should be considered in the construction and characterization of the devices, which have been reported infrequently for ST-DSCs. Window applications are often thought to benefit from neutral-color ST photovoltaic cells with good CRIs. Colorful ST devices, on the other hand, may be desirable in some circumstances because of their aesthetic functions. However, color customization of the active layer by composition and bandgap engineering may occasionally result in a reduction in device efficiency. As a result, optical approaches for tuning the color of ST solar cells have been proposed, such as employing dielectric mirrors or MDM electrodes (e.g., Ag/ITO/Ag). Although significant progress has been achieved in constructing efficient ST photovoltaic cells based on PSCs, DSCs, and OSCs, as discussed above, the devices' performance still lags well behind that of state-of-the-art opaque ones. The transparency and efficiency of ST solar cells are projected to be increased further by combining innovative materials for active layers and perfect electrodes that are transparent in the devices. Particular emphasis should be directed to innovative ST solar cell device architectures and fabrication processes, particularly light management strategies that can improve light absorption and control the color of the devices. As a result, for future work, we must optimize device performance and manufacturing procedures in the following areas. First, very transparent and electrically conductive substances for OPV/PSC top electrodes and DSC counter electrodes are urgently required. Although transparent electrodes composed of ultrathin metals (Pt, Ag, Au) and TCOs are utilized effectively in devices resulting in acceptable performance, the metal films' transparency and conductance cannot meet the requirements of the devices at the same time, and TCOs cannot be used as top electrodes. Metal electrode instability is also a disadvantage for some devices, particularly PSCs. As a result, we must design electrodes with transparency based on nanomaterials such as graphene, AgNWs, alternative carbon-based materials, or a combination of these. AgNW electrodes have already demonstrated a sheet resistance of less than 15 Ω sq^{-1} and transmittance of more than 90%, both of which are sufficient for high-performance solar cells. We still need to optimize the size as well as the aspect ratio of AgNWs, address device fabrication manufacturing conditions, and realize large-area uniform films having good conductivity and transparency. Suitable interfacial layers should be sought between the electrodes and the active layers in order to improve electron or hole transport from the active layers toward the electrodes. Because AgNWs are volatile in liquid electrolytes, they can only be employed in solid electrolytes for DSCs. Carbon-based electrodes, including conductive polymers, carbon nanotubes, and graphene, tend to be more stable and less expensive than AgNWs and can be readily coated using solution methods. The stated conductivity as well as transparency of carbon-based materials, however, are insufficient for high-efficiency ST photovoltaic cells. Given graphene's extremely high carrier mobilities, increasing carrier density by efficient doping procedures is a viable strategy for enhancing graphene conductivity. Future research should focus on various doping approaches such as modifying the surface of metallic nanoparticles, organic compounds, and acids. Second, device performance will be improved by revolutionary device structure and material innovations. Functional materials and nanostructures can be included in specific layers of devices with varying device structures (e.g., normal or inverted structures for PSCs and OSCs). (1) Photonic nanostructures (e.g., plasmonic metal nanoparticles) may be introduced in the active layers or interfacial layers of the devices to optimize their optical characteristics and increase their light absorption. The devices' transparency and color can be modified by modifying the nanostructures within the devices. (2) More appropriate OSC/PSC active materials and DSC dyes for ST photovoltaic cells should be developed. Material design is primarily concerned with the electrical and optical characteristics of materials. The color and transparency of devices may also be changed during device manufacture by modifying the active ingredients and compositions (e.g., the ratio of donor and acceptor in OPVs). The manufacture of neutral-color devices, which are likely to find applications in daylight harvesting and building integrated photovoltaics, should receive special attention. Third, the long-term stability of solar cells is required for applications in the real world, which should be given significantly more attention in future studies. When utilized as top electrodes, graphene and other materials made from carbon have demonstrated the potential to enhance device stability. Because graphene is impervious to gas and water, it is an excellent candidate for not just electrodes with transparency but also solar cell encapsulation layers [51]. As a result, carbon-derived transparent electrodes should be explored further for very stable devices. The reliability of both the interfacial and active layers of the devices, on the other hand, should be enhanced. A noteworthy example is the use of more reliable inorganic-based charge transport layers in PSCs, which result in significantly higher stability than organic layers. Fourth, large-area production of ST solar cells, which has received little attention to date, should be studied. We may improve on the fabrication procedures utilized in standard large-area OSCs, PSCs, and DSCs by concentrating on the utilization of roll-to-roll (R2R) or printing techniques on solution-processed ST photovoltaic cells. Inkjet printing, spray coating, gravure printing, doctor blading, and R2R processing are among the manufacturing processes available for large-area devices. R2R is a continuous, low-cost, solution-based manufacturing approach that is exclusively applicable to flexible devices. As a result, the R2R technique cannot be used to prepare devices on rigid substrates. Spray coating, as well as doctor blading, are simple methods for mass-producing multilayer devices on flexible and stiff substrates. Because the quality of each device layer is crucial to device performance, production conditions, such as solvent selection, precursor viscosity and temperature, and substrate surface qualities, should be carefully optimized at each stage. Because of their distinct features, developing ST solar cells would supplement the market of Si-based PVs, as they have applications in a wide range of niche markets. With the advancement of manufacturing processes, we will soon be able to utilize colorful ST devices combined with buildings, clothing, and even automobiles, which will not only produce power but also have eye-catching aesthetics.

Numerous research papers on the overall stability of QDSSCs have been published, but little is known regarding the chemical and physical mechanisms that induce cell deterioration or their link to the amounts of stress that produce the damage. In general, publications on stability assess the time-varying current density-voltage (J-V) properties of devices at room temperature or in an inert environment. To achieve significant progress in QDSSCs, reasonable strategies for increasing cell device J_{SC}, Voc, and FF values will be required. Some papers examine the influence of stability on chemical or spectroscopic approaches, providing experimental proof of chemical or molecular changes, although the majority just measure chemical parameters. Furthermore, most studies have published only positive stability data, whereas negative outcomes are rarely provided in order to hasten publication. As a result, all deterioration mechanisms and processes must be examined and explored in order to increase the overall stability of QDSSCs. The most important aspect of expanding the QDSSC experiment is providing chemical and physical information for QDSSC procedures to examine the tested material.

REFERENCES

1. F. Gonella and S. Spagnolo, "On sustainable PV–solar exploitation: An emergy analysis", in *Solar Cells and LightManagement*, Elsevier, 2020, pp. 481–507, doi: 10.1016/B978-0-08-102762-2.00014-8.

2. Howard T. Odum, *Environmental Accounting: Emergy and Environmental Decision Making.* Jhon Wiley Sons, Inc., 1996.
3. "World Economic Situation and Prospects", United Nations Report Online. Available: https://www.un.org/development/desa/dpad/wp-content/uploads/sites/45/publication/WESP2018_Full_Web
4. "Global Energy & CO2 Status Report", 2017. Online. . Available: https://www.iea.org/geco/
5. Global Energy Statistical Yearbook. 2018. Online. . Available: https://yearbook.enerdata.net/electricity/ electricity-domestic-consumption-data.html
6. "World Energy Outlook." Online. . Available: https://www.iea.org/weo2018/
7. "NASA, Global Climate Change. Carbondioxide." Online. . Available: https://climate.nasa.gov/vital-signs/carbon-dioxid
8. "UNFCCC 25th Anniversary - Climate Action is More Urgent than Ever." Online. . Available: https://unfccc.int
9. A. Einstein, "Über einen die Erzeugung und Verwandlung des Lichtes betreffenden heuristischen Gesichtspunkt", *Annals of Physics*, vol. 322, no. 6, pp. 132–148, 1905, doi: 10.1002/andp.19053220607.
10. R. A. Millikan, "A direct photoelectric determination of Planck's "h"," *Physical Review*, vol. 7, no. 3, pp. 355–388, Mar 1916, doi: 10.1103/PhysRev.7.355.
11. "Wikimedia Commons, Price History of Silicon PV Cells since 1977, 2015." Online. . Available: https://commons.wikimedia.org/wiki/File:Price_history_of_silicon_PV_cells_since_1977.svg
12. Y. Liang, Xu Zheng, Xia Jiangbin, Tsai Szu-Ting, Wu Yue, Li Gang, Ray Claire, Yu Luping, "For the bright future-bulk heterojunction polymer solar cells with power conversion efficiency of 7.4%", *Advanced Materials*, vol. 22, no. 20, pp. E135–E138, May 2010, doi: 10.1002/adma.200903528.
13. Wenchao Zhao, Deping Qian, Shaoqing Zhang, Sunsun Li, Olle Inganäs, Feng Gao, Jianhui Hou, "Fullerene-Free Polymer Solar Cells with over 11% Efficiency and Excellent Thermal Stability", *Advanced Materials*, vol. 28, no. 23, pp. 4734–4739, Jun 2016, doi: 10.1002/adma.201600281.
14. K. A. Mazzio, C. K. Luscombe, "The future of organic photovoltaics", *Chemical Society Reviews*, vol. 44, no. 1, pp. 78–90, 2015, doi: 10.1039/C4CS00227J.
15. M. C. Scharber, N. S. Sariciftci, "Efficiency of bulk-heterojunction organic solar cells", *Progress in Polymer Science*, vol. 38, no. 12, pp. 1929–1940, Dec 2013, doi: 10.1016/j.progpolymsci.2013.05.001.
16. Jingbi You, Letian Dou, Ken Yoshimura, Takehito Kato, Kenichiro Ohya, Tom Moriarty, Keith Emery, Chun-Chao Chen, Jing Gao, Gang Li, Yang Yang, "A polymer tandem solar cell with 10.6% power conversion efficiency", *Nature Communications*, vol. 4, no. 1, p. 1446, Feb 2013, doi: 10.1038/ncomms2411.
17. Chih-Yu Chang, Lijian Zuo, Hin-Lap Yip, Yongxi Li, Chang-Zhi Li, Chain-Shu Hsu, Yen-Ju Cheng, Hongzheng Chen, Alex K.-Y. Jen, "A versatile fluoro-containing low-bandgap polymer for efficient semitransparent and tandem polymer solar cells", *Advanced Functional Materials*, vol. 23, no. 40, pp. 5084–5090, Oct 2013, doi: 10.1002/adfm201301557.
18. K.-S. Chen, J.-F. Salinas, H.-L. Yip, L. Huo, J. Hou, A. K.-Y. Jen, "Semitransparent polymer solar cells with 6% PCE, 25% average visible

transmittance and a color rendering index close to 100 for power generating window applications", *Energy & Environmental Science*, vol. 5, no. 11, p. 9551, 2012, doi: 10.1039/c2ee22623e.

19. X. Ren, X. Li, W. C. H. Choy, "Optically enhanced semi-transparent organic solar cells through hybrid metal/nanoparticle/dielectric nanostructure", *Nano Energy*, vol. 17, pp. 187–195, Oct 2015, doi: 10.1016/j.nanoen.2015.08.014.

20. Sebastian Wilken, Verena Wilkens, Dorothea Scheunemann, Regina-Elisabeth Nowak, Karsten von Maydell, Jürgen Parisi, Holger Borchert, "Semitransparent polymer-based solar cells with aluminum-doped zinc oxide electrodes", *ACS Applied Materials & Interfaces*, vol. 7, no. 1, pp. 287–300, Jan 2015, doi: 10.1021/am5061917.

21. G. Li, C.-W. Chu, V. Shrotriya, J. Huang, Y. Yang, "Efficient inverted polymer solar cells", *Applied Physics Letters*, vol. 88, no. 25, p. 253503, Jun 2006, doi: 10.1063/1.2212270.

22. A. Bauer, T. Wahl, J. Hanisch, E. Ahlswede, "ZnO:Al cathode for highly efficient, semitransparent 4% organic solar cells utilizing TiO x and aluminum interlayers", *Applied Physics Letters*, vol. 100, no. 7, p. 073307, Feb 2012, doi: 10.1063/1.3685718.

23. J. Huang, G. Li, Y. Yang, "A semi-transparent plastic solar cell fabricated by a lamination process", *Advanced Materials*, vol. 20, no. 3, pp. 415–419, Feb 2008, doi: 10.1002/adma.200701101.

24. Jie Min, Carina Bronnbauer, Zhi-Guo Zhang, Chaohua Cui, Yuriy N. Luponosov, Ibrahim Ata, Peter Schweizer, Thomas Przybilla, Fei Guo, Tayebeh Ameri, Karen Forberich, Erdmann Spiecker, Peter Bäuerle, Sergei A. Ponomarenko, Yongfang Li, Christoph J. Brabec, "Fully solution-processed small molecule semitransparent solar cells: Optimization of transparent cathode architecture and four absorbing layers", *Advanced Functional Materials*, vol. 26, no. 25, pp. 4543–4550, Jul 2016, doi: 10.1002/adfm.201505411.

25. Abd. R. bin M. Yusoff, S. J. Lee, F. K. Shneider, W. J. da Silva, J. Jang, "High-performance semitransparent tandem solar cell of 8.02% conversion efficiency with solution-processed graphene mesh and laminated ag nanowire top electrodes", *Advanced Energy Materials*, vol. 4, no. 12, p. 1301989, Aug 2014, doi: 10.1002/aenm.201301989.

26. Yong Hyun Kim, Christoph Sachse, Alexander A. Zakhidov, Jan Meiss, Anvar A. Zakhidov, Lars Müller-Meskamp, Karl Leo, "Combined alternative electrodes for semi-transparent and ITO-free small molecule organic solar cells", *Organic Electronics*, vol. 13, no. 11, pp. 2422–2428, Nov 2012, doi: 10.1016/j.orgel.2012.06.034.

27. Jong Hyuk Yim, Sung-yoon Joe, Christina Pang, Kyung Moon Lee, Huiseong Jeong, Ji-Yong Park, Yeong Hwan Ahn, John C. de Mello, Soonil Lee, "Fully solution-processed semitransparent organic solar cells with a silver nanowire cathode and a conducting polymer anode", *ACS Nano*, vol. 8, no. 3, pp. 2857–2863, Mar 2014, doi: 10.1021/nn406672n.

28. Y. Zhou, H. Cheun, S. Choi, W. J. Potscavage, C. Fuentes-Hernandez, B. Kippelen, "Indium tin oxide-free and metal-free semitransparent organic solar cells", *Applied Physics Letters*, vol. 97, no. 15, p. 153304, Oct 2010, doi: 10.1063/1.3499299.

29. Z. Liu, P. You, S. Liu, F. Yan, "Neutral-color semitransparent organic solar cells with all- graphene electrodes", *ACS Nano*, vol. 9, no. 12, pp. 12026–12034, Dec 2015, doi: 10.1021/acsnano.5b04858.

30. Yu-Ying Lee, Kun-Hua Tu, Chen-Chieh Yu, Shao-Sian Li, Jeong-Yuan Hwang, Chih-Cheng Lin, Kuei- Hsien Chen, Li-Chyong Chen, Hsuen-Li Chen, Chun-Wei Chen, "Top laminated graphene electrode in a semitransparent polymer solar cell by simultaneous thermal annealing/releasing method", *ACS Nano*, vol. 5, no. 8, pp. 6564–6570, Aug 2011, doi: 10.1021/nn201940j.

31. Z. Liu, J. Li, Z.-H. Sun, G. Tai, S.-P. Lau, F. Yan, "The application of highly doped single-layer graphene as the top electrodes of semitransparent organic solar cells", *ACS Nano*, vol. 6, no. 1, pp. 810–818, Jan 2012, doi: 10.1021/nn204675r.

32. D. S. Hecht, L. Hu, G. Irvin, "Emerging transparent electrodes based on thin films of carbon nanotubes, graphene, metallic nanostructures", *Advanced Materials*, vol. 23, no. 13, pp. 1482–1513, Apr 2011, doi: 10.1002/adma.201003188.

33. C.-C. Chen, L. Dou, J. Gao, W.-H. Chang, G. Li, Y. Yang, "High-performance semi-transparent polymer solar cells possessing tandem structures", *Energy & Environmental Science*, vol. 6, no. 9, p. 2714, 2013, doi: 10.1039/c3ee40860d.

34. Fei Guo, Peter Kubis, Thomas Przybilla, Erdmann Spiecker, Andre Hollmann, Stefan Langner, Karen Forberich, Christoph J. Brabec, "Nanowire interconnects for printed large-area semitransparent organic photovoltaic modules", *Advanced Energy Materials*, vol. 5, no. 12, p. 1401779, Jun 2015, doi: 10.1002/aenm.201401779.

35. S. K. Hau, H.-L. Yip, J. Zou, A. K.-Y. Jen, "Indium tin oxide-free semi-transparent inverted polymer solar cells using conducting polymer as both bottom and top electrodes", *Organic Electronics*, vol. 10, no. 7, pp. 1401–1407, Nov 2009, doi: 10.1016/j.orgel.2009.06.019.

36. Fei Guo, Xiangdong Zhu, Karen Forberich, Johannes Krantz, Tobias Stubhan, Michael Salinas, Marcus Halik, Stefanie Spallek, Benjamin Butz, Erdmann Spiecker, Tayebeh Ameri, Ning Li, Peter Kubis, Dirk M. Guldi, Gebhard J. Matt, Christoph J. Brabec, "ITO-free and fully solution-processed semitransparent organic solar cells with high fill factors", *Advanced Energy Materials*, vol. 3, no. 8, pp. 1062–1067, Aug 2013, doi: 10.1002/aenm.201300100.

37. L. Meng, J. You, T.-F. Guo, Y. Yang, "Recent advances in the inverted planar structure of perovskite solar cells", *Accounts of Chemical Research*, vol. 49, no. 1, pp. 155–165, Jan 2016, doi: 10.1021/acs.accounts.5b00404.

38. Cristina Roldán-Carmona, Olga Malinkiewicz, Rafael Betancur, Giulia Longo, Cristina Momblona, Franklin Jaramillo, Luis Camacho, Henk J. Bolink, "High efficiency single-junction semitransparent perovskite solar cells", *Energy & Environmental Science*, vol. 7, no. 9, pp. 2968–2973, 2014, doi: 10.1039/C4EE01389A.

39. Enrico Della Gaspera, Yong Peng, Qicheng Hou, Leone Spiccia, Udo Bach, Jacek J. Jasieniak, Yi-Bing Cheng, "Ultra-thin high efficiency semitransparent perovskite solar cells", *Nano Energy*, vol. 13, pp. 249–257, Apr 2015, doi: 10.1016/j.nanoen.2015.02.028.

40. J. W. Jung, C.-C. Chueh, A. K.-Y. Jen, "High-performance semitransparent perovskite solar cells with 10% power conversion efficiency and 25% average visible transmittance based on transparent CuSCN as the hole-transporting material", *Advanced Energy Materials*, vol. 5, no. 17, p. 1500486, Sep 2015, doi: 10.1002/aenm.201500486.
41. C.-Y. Chang, Y.-C. Chang, W.-K. Huang, K.-T. Lee, A.-C. Cho, C.-C. Hsu, "Enhanced performance and stability of semitransparent perovskite solar cells using solution-processed thiol-functionalized cationic surfactant as cathode buffer layer", *Chemistry of Materials*, vol. 27, no. 20, pp. 7119–7127, Oct 2015, doi: 10.1021/acs.chemmater.5b03137.
42. H. S. Jung and N.-G. Park, "Solar cells: Perovskite solar cells: From materials to devices (small 1/2015)", *Small*, vol. 11, no. 1, pp. 2–2, Jan 2015, doi: 10.1002/smll.201570002.
43. J. H. Heo, H. J. Han, M. Lee, M. Song, D. H. Kim, S. H. Im, "Stable semi-transparent CH3NH3PbI3 planar sandwich solar cells", *Energy & Environmental Science*, vol. 8, no. 10, pp. 2922–2927, 2015, doi: 10.1039/C5EE01050K.
44. Philipp Löper, Soo-Jin Moon, Sílvia Martín de Nicolas, Bjoern Niesen, Martin Ledinsky, Sylvain Nicolay, Julien Bailat, Jun-Ho Yum, Stefaan De Wolfa, Christophe Ballif, "Organic–inorganic halide perovskite/crystalline silicon four-terminal tandem solar cells", *Physical Chemistry Chemical Physics*, vol. 17, no. 3, pp. 1619–1629, 2015, doi: 10.1039/C4CP03788J.
45. Kevin A. Bush, Colin D. Bailie, Ye Chen, Andrea R. Bowring, Wei Wang, Wen Ma, Tomas Leijtens, Farhad Moghadam, Michael D. McGehee, "Thermal and environmental stability of semi- transparent perovskite solar cells for tandems enabled by a solution-processed nanoparticle buffer layer and sputtered ITO electrode", *Advanced Materials*, vol. 28, no. 20, pp. 3937–3943, May 2016, doi: 10.1002/adma.201505279.
46. Fan Fu, Thomas Feurer, Timo Jäger, Enrico Avancini, Benjamin Bissig, Songhak Yoon, Stephan Buecheler, Ayodhya N. Tiwari, "Low-temperature-processed efficient semi-transparent planar perovskite solar cells for bifacial and tandem applications", *Nature Communications*, vol. 6, no. 1, p. 8932, Nov 2015, doi: 10.1038/ncomms9932.
47. Lukas Kranz, Antonio Abate, Thomas Feurer, Fan Fu, Enrico Avancini, Johannes Löckinger, Patrick Reinhard, Shaik M. Zakeeruddin, Michael Grätzel, Stephan Buecheler, Ayodhya N. Tiwari, "High-efficiency polycrystalline thin film tandem solar cells", *The Journal of Physical Chemistry Letters*, vol. 6, no. 14, pp. 2676–2681, Jul 2015, doi: 10.1021/acs.jpclett.5b01108.
48. C.-Y. Chang, K.-T. Lee, W.-K. Huang, H.-Y. Siao, Y.-C. Chang, "High-performance, air-stable, low-temperature processed semitransparent perovskite solar cells enabled by atomic layer deposition", *Chemistry of Materials*, vol. 27, no. 14, pp. 5122–5130, Jul 2015, doi: 10.1021/acs.chemmater.5b01933.
49. Xuezeng Dai, Ye Zhang, Heping Shen, Qiang Luo, Xingyue Zhao, Jianbao Li, Hong Lin, "Working from both sides: Composite metallic semitransparent top electrode for high performance perovskite solar cells", *ACS Applied Materials & Interfaces*, vol. 8, no. 7, pp. 4523–4531, Feb 2016, doi: 10.1021/acsami.5b10830.

50. Zhen Li, Sneha A. Kulkarni, Pablo P. Boix, Enzheng Shi, Anyuan Cao, Kunwu Fu, Sudip K. Batabyal, Jun Zhang, Qihua Xiong, Lydia Helena Wong, Nripan Mathews, Subodh G. Mhaisalkar, "Laminated carbon nanotube networks for metal electrode-free efficient perovskite solar cells", *ACS Nano*, vol. 8, no. 7, pp. 6797–6804, Jul 2014, doi: 10.1021/nn501096h.

51. P. You, Z. Liu, Q. Tai, S. Liu, F. Yan, "Efficient semitransparent perovskite solar cells with graphene electrodes", *Advanced Materials*, vol. 27, no. 24, pp. 3632–3638, Jun 2015, doi: 10.1002/adma.201501145.

52. Lingling Bu, Zonghao Liu, Meng Zhang, Wenhui Li, Aili Zhu, Fensha Cai, Zhixin Zhao, Yinhua Zhou, "Semitransparent fully air processed perovskite solar cells", *ACS Applied Materials & Interfaces*, vol. 7, no. 32, pp. 17776–17781, Aug 2015, doi: 10.1021/acsami.5b04040.

53. M. A. Green, A. Ho-Baillie, H. J. Snaith, "The emergence of perovskite solar cells", *Nature Photonics*, vol. 8, no. 7, pp. 506–514, Jul 2014, doi: 10.1038/nphoton.2014.134.

54. Anyi Mei, Xiong Li, Linfeng Liu, Zhiliang Ku, Tongfa Liu, Yaoguang Rong, Mi Xu, Min Hu, Jiangzhao Chen, Ying Yang, Michael Grätzel, Hongwei Han, "A hole-conductor–free, fully printable mesoscopic perovskite solar cell with high stability", *Science*, vol. 345, no. 6194, pp. 295–298, Jul 2014, doi: 10.1126/science.1254763.

55. Jérémie Werner, Guy Dubuis, Arnaud Walter, Philipp Löper, Soo-Jin Moon, Sylvain Nicolay, Monica Morales-Masis, Stefaan De Wolf, Bjoern Niesen, Christophe Ballif, "Sputtered rear electrode with broadband transparency for perovskite solar cells", *Solar Energy Materials and Solar Cells*, vol. 141, pp. 407–413, Oct 2015, doi: 10.1016/j.solmat.2015.06.024.

56. Hiroyuki Kanda, Abdullah Uzum, Ajay K. Baranwal, T. A. Nirmal Peiris, Tomokazu Umeyama, Hiroshi Imahori, Hiroshi Segawa, Tsutomu Miyasaka, Seigo Ito, "Analysis of sputtering damage on I –V curves for perovskite solar cells and simulation with reversed diode model", *The Journal of Physical Chemistry C*, vol. 120, no. 50, pp. 28441–28447, Dec 2016, doi: 10.1021/acs .jpcc.6b09219.

57. Jérémie Werner, Loris Barraud, Arnaud Walter, Matthias Bräuninger, Florent Sahli, Davide Sacchetto, Nicolas Tétreault, Bertrand Paviet-Salomon, Soo-Jin Moon, Christophe Allebé, Matthieu Despeisse, Sylvain Nicolay, Stefaan De Wolf, Bjoern Niesen, Christophe Ballif, "Efficient Near-Infrared- Transparent Perovskite Solar Cells Enabling Direct Comparison of 4-Terminal and Monolithic Perovskite/Silicon Tandem Cells", *ACS Energy Letters*, vol. 1, no. 2, pp. 474–480, Aug 2016, doi: 10.1021/acsenergylett.6b00254.

58. David P. Mcmeekin, Golnaz Sadoughi, Waqaas Rehman, Giles E. Eperon, Michael Saliba, Maximilian T. Hörantner, Amir Haghighirad, Nobuya Sakai, Lars Korte, Bernd Rech, Michael B. Johnston, Laura M. Herz, Henry J. Snaith, "A mixed-cation lead mixed-halide perovskite absorber for tandem solar cells", *Science*, vol. 351, no. 6269, pp. 151–155, Jan 2016, doi: 10.1126/science.aad5845.

59. Teodor Todorov, Talia Gershon, Oki Gunawan, Yun Seog Lee, Charles Sturdevant, Liang-Yi Chang, Supratik Guha, "Monolithic perovskite-CIGS tandem solar cells via in situ band gap engineering", *Advanced*

Energy Materials, vol. 5, no. 23, p. 1500799, Dec 2015, doi: 10.1002/aenm.201500799.

60. Jérémie Werner, Ching-Hsun Weng, Arnaud Walter, Luc Fesquet, Johannes Peter Seif, Stefaan De Wolf, Bjoern Niesen, Christophe Ballif, "Efficient monolithic perovskite/silicon tandem solar cell with cell area >1 cm 2", *The Journal of Physical Chemistry Letters*, vol. 7, no. 1, pp. 161–166, Jan 2016, doi: 10.1021/acs.jpclett.5b02686.

61. Chenxin Ran, Weiyin Gao, Jingrui Li, Jun Xi, Lu Li, Jinfei Dai, Yingguo Yang, Xingyu Gao, Hua Dong, Bo Jiao, Ioannis Spanopoulos, Christos D. Malliakas, Xun Hou, Mercouri G. Kanatzidis, Zhaoxin Wu, "Conjugated organic cations enable efficient self-healing FASnI3 solar cells", *Joule*, vol. 3, no. 12, pp. 3072–3087, Dec 2019, doi: 10.1016/j.joule.2019.08.023.

62. N. K. Tailor, P. Maity, M. I. Saidaminov, N. Pradhan, S. Satapathi, "Dark self-healing-mediated negative photoconductivity of a lead-free Cs 3 Bi 2 Cl 9 perovskite single crystal", *The Journal of Physical Chemistry Letters*, vol. 12, no. 9, pp. 2286–2292, Mar 2021, doi: 10.1021/acs.jpclett.1c00057.

63. Priya Bansal, B. Raj, "Memristor modeling and analysis for linear dopant drift kinetics", *Journal of Nanoengineering and Nanomanufacturing*, American Scientific Publishers, vol. 6, pp. 1–7, 2016.

64. Amandeep Singh, Mamta Khosla, B. Raj, "Circuit compatible model for electrostatic doped schottky barrier CNTFET", *Journal of Electronic Materials*, Springer, vol. 45, no. 12, pp. 4825–4835, 2016.

65. Ashima, D Vaithiyanathan, B. Raj", Performance analysis of charge plasma induced graded channel si nanotube", *Journal of Engineering Research (JER)*, EMSME Special Issue pp. 146–154, Aug 2021.

66. Abhishek Singh Tomar, Vijay Kumar Magraiya, B. Raj, "Scaling of Access and Data Transistor for High Performance DRAM Cell Design",*Quantum Matter*, vol. 2, pp. 412–416, Oct 2013.

67. Pawandeep Kaur, Avtar Singh Buttar, B. Raj, "A comprehensive analysis of nanoscale transistor based biosensor: A review", *Indian Journal of Pure and Applied Physics*, vol. 59, pp. 304–318, Apr 2021.

68. Divya Yadav, Balwant Raj, B. Raj, "Design and simulation of low power microcontroller for IoT applications", *Journal of Sensor Letters*, ASP, vol. 18, pp. 401–409, May 2020.

69. Shailendra Singh, B. Raj, "A 2-D analytical surface potential and drain current modeling of double-gate vertical t-shaped tunnel FET", *Journal of Computational Electronics*, Springer, vol. 19, pp. 1154–1163, Apr 2020.

70. Jeetendra Singh, B. Raj, "An accurate and generic window function for nonlinear memristor model", *Journal of Computational Electronics*, Springer, vol. 18, no. 2, pp. 640–647, Jun 2019.

71. Manjit Kaur, Neena Gupta, Sanjeev Kumar, B. Raj, Arun Kumar Singh, "Comparative RF and crosstalk analysis of carbon based nano interconnects", *IET Circuits, Devices & Systems*, vol. 15(6), pp. 493–503, Feb 2021.

72. Nehru Kandasamy, Firdous Ahmad, D. Ajitha, B. Raj, Nagarjuna Telagam "Quantum dot cellular automata based scan flip flop and boundary scan register", *IETE Journal of Research*, vol. 66, pp. 535–548, 2020.

73. S. K. Sharma, B. Raj, M. Khosla, "Enhanced photosensivity of highly spectrum selective cylindrical gate In1-xGaxAs nanowire MOSFET photodetector", *Modern Physics Letter-B*, vol. 33, no. 12, p. 1950144, 2019.

74. Jeetendra Singh, B. Raj, "Design and investigation of 7T2M NVSARM with enhanced stability and temperature impact on store/restore energy", *IEEE Transactions on Very Large Scale Integration Systems*, vol. 27, no. 6, pp. 1322–1328, Jun 2019.

75. Anil Kumar Bhardwaj, Sumeet Gupta, B. Raj, Amandeep Singh, "Impact of double gate geometry on the performance of carbon nanotube field effect transistor structures for low power digital design", *Computational and Theoretical Nanoscience*, ASP, vol. 16, pp. 1813–1820, 2019.

76. Neeraj Jain, B. Raj, "Thermal stability analysis and performance exploration of asymmetrical dual-k underlap spacer (ADKUS) SOI FinFET for security and privacy applications", *Indian Journal of Pure & Applied Physics (IJPAP)*, vol. 57, pp. 352–360, May 2019.

77. Amandeep Singh, Mamta Khosla, B. Raj, "Design and analysis of dynamically configurable electrostatic doped carbon nanotube tunnel FET", *Microelectronics Journal*, Elsevier, vol. 85, pp. 17–24, Mar 2019.

78. Neeraj Jain, Balwinder Raj, "Dual-k spacer region variation at the drain side of asymmetric SOI FinFET structure: Performance analysis towards the analog/RF design applications", *"Journal of Nanoelectronics and Optoelectronics"*, American Scientific Publishers, vol. 14, pp. 349–359, Mar 2019.

79. Jeetendra Singh, Sanjeev Sharma, B. Raj, Mamta Khosla, "Analysis of barrier layer thickness on performance of In1-xGaxAs based gate stack cylindrical gate nanowire MOSFET," *JNO*, ASP, vol. 13, pp. 1473–1477, Oct 2018.

80. Neeraj Jain, B. Raj, "Parasitic capacitance and resistance model development and optimization of raised source/drain SOI FinFET structure for analog circuit applications", *Journal of Nanoelectronics and Optoelectronins*, ASP, USA, vol. 13, pp. 531–539, Apr 2018.

81. Shradhya Singh, S. K. Vishvakarma, B. Raj, "Analytical modeling of split-gate junction-less transistor for a biosensor application", *Sensing and Biosensing*, Elsevier, vol. 18, pp. 31–36, Apr 2018.

82. Maisagalla Gopal, Balwinder Raj, "Low power 8T SRAM cell design for high stability video applications", *ITSI Transaction on Electrical and Electronics Engineering*, vol. 1, no. 5, pp. 91–97, 2013.

83. Balwinder Raj, Jatin Mitra, Deepak Kumar Bihani, Rangharajan V, A. K. Saxena, S. Dasgupta, "Analysis of noise margin, power and process variation for 32 nm FinFET based 6T SRAM cell", *Journal of Computer (JCP)*, Academy Publisher, Finland, vol. 5, no. 6, pp. 1–8, 2010.

84. Divya Sharma, Rajesh Mehra, B. Raj, "Comparative analysis of photovoltaic technologies for high efficiency solar cell design", *Superlattices and Microstructures*, Elsevier, vol. 153, pp. 106861, May 2021.

85. Neeraj Jain, B. Raj, "Analysis and performance exploration of high-k SOI FinFETs over the conventional low-k SOI FinFET toward analog/RF design", *Journal of Semiconductors (JoS)*, IOP Science, vol. 39, no. 12, p. 124002-1-7, Dec 2018.

86. Candy Goyal, Jagpal Singh Ubhi, B. Raj, "A reliable leakage reduction technique for approximate full adder with reduced ground bounce noise", *Journal of Mathematical Problems in Engineering*, Hindawi, vol. 2018, Article ID 3501041, p. 16, 15 Oct 2018.

87. Jeetendra Singh, B. Raj, Mamta Khosla, "Design and performance analysis of nano-scale memristor-based nonvolatile SRAM", *Journal of Sensor Letter"*, American Scientific Publishers, vol. 16, pp. 798–805, Oct 2018.

88. Girish Wadhwa, B. Raj, "Parametric variation analysis of charge-plasma-based dielectric modulated JLTFET for biosensor application" *IEEE Sensor Journal*, vol. 18, no. 15, pp. 6070–6077, 1 Aug 2018.

89. Jeetendra Singh, B. Raj, "Comparative analysis of memristor models for memories design", *JoS*, IoP, vol. 39, no. 7, pp. 074006-1-12, Jul 2018.

90. Divya Yadav, Shailesh Singh Chouhan, Santosh Kumar Vishvakarma, B. Raj, "Application specific microcontroller design for IoT based WSN", *Sensor Letter*, ASP, vol. 16, pp. 374–385, May 2018.

91. Gurmohan Singh, R. K. Sarin, B. Raj, "Fault-tolerant design and analysis of quantum-dot cellular automata based circuits", *IEEE/IET Circuits, Devices & Systems*, vol. 12, pp. 638–664, 2018.

92. Jeetendra Singh, B. Raj, "Modeling of mean barrier height levying various image forces of metal insulator metal structure to enhance the performance of conductive filament based memristor model", *IEEE Nanotechnology*, vol. 17, no. 2, pp. 268–267, Mar 2018.

93. Aakash Jain, Sanjeev Sharma, B. Raj, "Analysis of triple metal surrounding gate (TM-SG) III-V nanowire MOSFET for photosensing application", *Opto-Electronics Journal*, Elsevier, vol. 26, no. 2, pp. 141–148, May 2018.

94. Aakash Jain, Sanjeev Sharma, B. Raj, "Design and analysis of high sensitivity photosensor using cylindrical surrounding gate MOSFET for low power sensor applications", *Engineering Science and Technology, an International Journal*, Elsevier, vol. 19, no. 4, pp. 1864–1870, Dec 2016.

95. Amandeep Singh, Mamta Khosla, B. Raj, "Analysis of electrostatic doped schottky barrier carbon nanotube FET for low power applications", *Journal of Materials Science: Materials in Electronics*, Springer, vol. 28, pp. 1762–1768, 2017.

96. G. Saiphani Kumar, Amandeep Singh, B. Raj, "Design and analysis of gate all around CNTFET based SRAM cell design", *Journal of Computational Electronics*, Springer, vol. 17, no. 1, pp. 138–145, Mar 2018.

97. Gurinder pal Singh, B. S. Sohi, Balwinder Raj, "Material properties analysis of graphene base transistor (GBT) for VLSI analog circuits", *Indian Journal of Pure & Applied Physics (IJPAP)*, vol. 55, pp. 896–902, Dec 2017.

98. Amandeep Singh, Mamta Khosla, B. Raj, "Comparative analysis of carbon nanotube field effect transistor and nanowire transistor for low power circuit design", *Journal of Nanoelectronics and Optoelectronics*, American Scientific Publishers, USA, vol. 11, pp. 388–393, Jun 2016.

99. Sunil Kumar, B. Raj, "Estimation of stability and performance metric for inward access transistor based 6T SRAM cell design using n-type/p-type DMDG-GDOV TFET", *IEEE VLSI Circuits and Systems Letter*, vol. 3, no. 2, pp. 25–39, Jun 2017.

100. Shashikant Sharma, Anjan Kumar, Manisha Pattanaik, B. Raj, "Forward body biased multimode multi-threshold CMOS technique for ground bounce noise reduction in static CMOS adders", *International Journal of Information and Electronics Engineering*, vol. 3, no. 3, pp. 567–572, 2013.

101. Hamendra Singh, Pankaj Kumar, Balwinder Raj, "Performance analysis of majority gate SET based 1-bit full adder", *International Journal of Computer and Communication Engineering (IJCCE)*, IACSIT Press, Singapore, ISSN: 2010-3743, vol. 2, no. 4, 2013.

102. Anil Kumar Bhardwaj, Sumeet Gupta, B. Raj, "Investigation of parameters for schottky barrier (SB) height for schottky barrier based carbon nanotube field effect transistor device", *Journal of Nanoelectronics and Optoelectronics*, ASP, vol. 15, pp. 783–791, Jul 2020.

103. Priya Bansal, B. Raj, *"Memristor: A versatile nonlinear model for dopant drift and boundary issues"*, JCTN, American Scientific Publishers, vol. 14, no. 5, pp. 2319–2325, May 2017.

104. Neeraj Jain, B. Raj, "An analog and digital design perspective comprehensive approach on Fin-FET (Fin-Field Effect transistor) technology: A review", *Reviews in Advanced Sciences and Engineering (RASE)*, ASP, vol. 5, pp. 1–14, 2016.

105. Sanjeev Sharma, B. Raj, Mamta Khosla, "Subthreshold performance of In1-xGaxAs based dual metal with gate stack cylindrical/surrounding gate nanowire MOSFET for low power analog applications", *Journal of Nanoelectronics and Optoelectronics*, American Scientific Publishers, USA, vol. 12, pp. 171–176, 2017.

106. Balwinder Raj, A. K. Saxena, S. Dasgupta, "Analytical modeling for the estimation of leakage current and subthreshold swing factor of nanoscale double gate FinFET device", *Microelectronics International, UK*, vol. 26, pp. 53–63, 2009.

107. Shailendra Singh, Girish Wadhwa, Balwinder Raj, "An analytical modeling for dual source vertical tunnel field effect transistor", *International Journal of Recent Technology and Engineering (IJRTE)*, vol. 8, no. 2, pp. 603–608, Jul 2019.

108. Shailendra Singh, B. Raj, "Design and analysis of hetrojunction vertical T-shaped tunnel field effect transistor", *Journal of Electronics Material*, Springer, vol. 48, no. 10, pp. 6253–6260, Oct 2019.

109. Candy Goyal, Jagpal Singh Ubhi, B. Raj, "A low leakage CNTFET based inexact full adder for low power image processing applications", *International Journal of Circuit Theory and Applications*, Wiley, vol. 47, no. 9, pp. 1446–1458, Sep 2019.

110. B. Raj, A. K. Saxena, S. Dasgupta, "A compact drain current and threshold voltage quantum mechanical analytical modeling for FinFETs" *Journal of Nanoelectronics and Optoelectronics (JNO)*, USA, vol. 3, no. 2, pp. 163–170, 2008.

111. Girish Wadhwa, B. Raj, "An analytical modeling of charge plasma based tunnel field effect transistor with impacts of gate underlap region" *Superlattices and Microstructures*, Elsevier, vol. 142, p. 106512, Jun 2020.

112. Shailendra Singh, B. Raj, "Modeling and simulation analysis of SiGe hetrojunction double gatevertical t-shaped tunnel FET", *Superlattices and Microstructures*, Elsevier, vol. 142, p. 106496, Jun 2020.

113. Amandeep Singh, Dinesh Kumar Saini, Dinesh Agarwal, Sajal Aggarwal, Mamta Khosla, B. Raj, "Modeling and simulation of carbon nanotube field effect transistor and its circuit application", *Journal of Semiconductors (JoS)*, IOP Science, vol. 37, p. 074001-6, Jul 2016.

114. Neeraj Jain, Balwinder Raj, "Device and circuit co-design perspective comprehensive approach on FinFET technology: A review", *Journal of Electron Devices*, vol. 23, no. 1, pp. 1890–1901, 2016.

115. Sunil Kumar, B. Raj, "Analysis of ION and ambipolar current for dual-material gate-drain overlapped DG-TFET", *Journal of Nanoelectronics and Optoelectronics*, American Scientific Publishers, USA, vol. 11, pp. 323–333, Jun 2016.

116. Naveed Anjum, Tarun Bali, B. Raj, "Design and simulation of handwritten multiscript character recognition", *International Journal of Advanced Research in Computer and Communication Engineering*, vol. 2, no. 7, pp. 2544–2549, Jul 2013.

117. Sanjeev Sharma, B. Raj, Mamta Khosla, "A gaussian approach for analytical subthreshold current model of cylindrical nanowire FET with quantum mechanical effects",*Microelectronics Journal*, Elsevier, vol. 53, pp. 65–72, Apr 2016.

118. Karmjit Singh, B. Raj, "Performance and analysis of temperature dependent multi-walled carbon nanotubes as global interconnects at different technology nodes", *Journal of Computational Electronics*, Springer, vol. 14, no. 2, pp. 469–476, Jun 2015.

119. Sunil Kumar, B. Raj, "Compact channel potential analytical modeling of DG-TFET based on evanescent–mode approach", *Journal of Computational Electronics*, Springer, vol. 14, no. 2, pp. 820–827, Jul 2015.

120. Karmjit Singh, B. Raj, "Temperature dependent modeling and performance evaluation of multi-walled CNT and single-walled CNT as global interconnects", *Journal of Electronic Materials*, Springer, vol. 44, no. 12, pp. 4825–4835, Dec 2015.

121. V. K. Sharma, M. Pattanaik, B. Raj, "INDEP approach for leakage reduction in nanoscale CMOS circuits", *International Journal of Electronics*, Taylor & Francis, vol. 102, no. 2, pp. 200–215, 2014.

122. Karmjit Singh, B. Raj, "Influence of temperature on MWCNT bundle, SWCNT bundle and copper interconnects for nanoscaled technology nodes", *Journal of Materials Science: Materials in Electronics*, Springer, vol. 26, no. 8, pp. 6134–6142, 2015.

123. Naveed Anjum, Tarun Bali, B. Raj, "Design and simulation of handwritten gurumukhi and devanagri numerical recognition", *International Journal of Computer Applications*, Foundation of Computer Science, New York, USA, vol. 73, no. 12, pp. 16–21, 2013.

124. S. Khandelwal, V. Gupta, B. Raj, R. D. Gupta, "Process variability aware low leakage reliable nano scale DG-FinFET SRAM cell design technique", *Journal of Nanoelectronics and Optoelectronics*, vol. 10, no. 6, pp. 810–817, Dec 2015.

125. Shradhya Singh, Shashi Bala, Balwant Raj, B. Raj, "Improved sensitivity of dielectric modulated junctionless transistor for nanoscale biosensor design", *Sensor Letter*, ASP, vol. 18, pp. 328–333, Apl 2020.

126. Vivek Kumar, Santosh Kumar Vishvakarma, B. Raj, "Design and performance analysis of ASIC for IoT applications", *Sensor Letter*, ASP, vol. 18, pp. 31–38, Jan 2020.

127. Akanksha Jaiswal, R. K. Sarin, B. Raj, Shikha Sukhija, "A novel circular slotted microstrip-fed patch antenna with three triangle shape defected ground structure for multiband applications", *Advanced Electromagnetic (AEM)*, vol. 7, no. 3, pp. 56–63, Aug 2018.

128. Girish Wadhwa, B. Raj, "Label free detection of biomolecules using charge-plasma-based gate underlap dielectric modulated junctionless TFET", *Journal of Electronic Materials (JEMS)*, Springer, vol. 47, no. 8, pp. 4683–4693, Aug 2018.

129. V. K. Sharma, M. Pattanaik, B. Raj, "ONOFIC approach: Low power high speed nanoscale VLSI circuits design", *International Journal of Electronics*, Taylor & Francis, vol. 101, no. 1, pp. 61–73, 2014.

130. S. Khandelwal, Balwinder Raj, R. D. Gupta, "FinFET based 6T SRAM cell design: Analysis of performance metric, process variation and temperature effect", *Journal of Computational and Theoretical Nanoscience*, ASP, USA, vol. 12, pp. 2500–2506, 2015.

131. Sumit Singh, Shekhar Yadav, Jagdeep Rahul, Anurag Srivastava, B. Raj, "Impact of HfO2 in graded channel dual insulator double gate MOSFET", *Journal of Computational and Theoretical Nanoscience*, American Scientific Publishers, vol. 12, no. 6, pp. 950–953, Apr 2015.

132. Vijay Kumar Sharma, Manisha Pattanaik, B. Raj, "PVT variations aware low leakage INDEP approach for nanoscale CMOS circuits, *Microelectronics Reliability*, Elsevier, vol. 54, pp. 90–99, 2014.

133. B. Raj, A. K. Saxena, S. Dasgupta, "Quantum mechanical analytical modeling of nanoscale DG FinFET: Evaluation of potential, threshold voltage and source/drain resistance", *Journal of Material Science in Semiconductor Processing*, Elsevier, vol. 16, no. 4, pp. 1131–1137, 2013.

134. Maisagalla Gopal, Siva Sankar D Prasad, Balwinder Raj, "8T SRAM cell design for dynamic and leakage power reduction", *International Journal of Computer Applications*, Foundation of Computer Science, New York, USA, vol. 71, no. 9, pp. 43–48, Jun 2013.

135. Manisha Pattanaik, B. Raj, Shashikant Sharma, Anjan Kumar, "Diode based triode multi-threshold CMOS technique for ground bounce noise reduction in static CMOS adders", *Advanced Materials Research*, Trans Tech Publications, Switzerland, vol. 548, pp. 885–889, 2012.

136. Balwinder Raj, A. K. Saxena, S. Dasgupta, "Nanoscale FinFET based SRAM cell design: Analysis of performance metric, process variation, underlapped FinFET and temperature effect", *IEEE Circuits and System Magazine*, vol. 11, no. 2, pp. 38–50, 2011.

137. V. K. Sharma, M. Pattanaik, Balwinder Raj, "Leakage current ONOFIC approach for deep submicron VLSI circuit design", *International Journal of Electrical, Computer, Electronics and Communication Engineering, World Academy of Sciences, Engineering and Technology*, vol. 7, no. 4, pp. 239–244, 2013.

138. Tulika Chawla, Mamta Khosla, B. Raj, "Design and simulation of triple metal double-gate germanium on insulator vertical tunnel field effect transistor",*Microelectronics Journal*, Elsevier, vol. 114, p. 105125, Aug 2021.

139. Parminder Kaur, Sandeep Singh Gill, B. Raj, "Comparative analysis of OFETs materials and devices for sensor applications", *Journal of Silicon*, Springer, vol. 14, pp. 4463–4471, 2022.

140. Sanjeev Kumar Sharma, Parveen Kumar, Balwant Raj, B. Raj, "In1–xGaxAs double metal gate-stacking cylindrical nanowire MOSFET for highly sensitive photo detector", *Journal of Silicon*, Springer, vol. 14, pp. 3535–3541, 2022.

141. B. Raj, A. K. Saxena, S. Dasgupta, "Analytical modeling of quasi planar nanoscale double gate FinFET with source/drain resistance and field dependent carrier mobility: A quantum mechanical study", *Journal of Computer (JCP)*, Academy Publisher, Finland, vol. 4, no. 9, pp. 1–8, 2009.

142. S. Bhushan, S. Khandelwal, B. Raj, "Analyzing different mode FinFET based memory cell at different power supply for leakage reduction", *Seventh International Conference on Bio-Inspired Computing: Theories and Application, (BIC-TA 2012) Advances in Intelligent Systems and Computing*, vol. 202, pp. 89–100, 2013.

143. Jeetendra Singh, B. Raj, "Temperature dependent analytical modeling and simulations of nanoscale memristor", *Journal: Engineering Science and Technology, an International Journal*, Elsevier, vol. 21, pp. 862–868, Oct 2018.

144. Gurmohan Singh, R. K. Sarin, B. Raj, "Design and performance analysis of a new efficient coplanar quantum-dot cellular automata adder", *Indian Journal of Pure & Applied Physics (IJPAP)*, vol. 55, pp. 97–103, Feb 2017.

145. Amandeep Singh, Mamta Khosla, Balwinder Raj, "Design and analysis of electrostatic doped schottky barrier CNTFET based low power SRAM", *International Journal of Electronics and Communications, (AEÜ)*, Elsevier, vol. 80, pp. 67–72, 2017.

146. Parminder Kaur, Vikas Pandey, B. Raj, "Comparative study of efficient design, control and monitoring of solar power using IoT", *Sensor Letter*, ASP, vol. 18, pp. 419–426, May 2020.

147. Anil Kumar Bhardwaj, Sumeet Gupta, B. Raj, "Development & analysis of compact model for double gate schottky barrier CNTFET, "*Journal of Nanoelectronics and Optoelectronics*, ASP, vol. 15, pp. 1199–1208, Aug 2020.

148. Shailendra Singh, Girish Wadhwa, Balwinder Raj, "Design and analysis of dual source vertical tunnel field effect transistor for high performance, *Transactions on Electrical and Electronics Materials*, Springer, vol. 21, pp. 74–82, Oct 2019.

149. Manjit Kaur, Neena Gupta, Sanjeev Kumar, B. Raj, Arun Kumar Singh, "RF performance analysis of intercalated graphene nanoribbon based global level interconnects", *Journal of Computational Electronics*, Springer, vol. 19, pp. 1002–1013, Jun 2020.

150. Girish Wadhwa, B. Raj, "Design and performance analysis of junctionless TFET biosensor for high sensitivity", *IEEE Nanotechnology*, vol. 18, pp. 567–574, 2019.

151. Jeetendra Singh, B. Raj, "Enhanced nonlinear memristor model encapsulating stochastic dopant drift", *JNO*, ASP, vol. 14, pp. 958–963, 2019.

152. Girish Wadhwa, Priyanka Kamboj, Jeetendra Singh, B. Raj, "Design and investigation of junctionless DGTFET for biological molecule recognition", *Transactions on Electrical and Electronic Materials*, Springer, vol. 22, pp. 282–289, 2021.

153. Tulika Chawla, Mamta Khosla, B. Raj, "Optimization of double-gate dual material GeOI-vertical TFET for VLSI circuit design",*IEEE VLSI Circuits and Systems Letter*, vol. 6, no. 2, pp. 13–25, Aug 2020.

154. Sachin Kumar Verma, Shailendra Singh, Girish Wadhwa, B. Raj, "Detection of biomolecules using charge-plasma based gate underlap dielectric modulated

dopingless TFET", *Transactions on Electrical and Electronic Materials (TEEM)*, Springer, vol. 21, pp. 528–535, Jun 2020.

155. Neeraj Jain, B. Raj, "Impact of underlap spacer region variation on Electrostatic and analog/RF performance of symmetrical high-k SOI FinFET at 20 nm channel length", *Journal of Semiconductors (JoS)*, IOP Science, vol. 38, no. 12, p. 122002, Dec 2017.

156. Shailendra Singh, B. Raj", Analytical modeling and simulation analysis of T-shaped III-V heterojunction vertical T-FET", *Superlattices and Microstructures*, Elsevier, vol. 147, p. 106717, Nov 2020.

157. Gurmohan Singh, R. K. Sarin, B. Raj, "Design and analysis of area efficient QCA based reversible logic gates", *Journal of Microprocessors and Microsystems*, Elsevier, vol. 52, pp. 59–68, May 2017.

158. Amandeep Singh, Mamta Khosla, B. Raj, "Compact model for ballistic single wall CNTFET under quantum capacitance limit", *Journal of Semiconductors (JoS)*, IOP Science, vol. 37, pp. 104001–8, Oct 2016.

159. Sonal Singh, Mamta Khosla, Girish Wadhwa, B. Raj, "Design and analysis of double-gate junctionless vertical TFET for gas sensing applications", *Applied Physics A*, Springer, vol. 127, no. 16, 2 Jan 2021.

160. Inderjit Singh, B. Raj, Mamta Khosla, B. Rajesh Kumar Kaushik, "Potential MRAM technologies for low power SoCs", *SPIN World Scientific Publisher, SCIE*, vol. 10, no. 4, p. 2050027, Dec 2020.

161. Shailendra Singh, B. Raj, "Parametric variation analysis on hetero-junction vertical t-shape TFET for supressing ambipolar conduction", *Indian Journal of Pure and Applied Physics*, vol. 58, pp. 478–485, Jun 2020.

162. Anjana Bhardwaj, Pradeep Kumar, B. Raj, Sunny Anand, "Design and performance optimization of doping-less vertical nanowire TFET using gate stack technique", *Journal of Electronic Materials (JEMS)*, Springer, vol. 41, no. 7, pp. 4005–4013, 2022.

163. Jeetendra Singh, B. Raj, "Tunnel current model of asymmetric MIM structure levying various image forces to analyze the characteristics of filamentary memristor", *Applied Physics A*, Springer, vol. 125, no. 3, p. 203.1, 11 Feb 2019.

164. Candy Goyal, Jagpal Singh Ubhi, B. Raj, Low leakage zero ground noise nanoscale full adder using source biasing technique, "*Journal of Nanoelectronics and Optoelectronics*", American Scientific Publishers, vol. 14, pp. 360–370, Mar 2019.

165. Gurmohan Singh, R. K. Sarin, B. Raj, "A novel robust exclusive-OR function implementation in QCA nanotechnology with energy dissipation analysis", *Journal of Computational Electronics*, Springer, vol. 15, no. 2, pp. 455–465, Jun 2016.

166. Tanu Wadhera, Deepti Kakkar, Girish wadhwa, B. Raj, " Recent advances and progress in development of the field effect transistor biosensor: A review" *Journal of Electronic Materials*, Springer, vol. 48, no. 12, pp. 7635–7646, Dec 2019.

167. Girish Wadhwa, Priyanka Kamboj, Balwinder Raj" Design optimisation of junctionless TFET biosensor for high sensitivity", *Advances in Natural Sciences: Nanoscience and Nanotechnology*, vol. 10, p. 045001, 2019.

Chapter 2

Design and parametric analysis of copper iodide HTL-based perovskite solar cell

Srishtee Chaudhary, Rajesh Mehra, and Balwinder Raj

2.1 INTRODUCTION

For many decades, growing energy demand has been a great concern. With the extinction of some resources and limited availability of others, concerns about energy have led to controversy. Resources belong to all, but resources are not accessible to all. Nature has provided the best and most infinite sources of energy for human existence, but over time, the best has been converted to the worst, and what was once thought infinite has decreased to a finite amount of time. Researchers cannot just sit idly by and wait for the disappearance of resources. Solar power would provide an infinite source of energy, as sunlight produces more than 100 times the energy of all known fossil fuel reserves on the planet.

The responsibility of researchers is to find alternatives for future sustainable survival. A clean and green source that is in abundance is solar energy. Solar cells are an excellent source for solar energy utilization [1]. The various possible ways to utilize it are being explored through solar device biases for efficient outcomes. The structural proposals are investigated and analyzed for effective results. Even though solar energy is the most freely available resource, researchers are still struggling to achieve enhanced efficiency. Environmental degradation issues, the toxicity of materials, and the stability of the device impact efficiency. All these issues need to be individually or simultaneously addressed in order to achieve great design efficiency [2].

Solar cell technologies need to be economical, stable, and non-toxic. Solar cells, often known as photovoltaic cells, are devices that turn solar energy into electricity by utilizing the photovoltaic effect. As shown in Figure 2.1, the P-N junction diode forms the basic working concept of a solar cell. When sunlight reaches the conducting electrode and the junction (P-N), an electron and a hole are generated by the photons produced. The created electron and hole generate a circuit current. There are various factors involved in analyzing the enhanced performance of solar cells [3]. The most important is how much solar energy is capable of reaching the absorber that is then converted into electricity. Further, different materials will theoretically procure different efficiencies because they possess their own unique

DOI: 10.1201/9781003487692-2

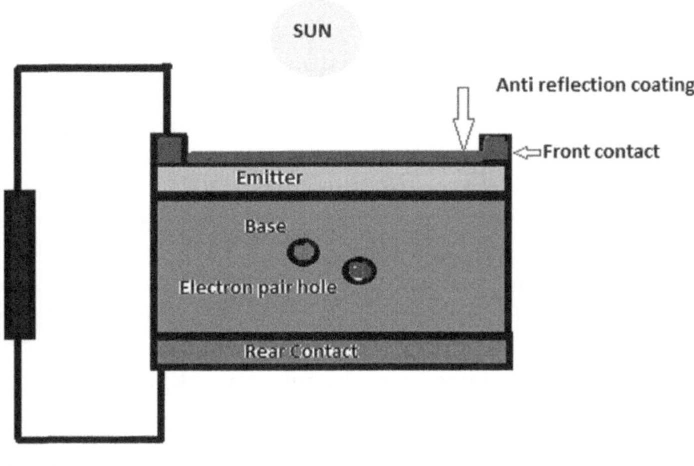

Figure 2.1 Basic structure of a solar cell.

Table 2.1   Technology-based efficiencies of solar cells [5]

S.No.	Solar cell technology type	Efficiency documented
1	Dye-sensitized solar	11.9
2	Si crystalline	25.6
3	Si multi-crystalline	21.3
4	Organic solar cell	11.2
5	Cadmium telluride (Cd Te)	21.0
6	C.I.G.S (copper indium gallium selenide)	21.0
7	Perovskite	19.7
8	Multi-junction	38.8

band gap and absorption coefficients. A crystalline structure with suitable doping provides an enhanced performance [4]. Figure 2.1 shows the basic workings of a solar cell, and Table 2.1 gives a comparative analysis of solar efficiencies.

2.2 PEROVSKITES

The most important advancement in photovoltaics during the past ten years has been the creation of halide perovskite solar cells (PSCs), both organic and inorganic. Currently, perovskite, being a resourceful variant in the field, offers significant possibilities. Perovskite was named after

Russian mineralogist Lev A. Perovski. This material has gained popularity over several decades [6]. Early investigations showed a 3.8% power conversion efficiency (PCE) using dye-sensitized solar cells along with a thin coating of perovskite on the electron collector mesoporous TiO_2. The usual name for substances with the general chemical formula AMX_3 is perovskite. The robust solid-state solar cells established on $CH_3NH_3PbI_3$ perovskite have been reported since mid-2012, and their PCEs have already crossed 15%, outperforming all other solution-processed solar cell technologies [23–30].

The general formula of ABX_3 indicates that A's cation occupies a cube-octahedral site and B's cation occupies an octahedral site. The X denotation can be oxygen, carbon, nitrogen, or halogen. When using an O_2 anion, then A is divalent and B is tetravalent. Simple perovskites, such as $MAPbI_3$ and $FAPbI_3$, are compositions of perovskite with a single ion constituting each of the A, B, and X sites [31–36]. These compositions have undergone extensive research. For a perovskite structure, the tolerance factor rule given below must be followed:

$$.81 < TF = \frac{Ra + Rx}{\sqrt{2}(Rb + Rx)} < 1.11 \tag{1.1}$$

$CaTiO_3$ and $SrTiO_3$ are typical inorganic perovskites. $CaTiO_3$ perovskite is a calcium titanium oxide mineral, and $SrTiO_3$ is strontium titanate. The materials possess extraordinary phenomena, such as superconductivity, ferroelectricity, magnetoelectricity, anti-ferromagnetism, magnetoresistance, and anti-ferroelectricity [37–43]. Table 2.3 depicts the properties of inorganic perovskites, and Figure 2.2 shows the basic structure of perovskite.

Considering the issue of corrosive liquid electrolytes in perovskite DSSC, Spiro-OMeTAD was created to operate as a solid-state electrolyte as a substance for transporting holes (HTM). The one-step method for perovskite

Table 2.2 AMX$_3$ structure

Large organic or inorganic cation – "A"
Smaller metal cation – "M"
An anion in the halide series – "X$_3$"

Table 2.3 Properties of inorganic perovskites

Formula	CaTiO$_3$	SrTiO$_3$
Molar mass	135.943 g/mol	183.49 g/mol
Density	3.98 g/cm³	4.81 g/cm³
Melting point	1,975 °C	2,080 °C
Band gap eV	3–3.5	1.77

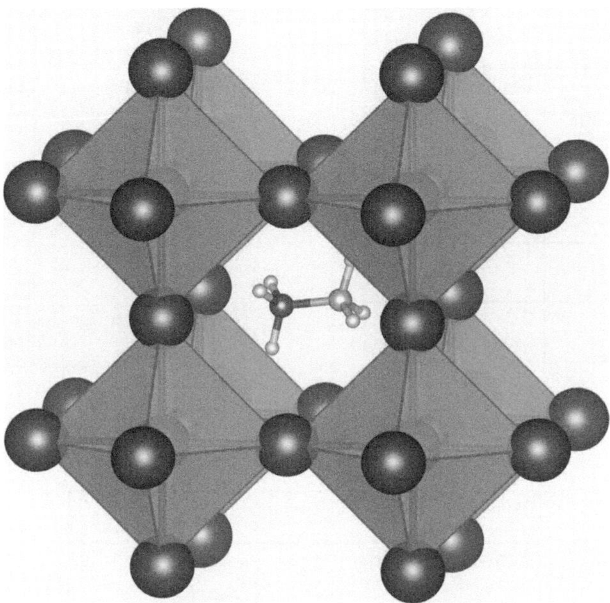

Figure 2.2 Basic structure of perovskite [6].

with a solid-state electrolyte that can be processed in a solution consistently reached a PCE of 9.7% with excellent consistency [5].

The effectiveness of the device increased to 17.9% by calibrating the thickness ratio for perovskite-infiltrated TiO_2 [44–52]. By changing the alignment of bands for hole transport material/electron transport material in the device structure, a PCE of 19.3% was achieved in perovskites [7]. However, TiO_2 disadvantages, such as TiO_2 deterioration under UV irradiation and TiO_2 photocatalytic characteristics, reduce PSC functionality. This led to the suggestion that a non-oxide substance should be employed as an electron material [8]. Metal sulfides were effective in matching the energy levels of the perovskite absorber and fluorine-doped tin oxide (FTO) layers.

Figure 2.3 depicts the basic perovskites with their respective bandgaps. Through standard material synthesis and simple device engineering, the PCE increased from 3.8% to 22.1% in a short time. The outcome of the resulting cell, depending upon the charge, is similarly influenced by transport layers (hole and electron) [53–60]. The majority of current research is based on enhancing the PSC's already-efficient device properties in order to increase PCE [8]. In a classical cell configuration, an active perovskite layer substituted as an absorber layer with a thickness of a few hundred nanometers is positioned between transport layers. The transporting carrier layer subsequently removes the excitons, and the charges are then collected at the suitable electrodes [9].

A Site	
Perovskite Material	Band Gap Eg (eV)
FAPbI3	1.4
MAPbI3	1.5
EAPblj	2.2

X Site	
Perovskite Material	Band Gap Eg (eV)
MAPbClj	3.11
MAPbBr3	2.2
MAPbljBr	1.8

B Site	
Perovskite Material	Band Gap Eg (eV)
MASnL	1.1
MASlio.9Pbo.jI3I3	1.18
MAPblj	1.5
MASlio.3Pbo.7I3	1.31
MASno.5Pbo.5I3	1.28

Figure 2.3 Basic perovskite with bandgaps.

2.3 HOLE TRANSPORT MATERIALS AND ELECTRON TRANSPORT MATERIALS

Hole transport materials (HTMs) and electron transport materials (ETMs) are also undergoing different adjustments and enhancements to better improve the device's performance. To solve the drawbacks, alternative HTMs that are both economical and reliable in all environmental conditions must be developed [61–68]. To this end, effective inorganic hole transporters are readily available, inexpensive, non-toxic, and energy-efficient. Some of the materials used are CuO, CuI, NiO, Cu_2O, $CuCrO_2$, and $CuGaO_2$. In 2012, a breakthrough in efficiency and stability was obtained by using $CH_3NH_3PbI_3$ and $CH_3NH_3PbI_{3-x}Cl_x$ as light absorbers along with a solid-state hole transporter (Spiro-OMeTAD) [10–11]. However, due to the expensive commercial pricing of Spiro-OMeTAD and its proclivity to be uncontrollably doped by O_2, alternatives must be developed. Figure 2.4 shows some of the hole transport layers (HTLs) with their respective chemical names.

Increasing the absorber layer efficiency is attained by combining an appropriate hole layer material, which improves the withdrawal of holes, with an ETM that extracts the photoelectrons produced by the perovskite. The nature of these layers is also crucial for the real-time performance and environmental stability of PSCs [12–13].

Considering $MASnI_3$ as an absorber, TiO_2 as electron material, and cuprous thiocyanate as hole material, the outcome reported is 28.32% efficiency [14]. In comparison, with ZnO as electron material and CuI as hole material for $MASnI_3$-based perovskite cells, one study showed an efficiency of 24.82% [15]. Figure 2.5 shows some of the electron transport layers (ETLs) with their respective chemical names.

Figure 2.6 summarizes some of the back metal contact work functions. Further effort is needed to reduce the bandgap of perovskite in order to enhance spectral responsiveness and thus efficiency [69–75]. Solar cells based on $CH_3NH_3PbI_3$ do not systematically harvest photons close to the optical absorption start (600–780 nm) and give outcomes in photocurrents that do not approach the theoretical maximum [16].

HTLs	Chemical Names of HTLs
Cu_2O	Copper(I) oxide
CuSCN	Cuprous thiocyanate
$CuSbS_2$	Copper antimony sulphide
P3HT	Regioregular poly(3-hexylthiophene-2,5-diyl)
PEDOT:PSS	Poly(3,4-ethylenedioxythiophene)-poly(styrenesulfonate)
Spiro-OMeTAD	2,2',7,7'-Tetrakis[N,N-di(4-methoxyphenyl)amino]-9,9'-spirobifluorene
CuI	Copper(I) iodide
CuO	Copper(II) oxide
V_2O_5	Vanadium pentoxide
NiO	Nickel oxide
$CuCrO_2$	Mcconnellite
$CuGaO_2$	Copper Gallium Oxide
MoOx	Molybdenum Oxides

Figure 2.4 Basic HTLs.

ETL	Chemical Names of ETLs
TiO_2	Titanium dioxide
PCBM	[6,6]-Phenyl C61 butyric acid methyl ester
ZnO	Zinc oxide
C60	Buckminsterfullerene
IGZO	Indium Gallium Zinc Oxide
SnO_2	Tin(IV) oxide
WS_2	Tungsten di-sulfide
CeO_2	Cerium di-oxide
Zn_2SnO_4	Zinc Stannate
ZnS	Zinc sulphide

Figure 2.5 Basic ETLs.

Back Metal Contact	Work function (eV)
Cu	4.65
Ag	4.26
Fe	4.81
C	5
Au	5.1
W	5.22
Ni	5.5
Pd	5.6
Pt	5.7
Se	5.9

Figure 2.6 Work function of back metal contacts.

2.4 DESIGN METHODOLOGY

The outcome of solar cells can vary depending on the material properties of the various solar cell layers, and numerical simulation has been shown to be a useful tool for understanding this. The objective is to study efficiency prospects by varying the morphological parameters of the device [76–81].

Achieving high efficiency and stable perovskites is a challenge for researchers in this field. Because device architectures have distinct principles governing their performance, ensuring that specific parameters are integrated into the operating mechanisms will enable the best method to be determined. Enhanced efficiency is the prime aim when considering perovskite structures [82–88].

The PSC structure proposed in the research is with intrinsic perovskite utilized as an absorber layer. The basic three input layers utilized to construct a model used Spiro-OMeTAD as the HTM, intrinsic perovskite as the absorber layer, and WS_2 as the ETL [17]. SCAPS, a simulation program, numerically resolves three critical semiconductor equations: the Poisson equation, the continuity equation for holes, and the continuity equation for electrons [89–94].

Different halogens have been used to replace iodine in perovskites, but their role is yet unknown. Cl- and Br cover most of the visible spectral range and allow uninterrupted tweaking of the optical band gap. Unfortunately, the high band gap of $CH_3NH_3PbBr_3$ prevents light intake with wavelengths below 550 nm, resulting in a significant reduction in photocurrent [18].

Because of the increased carrier movement across the junction of the hetero-structure in mixed halide perovskites as a result of chloride substitution, photovoltaic performance is increased [95–97]. Halide substitution causes the band gap to narrow, which then causes electroluminescence to shift from the blue to the green part of the spectrum.

On the SCAPS 3.3.07 platform, the above-mentioned PSC design is simulated. The device's structure is Contact (Front)/TCO/Buffer layer/ Defect layer1/Absorber layer (CH3NH3PbI3-xClx)/Defect layer2/HTL (CuI)/Contact (Back). In the proposed cell, CuI of 3.1 eV direct bandgap is utilized. Each layer's typical thickness is fixed. A perovskite absorber layer of 700 nm thickness, in particular, is examined because it assures maximal radiation absorption with minimal recombination losses [19]. Direct band gap semiconductors provide stronger photon absorption on a smaller semiconductor as well as more photon emission through EH recombination. Figure 2.7 shows the device's simulated structure on SCAPS. The uninterrupted band gap aids electron mobility. The work function is as flat bands. The front and back contact is proposed with thermionic emission/surface recombination velocity of 10^5cm/s for electrons and 10^{10}cm/s for holes [98–105]. The parameters considered in the proposed design of PSCs are analyzed and gathered systematically for efficient performance.

A perovskite cell is shown in a methodical layered shape in Figure 2.8. The criteria for selecting an acceptable architecture are part of the solar cell design framework concerning single or multiple junction designs. For PSCs, the layers that are involved must be carefully chosen. The critical components for an effective device formulation are the materials utilized for certain layers and the settings for those layers [106–111].

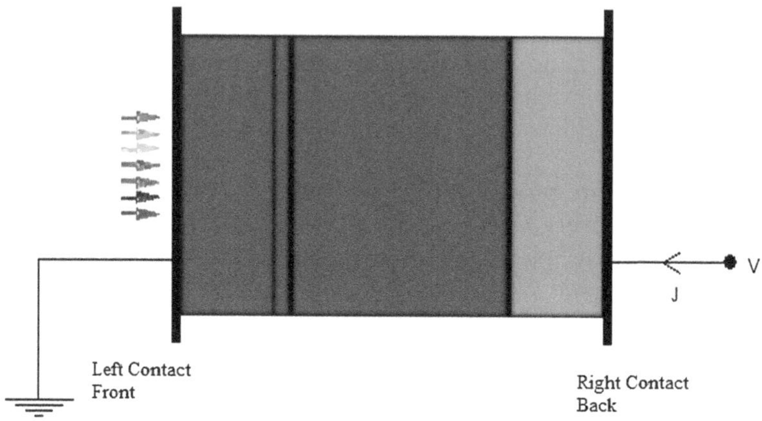

Figure 2.7 Simulated PSC.

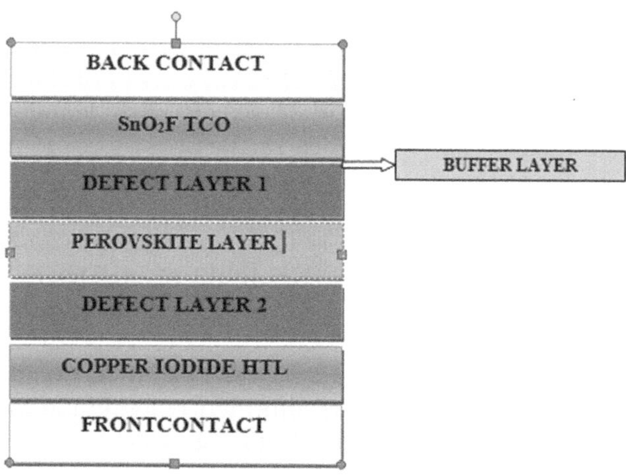

Figure 2.8 Schematic of proposed PSC.

The conduction band effective density is kept at $2.2 \times 10^{18} cm^3$, and the valence band effective density is kept a $1.8 \times 10^{18} cm^3$ for all the respective layers [21]. The defect type considered for simulation is neutral while the energetic distribution is single. The "neutral" defect is an idealized version of a defect that contributes to Shockley-Read-Hall recombination but not to the space charge [22]. Just the product of σ (Stefan-Boltzmann constant) and Nt impacts the dc and ac solutions in the event of a "neutral" fault. Figures 2.9 and 2.10 show the respective energy band diagram and current density for the proposed design. Table 2.4 summarizes the parameters proposed for the design of solar cell devices for efficient outcomes.

Band Diagram

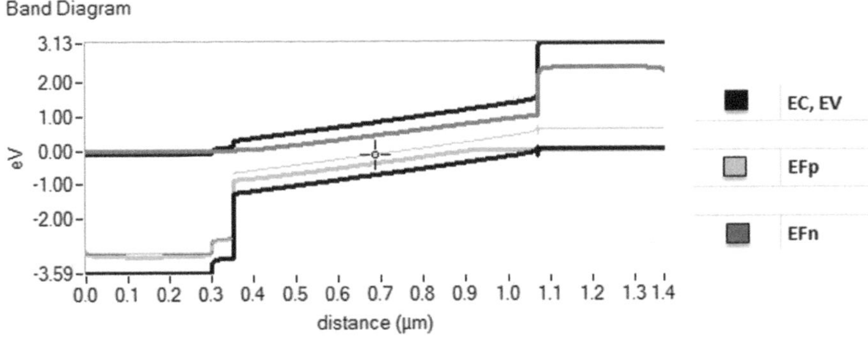

Figure 2.9 Energy band diagram for the design.

Current Density

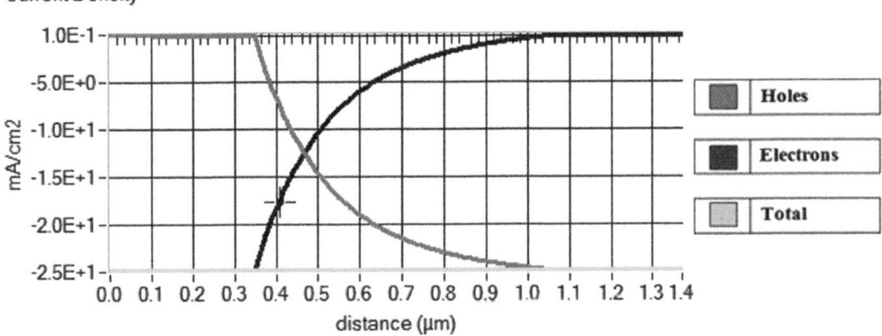

Figure 2.10 Current density diagram.

Table 2.4 Solar cell device parameters

Parameter	SnO_2F TCO	Buffer layer	Defect layer1	$CH_3NH_3PbI_3$ absorber layer	Defect layer2	CuI HTL [20]
Layer thickness (nm)	300	50	10	700	10	300
Eg (eV)	3.5	3.25	1.55	1.55	1.55	3.1
Electron affinity (eV)	4.0	4.08	3.9	3.90	3.9	2.6
(ε_r)	9.0	9.0	6.5	6.5	6.5	6.5
E_M (cm²/Vs) e⁻ mobility	20	340	2	2	2	44
H_M (cm²/Vs) h⁻ mobility	10	50	2	2	2	44
N_D (cm⁻³)	6×10¹⁹	1×10¹⁷	1×10¹⁴	1×10¹⁴	1×10¹⁶	-
N_A (cm⁻³)	-	-	-	-	-	5×10¹⁸
N_T (cm⁻³)	1×10¹⁶	1×10¹⁹	1×10¹¹	1×10¹⁰	1×10¹¹	1×10¹⁸

2.5 RESULTS AND DISCUSSION

2.5.1 Impact of capture cross-section

In physics, the concept of capture cross-section refers to the interaction of elementary particles with one another. Cross-section is the likelihood that two particles will definitely interact with each other in a specific scenario when one particle approaches another particle. The capture cross-section can be rationalized as the area within which a reaction will occur in a straightforward sense [112–118]. As a result, the unit of cross-section is the unit of area. Many factors influence the cross-section, including the type of particle contact, the target particle, the composition, and the type of particle being propelled. The capture cross-section is determined by choosing a defect (level). For the proposed device structure, the defect type is considered neutral with optimal values of 1×10^{-15} cm^2, 1×10^{-16} cm^2, and 2×10^{-16} cm^2 capture cross-section. The capture cross-section value is kept the same for both holes and electrons. The capture cross-section value of 1×10^{-16} cm^2 impacted the device with a proficient result of Voc 1.8V, Jsc 23.41mA/ cm^2, fill factor of 87.56%, and PCE 36.36% [119–127].

Table 2.5 depicts the impact of capture cross-section on the performance parameters of the device for various values. Figure 2.11 shows the graphical representation of the same outcomes.

2.5.2 Electron affinity of HTM and ETM

The layered structure is a blend of different materials. The electron transfer between these materials is measured through electron affinity. In general, it describes how convenient it is to add or extract an electron. The change in affinity will lead to a change in the conduction band and valence band offset of the P-N junction. The electron affinity of HTM ranges from 1 to 3 eV. Voc affixed to virtual 0 V due to the greater band offsets at the minimum value of electron affinity, whereas Jsc has a considerable reduction at the peak value. An effective band offset can be easily modified by altering the affinities and band gaps. ETM affinity varies depending on the percentage

Table 2.5 Variation in capture cross-section vs performance parameters

Capture cross-section	Voc	Jsc	Fill factor	PCE
1×10^{-15} cm^2	1.67	24.92	85.88	35.83
1×10^{-16} cm^2	1.8	23.41	87.56	36.96
2×10^{-15} cm^2	1.67	23.39	86.06	33.66
2×10^{-16} cm^2	1.76	23.41	87.64	36.16

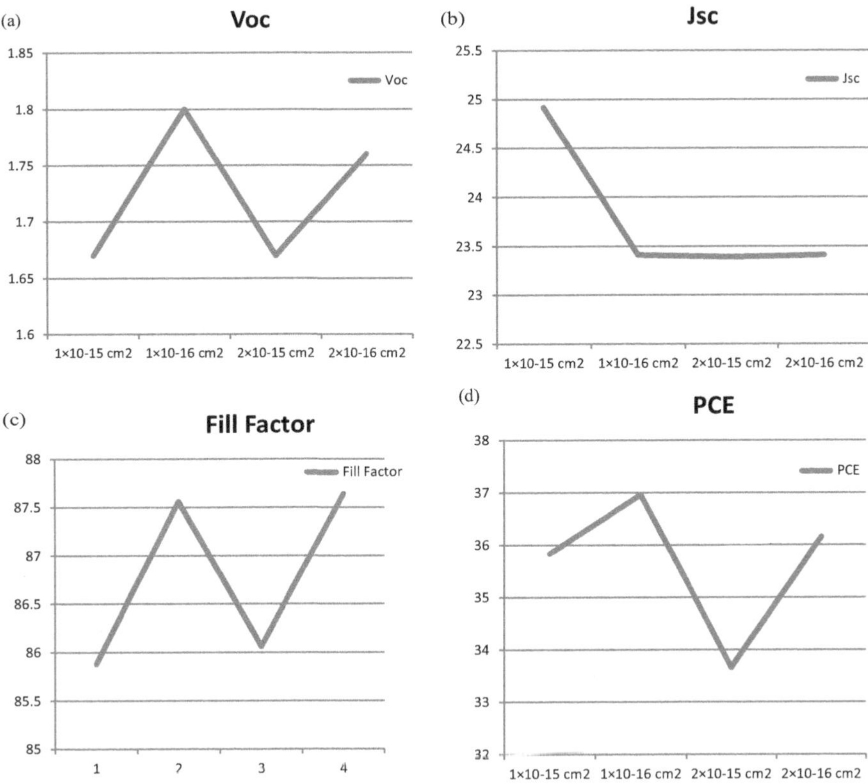

Figure 2.11 Capture cross-section vs performance parameters (a) Voc, (b) Jsc, (c) fill factor, and (d) efficiency.

of oxide in the alloy. As the concentration in the alloy rises, the PSC's performance continues to improve. This is because when the offset lowers, the barrier height drops.

The outcomes of the proposed device cell for varying electron affinity to the value 3.9 eV for the transparent conducting oxide (TCO) layer exhibited PCE at 36.44%, Voc at 1.67, Jsc at 25.34 mA/cm^2, and FF at 85.92%, with all other values the same as in Table 2.4. Considering an absorber layer electron affinity value at 4.08 eV and keeping all other parameters the same, a remarkable PCE of 36.44%, Voc of 1.67, Jsc of 25.34 mA/cm^2, and FF of 85.89% is achieved. The electron affinity set to 2.55eV for CuI gave a PCE of 35.83%, Voc of 1.67, Jsc of 24.92 mA/cm^2, and FF of 85.88%. The variation in electron affinity for the CuI layer decreased the efficiency from 36.44% to 35.83%. Table 2.6 shows the performance parameters with respect to electron affinity for TCO, CuI, and absorber perovskites.

Table 2.6 Variation in electron affinity vs performance parameters

Electron affinity (eV)	TCO 3.9 eV	Absorber layer 4.08 eV	CuI 2.55 eV
Voc (V)	1.67	1.67	1.67
Jsc (mA/cm^2)	25.34	25.34	24.92
Fill factor	85.92	85.89	85.88
Efficiency %	36.44	36.47	35.83

Table 2.7 Doping concentration for TCO and buffer layer

Performance parameters	TCO NA (6.9×10^{19})	Buffer layer ND (5.9× 10^{17})
Voc (V)	1.67	1.67
Jsc (mA/cm^2)	25.34	25.32
Fill factor	85.92	85.98
Efficiency %	36.44	36.47

2.5.3 Effect of doping concentration of HTM and ETM

The performance of a cell can be enhanced by properly doping its contributing components. As the rise in doping concentration shows an increase in the built-in electric field, acceptor density (NA) and donor density (ND) in hole transport/electron transport are responsible for the functional transport of holes and electrons under the impact of the electric field.

Table 2.7 shows the effect of doping concentration for the TCO and buffer layer. In the simulation (NA), the influence of doping concentration ranges from 3.9×10^{18} cm^3 to 7.9×10^{19} cm^3. PCE is low, with lower values of NA because of the existence of significant series resistance. A fill factor of 85.92 was attained by optimizing NA (6.9×10^{19}) for TCO, and ND (5.9× 10^{17}) for the buffer layer.

2.5.4 Effect of operating temperature

The working temperature is critical to the device's performance. Solar panels are typically installed outside and work at temperatures greater than 300 K. Temperature augmentations have been shown to enhance strain and stress in structures. As a result, there are more interfacial flaws, disorganization, and poor connection between layers. We investigated the effect of operation temperature on the cell parameters. By holding all other parameters constant, the simulated working temperature was adjusted from 300 K to 440 K. At a low temperature of 300 K, the highest efficiency of 36.44% was obtained. The increase in temperature also impacts the hole

Table 2.8 Variation in temperature vs performance parameters

Performance parameters	Temperature 300 K	Temperature 350 K	Temperature 410 K	Temperature 440 K
Voc (V)	1.67	1.68	1.69	1.69
Jsc (mA/cm²)	25.34	24.9	24.93	24.93
Fill factor	85.92	84.80	83.43	82.69
Efficiency %	36.44	35.60	35.19	34.94

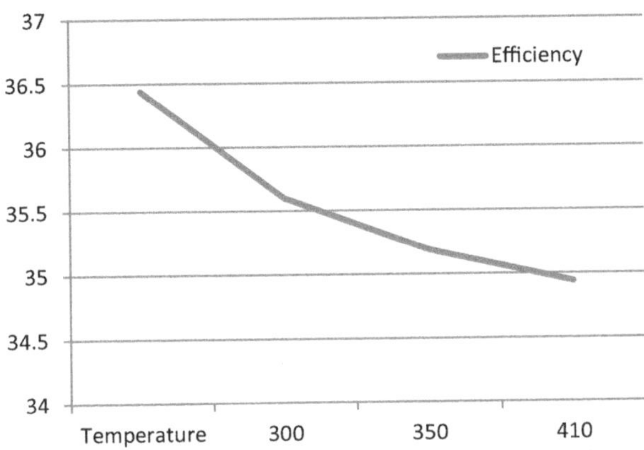

Figure 2.12 Temperature vs efficiency.

and electron mobilities as well as the concentration of the carrier, resulting in a decrease in PSC efficiency. Table 2.8 shows the performance outcome with respect to variation in temperature.

2.6 CONCLUSION

The potential of mixed layered structure SnO_2F perovskites with CuI as HTM is theoretically investigated using a modeling tool to explore their potential for application as an alternative to the options currently available. To evaluate important factors that define overall performance, the device's performance measurements were carefully reviewed. The best electron affinities for HTM and ETM can lessen recombination at the interfaces. The impact of doping profiles on performance is also significant. Many factors influencing cells have been examined. Parameters include the layer thickness, the density of absorber defects, and the operating temperature. We discovered that all of these variables have an effect on the electrical characteristics of PSCs. An improved absorber layer thickness (700 nm)

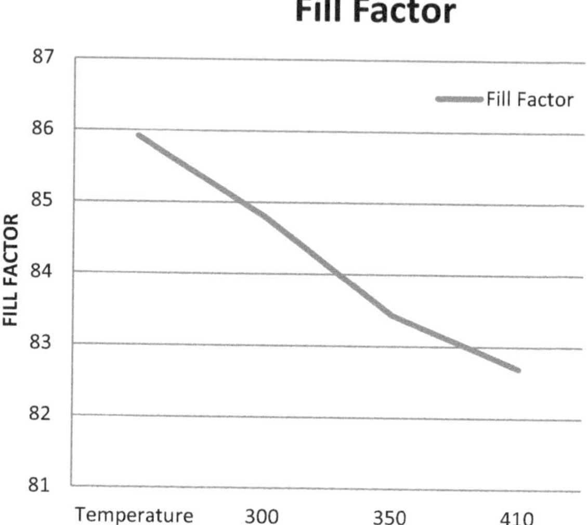

Figure 2.13 Temperature vs fill factor.

Table 2.9 Performance results

Performance	Results
Voc (V)	1.67
Jsc (mA/cm^2)	25.33
Fill factor	85.85
Efficiency %	36.44

could consume more luminescence and give a PCE of 36.44%. CuI HTM with absorber perovskite is proficient enough to replace expensive Spiro-OMeTAD and has significant potential as a future replacement, according to the findings. Parameter optimization can boost the PSC performance even further. Table 2.9 shows the performance results for the proposed solar cell design.

REFERENCES

1. Longbin Qiu, Luis K Ono, Yabing Qi, "Advances and Challenges to the Commercialization of Organic-Inorganic Halide Perovskite Solar Cell Technology", *Journal of Materials Today Energy*, Elsevier, Vol. 7, pp. 169–189, 2017.
2. BJ Kim, DH Kim, SL Kwon, SY Park, Z Li, K Zhu, HS Jung, "Selective Dissolution of Halide Perovskites as a Step Towards Recycling Solar Cells", *Journal of Nature Communications*, Vol. 7, pp. 1–9, 2016.

3. P Kumar, "*Organic Solar Cells-Device Physics, Processing, Degradation, and Prevention*", 1st edn. CRC Press Taylor & Francis Group, New York, 2017.

4. MM Lee, J Teuscher, T Miyasaka, TN Murakami, HJ Snaith, "Efficient Hybrid Solar Cells Based on Meso-Superstructured Organometal Halide Perovskites", *Science*, Vol. 338(6107), pp. 643–647, 2 Nov 2012. http://dx.doi.org/10.1126/science.1228604

5. Srishtee Chaudhary and Rajesh Mehra, "Analytical Review of Solar Cell as Globalized Approach", *Advances in Renewable Energy and Sustainable Environment*, pp. 201–209, Jan 2021. http://dx.doi.org/10.1007/978-981-15-5313-4_19

6. CS Ponseca Jr., TJ Savenije, M Abdellah, K Zheng, A Yartsev, T Pascher, T Harlang, P Chabera, T Pullerits, A Stepanov, JP Wolf, V Sundstrom, "Organometal Halide Perovskite Solar Cell Materials Rationalized: Ultrafast Charge Generation, High and Microsecond-Long Balanced Mobilities, and Slow Recombination", *Journal of American Chemical Society*, Vol. 136, pp. 5189–5192, 2014.

7. J Heo, S Im, J Noh, "Efficient Inorganic–Organic Hybrid Heterojunction Solar Cells Containing Perovskite Compound and Polymeric Hole Conductors", *Nature Photon*, Vol. 7, pp. 486–491, 2013. https://doi.org/10.1038/nphoton.2013.80

8. NJ Jeon, JH Noh, YC Kim, WS Yang, S Ryu, SI Seok, "*Solvent Engineering for High-Performance Inorganic–Organic Hybrid Perovskite Solar Cells*", *Nature Materials*, Vol. 13(9), pp. 897–903, 2014.

9. H Zhou, et al., "Interface Engineering of Highly Efficient Perovskite Solar Cells", *Science*, Vol. 345(6196), p. 5426, 2014.

10. HS Kim, CR Lee, JH Im, KB Lee, T Moehl, A Marchioro, SJ Moon, R Humphry-Baker, JH Yum, JE Moser, M Gratzel, NG Park, "Lead Iodide Perovskite Sensitized All-Solid-State Submicron Thin Film Mesoscopic Solar Cell with Efficiency Exceeding 9%", *Scientific Reports*, Vol. 2, p. 591, 2012.

11. TC Sum, N Mathews, "Advancements in Perovskite Solar Cells: Photophysics Behind the Photovoltaics", *Energy & Environmental Science*, Vol. 7(8), pp. 2518–2534, 2014.

12. Assadi M Khalaji, S Bakhoda, R Saidur, H Hanaei, "Recent Progress in Perovskite Solar Cells", *Renewable and Sustainable Energy Reviews*, 2018.

13. Seelam Prasanthkumar, Lingamallu Giribabu, "Recent Advances in Perovskite-Based Solar Cells", *Current Science*, 2016.

14. Neha Thakur, Rajesh Mehra, Chandni Devi, "Efficient Design of Perovskite Solar Cell Using Parametric Grading of Mixed Halide Perovskite and Copper Iodide", *Journal of Electronic Materials*, 2018.

15. Usha Mandadapu1 S Victor Vedanayakam, K Thyagarajan, M Raja Reddy, BJ Babu, "Design and Simulation of High Efficiency Tin Halide Perovskite Solar Cell", *International Journal of Renewable Energy Research*, Vol. 7(4), pp. 1603–1614, 2017.

16. GE Eperon, SD Stranks, C Menelaou, MB Johnston, LM Herz, HJ Snaith, "Formamidinium Lead Trihalide: A Broadly Tunable Perovskite for Efficient Planar Heterojunction Solar Cells", *Energy and Environmental Science*, Vol. 2014(3), pp. 982–988, 2014.

17. K Sobayell, KS Akhtaruzzaman, MT Rahman, Zeyad A Ferdaous, Hamad F Al-Mutairi Alharbi, Nabeel H Alharthi, Mohammad R Karim, S Hasmady,

N Amin, "A Comprehensive Defect Study of Tungsten Disulfide (WS2) as Electron Transport Layer in Perovskite Solar Cells by Numerical Simulation", *Results in Physics*, 2018. http://dx.doi.org/10.1016/j.rinp.2018.12.049

18. F Liu, J Zhu, J Wei, Y Li, M Lv, S Yang, "Numerical Simulation: Toward the Design of High-Efficiency Planar Perovskite Solar Cells", *Applied Physics Letter*, Vol. 104(25), p. 253508, 2014.

19. L Peng, "Device Simulation of Solid-State Perovskite Solar Cells", *31st European Photovoltaic Solar Energy Conference and Exhibition*, 2015.

20. Srishtee Chaudhary, Rajesh Mehra, "35.83% Efficient Non-Toxic Perovskite Solar Cell using Copper Iodide and Tin-oxide", *2020 International Conference on Computation, Automation and Knowledge Management (ICCAKM)*, Dubai, United Arab Emirates, 2020, pp. 258–262, http://doi.org/10.1109/ICCAKM46823.2020.9051498

21. Srishtee Chaudhary, Rajesh Mehra, "Perovskite Solar Cell Using HTLs Copper Iodide and Spiro-OMeTAD Comparative Analysis in Terms of Efficiency and Resource Utilization", *Applied Optics*, Vol. 61, pp. 101–107, 2022. https://opg.optica.org/ao/abstract.cfm?URI=ao-61-1-101

22. S Ahmed, F Jannat, MAK Khan, MA Alim, "Numerical Development of Ecofriendly Cs2TiBr6 Based Perovskite Solar Cell with All-Inorganic Charge Transport Materials Via SCAPS-1D", *Optik*, Vol. 225, p. 165765, 2021.

23. Anjana Bhardwaj, Pradeep Kumar, B Raj, Sunny Anand, "Design and Performance Optimization of doping-less Vertical Nanowire TFET using Gate Stack Technique", *Journal of Electronic Materials (JEMS)*, Springer, Vol. 41(7), pp. 4005–4013, 2022.

24. Jeetendra Singh, B Raj, "Tunnel Current Model of Asymmetric MIM Structure Levying Various Image Forces to Analyze the Characteristics of Filamentary Memristor", *Applied Physics A*, Springer, Vol. 125(3), p. 203.1, Feb 2019.

25. Candy Goyal, Jagpal Singh Ubhi, B Raj, "Low Leakage Zero Ground Noise Nanoscale Full Adder using Source Biasing Technique", *Journal of Nanoelectronics and Optoelectronics*, American Scientific Publishers, Vol. 14, pp. 360–370, Mar 2019.

26. Gurmohan Singh, RK Sarin, B Raj, "A Novel Robust Exclusive-OR Function Implementation in QCA Nanotechnology with Energy Dissipation Analysis", *Journal of Computational Electronics*, Springer, Vol. 15(2), pp. 455–465, Jun 2016.

27. Tanu Wadhera, Deepti Kakkar, Girish Wadhwa, B Raj, "Recent Advances and Progress in Development of the Field Effect Transistor Biosensor: A Review", *Journal of Electronic Materials*, Springer, Vol. 48(12), pp. 7635–7646, Dec 2019.

28. Girish Wadhwa, Priyanka Kamboj, Balwinder Raj, "Design Optimisation of Junctionless TFET Biosensor for High Sensitivity", *Advances in Natural Sciences: Nanoscience and Nanotechnology*, Vol. 10(7), p. 045001, 2019.

29. Priya Bansal, B Raj, "Memristor Modeling and Analysis for Linear Dopant Drift Kinetics", *Journal of Nanoengineering and Nanomanufacturing*, American Scientific Publishers, Vol. 6, pp. 1–7, 2016.

30. Amandeep Singh, Mamta Khosla, B Raj, "Circuit Compatible Model for Electrostatic Doped Schottky Barrier CNTFET", *Journal of Electronic Materials*, Springer, Vol. 45(12), pp. 4825–4835, 2016.

31. D Vaithiyanathan Ashima, B Raj, "Performance Analysis of Charge Plasma induced Graded Channel Si nanotube", *Journal of Engineering Research (JER)*, EMSME Special Issue pp. 146–154 Aug 2021.

32. Abhishek Singh Tomar, Vijay Kumar Magraiya, B Raj, "Scaling of Access and Data Transistor for High Performance DRAM Cell Design", *Quantum Matter*, Vol. 2, pp. 412–416 Oct 2013.

33. Jain, Neeraj, B Raj, "Parasitic Capacitance and Resistance Model Development and Optimization of Raised Source/Drain SOI FinFET Structure for Analog Circuit Applications", *Journal of Nanoelectronics and Optoelectronins*, ASP, USA, Vol. 13, pp. 531–539, Apr 2018.

34. Shradhya Singh, S. K. Vishvakarma, B Raj, "Analytical Modeling of Split-Gate Junction-Less Transistor for a Biosensor Application", *Sensing and Bio-sensing*, Elsevier, Vol. 18, pp. 31–36, Apr 2018.

35. Maisagalla Gopal, Balwinder Raj, "Low Power 8T SRAM Cell Design for High Stability Video Applications", *ITSI Transaction on Electrical and Electronics Engineering*, Vol. 1(5), pp. 91–97, 2013.

36. Balwinder Raj, Jatin Mitra, Deepak Kumar Bihani, V Rangharajan, AK Saxena, S Dasgupta, "Analysis of Noise Margin, Power and Process Variation for 32 nm FinFET Based 6T SRAM Cell", *Journal of Computer (JCP)*, Academy Publisher, Finland, Vol. 5(6), pp. 1–8, 2010.

37. Divya Sharma, Rajesh Mehra, B Raj, "Comparative Analysis of Photovoltaic Technologies for High Efficiency Solar Cell Design", *Superlattices and Microstructures*, Elsevier, Vol. 153, . 106861, May 2021.

38. Pawandeep Kaur, Avtar Singh Buttar, B Raj, "A Comprehensive Analysis of Nanoscale Transistor Based Biosensor: A Review", *Indian Journal of Pure and Applied Physics*, Vol. 59, pp. 304–318, Apr 2021.

39. Divya Yadav, Balwant Raj, B Raj, "Design and Simulation of Low Power Microcontroller for IoT Applications", *Journal of Sensor Letters*, ASP, Vol. 18, pp. 401–409, May 2020.

40. Shailendra Singh, B Raj, "A 2-D Analytical Surface Potential and Drain current Modeling of Double-Gate Vertical t-shaped Tunnel FET", *Journal of Computational Electronics*, Springer, Vol. 19, pp. 1154–1163, Apl 2020.

41. Jeetendra Singh, B Raj, "An Accurate and Generic Window function for Non-linear Memristor Model", *Journal of Computational Electronics*, Springer, Vol. 18(2), pp. 640–647, Jun 2019.

42. Manjit Kaur, Neena Gupta, Sanjeev Kumar, B Raj, Arun Kumar Singh, "Comparative RF and Crosstalk Analysis of Carbon Based Nano Interconnects", *IET Circuits, Devices & Systems*, Vol. 15(6), pp. 493–503, Feb 2021.

43. Nehru Kandasamy, Firdous Ahmad, D Ajitha, B Raj, Nagarjuna Telagam "Quantum Dot Cellular Automata Based Scan Flip Flop and Boundary Scan Register", *IETE Journal of Research*, Vol. 66, pp. 535–548, 2020.

44. SK Sharma, B Raj, M Khosla, "Enhanced Photosensivity of Highly Spectrum Selective Cylindrical Gate In1-xGaxAs Nanowire MOSFET Photodetector", *Modern Physics Letter-B*, Vol. 33(12), p. 1950144, 2019.

45. Jeetendra Singh, B Raj, "Design and Investigation of 7T2M NVSARM with Enhanced Stability and Temperature Impact on Store/Restore Energy", *IEEE Transactions on Very Large Scale Integration Systems*, Vol. 27(6), pp. 1322–1328, Jun 2019.

46. Anil Kumar Bhardwaj, Sumeet Gupta, B Raj, Amandeep Singh, "Impact of Double Gate Geometry on the Performance of Carbon Nanotube Field Effect Transistor Structures for Low Power Digital Design", *Computational and Theoretical Nanoscience*, ASP, Vol. 16, pp. 1813–1820, 2019.

47. Neeraj Jain, B Raj, "Thermal Stability Analysis and Performance Exploration of Asymmetrical Dual-k underlap Spacer (ADKUS) SOI FinFET for Security and Privacy Applications", *Indian Journal of Pure & Applied Physics (IJPAP)*, Vol. 57, pp. 352–360, May 2019.

48. Amandeep Singh, Mamta Khosla, B Raj, "Design and Analysis of Dynamically Configurable Electrostatic Doped Carbon Nanotube Tunnel FET", *Microelectronics Journal*, Elsevier, Vol. 85, pp. 17–24, Mar 2019.

49. Neeraj Jain, Balwinder Raj, "Dual-k Spacer Region Variation at the Drain Side of Asymmetric SOI FinFET Structure: Performance Analysis towards the Analog/RF Design Applications", *Journal of Nanoelectronics and Optoelectronics*", American Scientific Publishers, Vol. 14, pp. 349–359, Mar 2019.

50. Jeetendra Singh, Sanjeev Sharma, B Raj, Mamta Khosla, "Analysis of Barrier Layer Thickness on Performance of In1-xGaxAs Based Gate Stack Cylindrical Gate Nanowire MOSFET", *JNO*, ASP, Vol. 13, pp. 1473–1477, Oct 2018.

51. Neeraj Jain, B Raj, "Analysis and Performance Exploration of High-k SOI FinFETs Over the Conventional Low-k SOI FinFET toward Analog/RF Design", *Journal of Semiconductors (JoS)*, IOP Science, Vol. 39(12), pp. 124002-1-7, Dec 2018.

52. Candy Goyal, Jagpal Singh Ubhi, B Raj, A Reliable Leakage Reduction Technique for Approximate Full Adder with Reduced Ground Bounce Noise", *Journal of Mathematical Problems in Engineering*, Hindawi, Vol. 2018, Article ID 3501041, p. 16, 15 Oct 2018.

53. Jeetendra Singh, B Raj, Mamta Khosla, "Design and Performance Analysis of Nano-scale Memristor-based Nonvolatile SRAM", *Journal of Sensor Letter*", American Scientific Publishers, Vol. 16, pp. 798–805, Oct 2018.

54. Girish Wadhwa, B Raj, "Parametric Variation Analysis of Charge-Plasma-based Dielectric Modulated JLTFET for Biosensor Application", *IEEE Sensor Journal*, Vol. 18(15), pp. 6070–6077, 1 Aug 2018.

55. Jeetendra Singh, B Raj, "Comparative Analysis of Memristor Models for Memories Design", *JoS*, IOP, Vol. 39(7), pp. 074006-1-12, July 2018.

56. Divya Yadav, Shailesh Singh Chouhan, Santosh Kumar Vishvakarma, B Raj, "Application Specific Microcontroller Design for IoT based WSN", *Sensor Letter*, ASP, Vol. 16, pp. 374–385, May 2018.

57. Gurmohan Singh, RK Sarin, B Raj, "Fault-Tolerant Design and Analysis of Quantum-Dot Cellular Automata Based Circuits", *IEEE/IET Circuits, Devices & Systems*, Vol. 12, pp. 638–664, 2018.

58. Jeetendra Singh, B Raj, "Modeling of Mean Barrier Height Levying Various Image Forces of Metal Insulator Metal Structure to Enhance the Performance of Conductive Filament Based Memristor Model", *IEEE Nanotechnology*, Vol. 17(2), pp. 268–267, Mar 2018.

59. Aakash Jain, Sanjeev Sharma, B Raj, "Analysis of Triple Metal Surrounding Gate (TM-SG) III-V Nanowire MOSFET for Photosensing Application", *Opto-Electronics Journal*, Elsevier, Vol. 26(2), pp. 141–148, May 2018.

60. Aakash Jain, Sanjeev Sharma, B Raj, "Design and Analysis of High Sensitivity Photosensor Using Cylindrical Surrounding Gate MOSFET For Low Power Sensor Applications", *Engineering Science and Technology, an International Journal*, Elsevier, Vol. 19(4), pp. 1864–1870, Dec 2016.

61. Amandeep Singh, Mamta Khosla, B Raj, "Analysis of Electrostatic Doped Schottky Barrier Carbon Nanotube FET for Low Power Applications", *Journal of Materials Science: Materials in Electronics*, Springer, Vol. 28, pp. 1762–1768, 2017.

62. G Saiphani Kumar, Amandeep Singh, B Raj, "Design and Analysis of Gate All Around CNTFET Based SRAM Cell Design", *Journal of Computational Electronics*, Springer, Vol. 17(1), pp. 138–145, Mar 2018.

63. Gurinder Pal Singh, BS Sohi, Balwinder Raj, "Material Properties Analysis of Graphene Base Transistor (GBT) for VLSI Analog Circuits", *Indian Journal of Pure & Applied Physics (IJPAP)*, Vol. 55, pp. 896–902, Dec 2017.

64. Amandeep Singh, Mamta Khosla, B Raj, "Comparative Analysis of Carbon Nanotube Field Effect Transistor and Nanowire Transistor for Low Power Circuit Design", *Journal of Nanoelectronics and Optoelectronics*, American Scientific Publishers, USA, Vol. 11, pp. 388–393, Jun 2016.

65. Sunil Kumar, B Raj, "Estimation of Stability and Performance metric for Inward Access Transistor Based 6T SRAM Cell Design Using n-type/p-type DMDG-GDOV TFET", *IEEE VLSI Circuits and Systems Letter*, Vol. 3(2), pp. 25–39, Jun 2017.

66. Shashikant Sharma, Anjan Kumar, Manisha Pattanaik, B Raj, "Forward Body Biased Multimode Multi-Threshold CMOS Technique for Ground Bounce Noise Reduction in Static CMOS Adders", *International Journal of Information and Electronics Engineering*, Vol. 3(3), pp. 567–572, 2013.

67. Hamendra Singh, Pankaj Kumar, Balwinder Raj, "Performance Analysis of Majority Gate SET Based 1-bit Full Adder", *International Journal of Computer and Communication* Engineering *(IJCCE)*, IACSIT Press, Singapore, ISSN: 2010-3743, Vol. 2(4), 2013.

68. Anil Kumar Bhardwaj, Sumeet Gupta, B Raj, "Investigation of Parameters for Schottky Barrier (SB) Height for Schottky Barrier Based Carbon Nanotube Field Effect Transistor Device", *Journal of Nanoelectronics and Optoelectronics*, ASP, Vol. 15, pp. 783–791, Jul 2020.

69. Priya Bansal, B Raj, "Memristor: A Versatile Nonlinear Model for Dopant Drift and Boundary Issues", *JCTN*, American Scientific Publishers, Vol. 14 (5), pp. 2319–2325, May 2017.

70. Neeraj Jain, B Raj, "An Analog and Digital Design Perspective Comprehensive Approach on Fin-FET (Fin-Field Effect transistor) Technology - A Review", *Reviews in Advanced Sciences and Engineering (RASE)*, ASP, Vol. 5, pp. 1–14, 2016.

71. Sanjeev Sharma, B Raj, Mamta Khosla, "Subthreshold Performance of In1-xGaxAs Based Dual Metal with Gate Stack Cylindrical/Surrounding Gate Nanowire MOSFET for Low Power Analog Applications", *Journal of Nanoelectronics and Optoelectronics*, American Scientific Publishers, USA, Vol. 12, pp. 171–176, 2017.

72. Balwinder Raj, AK Saxena, S Dasgupta, "Analytical Modeling for the Estimation of Leakage Current and Subthreshold Swing Factor of Nanoscale

Double Gate FinFET Device", *Microelectronics International, UK*, Vol. 26, pp. 53–63, 2009.

73. Shailendra Singh, Girish Wadhwa, Balwinder Raj, "An Analytical Modeling for Dual Source Vertical Tunnel Field Effect Transistor", *International Journal of Recent Technology and Engineering (IJRTE)*, Vol. 8(2), pp. 603–608, Jul 2019.

74. Shailendra Singh, B Raj, "Design and Analysis of Hetrojunction Vertical T-shaped Tunnel Field Effect Transistor", *Journal of Electronics Material*, Springer, Vol. 48(10), pp. 6253–6260, Oct 2019.

75. Candy Goyal, Jagpal Singh Ubhi, B Raj, "A Low Leakage CNTFET Based Inexact Full Adder for Low Power Image Processing Applications", *International Journal of Circuit Theory and Applications*, Wiley, Vol. 47(9), pp. 1446–1458, Sep 2019.

76. B Raj, A. K. Saxena, S Dasgupta, "A Compact Drain Current and Threshold Voltage Quantum Mechanical Analytical Modeling for FinFETs", *Journal of Nanoelectronics and Optoelectronics (JNO)*, USA, Vol. 3(2), pp. 163–170, 2008.

77. Girish Wadhwa, B Raj, "An Analytical Modeling of Charge Plasma based Tunnel Field Effect Transistor with Impacts of Gate underlap Region", *Superlattices and Microstructures*, Elsevier, Vol. 142, pp. 106512, Jun 2020.

78. Shailendra Singh, B Raj, "Modeling and Simulation Analysis of SiGe Hetrojunction Double GateVertical t-shaped Tunnel FET", *Superlattices and Microstructures*, Elsevier, Vol. 142, pp. 106496, Jun 2020.

79. Amandeep Singh, Dinesh Kumar Saini, Dinesh Agarwal, Sajal Aggarwal, Mamta Khosla, B Raj, "Modeling and Simulation of Carbon Nanotube Field Effect Transistor and Its Circuit Application", *Journal of Semiconductors (JoS)*, IOP Science, Vol. 37, pp. 074001-6, July 2016.

80. Neeraj Jain, Balwinder Raj, "Device and Circuit Co-Design Perspective Comprehensive Approach on FinFET Technology - A Review", *Journal of Electron Devices*, Vol. 23(1), pp. 1890–1901, 2016.

81. Sunil Kumar, B Raj, "Analysis of I_{ON} and Ambipolar Current for Dual-Material Gate-drain Overlapped DG-TFET", *Journal of Nanoelectronics and Optoelectronics*, American Scientific Publishers, USA, Vol. 11, pp. 323–333, Jun 2016.

82. Naveed Anjum, Tarun Bali, B Raj, "Design and Simulation of Handwritten Multiscript Character Recognition", *International Journal of Advanced Research in Computer and Communication Engineering*, Vol. 2(7), pp. 2544–2549, Jul 2013.

83. Sanjeev Sharma, B Raj, Mamta Khosla, "A Gaussian Approach for Analytical Subthreshold Current Model of Cylindrical Nanowire FET with Quantum Mechanical Effects", *Microelectronics Journal*, Elsevier, Vol. 53, pp. 65–72, Apr 2016.

84. Karmjit Singh, B Raj, "Performance and Analysis of Temperature Dependent Multi-walled Carbon Nanotubes as Global Interconnects at Different Technology Nodes", *Journal of Computational Electronics*, Springer, Vol. 14(2), pp. 469–476, Jun 2015.

85. Sunil Kumar, B Raj, "Compact Channel Potential Analytical Modeling of DG-TFET Based on Evanescent–mode Approach", *Journal of Computational Electronics*, Springer Vol. 14(2), pp. 820–827, Jul 2015.

86. Karmjit Singh, B Raj, "Temperature Dependent Modeling and Performance Evaluation of Multi-Walled CNT and Single-Walled CNT as Global Interconnects", *Journal of Electronic Materials*, Springer, Vol. 44(12), pp. 4825–4835, Dec 2015.

87. VK Sharma, M Pattanaik, B Raj, "INDEP Approach for Leakage Reduction in Nanoscale CMOS Circuits", *International Journal of Electronics*, Taylor & Francis, Vol. 102(2), pp. 200–215, 2014.

88. Karmjit Singh, B Raj, "Influence of Temperature on MWCNT Bundle, SWCNT Bundle and Copper Interconnects for Nanoscaled Technology Nodes", *Journal of Materials Science: Materials in Electronics*, Springer, Vol. 26(8), pp. 6134–6142, 2015.

89. Naveed Anjum, Tarun Bali, B Raj, "Design and Simulation of Handwritten Gurumukhi and Devanagri Numerical Recognition", *International Journal of Computer Applications*, Foundation of Computer Science, New York, USA, Vol. 73(12), pp. 16–21, 2013.

90. S Khandelwal, V Gupta, B Raj, RD Gupta, "Process Variability Aware Low Leakage Reliable Nano Scale DG-FinFET SRAM Cell Design Technique", *Journal of Nanoelectronics and Optoelectronics*, Vol. 10(6), pp. 810–817, Dec 2015.

91. VK Sharma, M Pattanaik, B Raj, "ONOFIC Approach: Low Power High Speed Nanoscale VLSI Circuits Design", *International Journal of Electronics*, Taylor & Francis, Vol. 101(1), pp. 61–73, 2014.

92. S Khandelwal, Balwinder Raj, RD Gupta, "FinFET Based 6T SRAM Cell Design: Analysis of Performance Metric, Process Variation and Temperature Effect", *Journal of Computational and Theoretical Nanoscience*, ASP, USA, Vol. 12, pp. 2500–2506, 2015.

93. Sumit Singh, Shekhar Yadav, Jagdeep Rahul, Anurag Srivastava, B Raj, "Impact of HfO_2 in Graded Channel Dual Insulator Double Gate MOSFET", *Journal of Computational and Theoretical Nanoscience*, American Scientific Publishers, Vol. 12(6), pp. 950–953, Apr 2015.

94. Vijay Kumar Sharma, Manisha Pattanaik, B Raj, "PVT Variations Aware Low Leakage INDEP Approach for Nanoscale CMOS Circuits", *Microelectronics Reliability*, Elsevier, Vol. 54, pp. 90–99, 2014.

95. B Raj, AK Saxena, S Dasgupta, "Quantum Mechanical Analytical Modeling of Nanoscale DG FinFET: Evaluation of Potential, Threshold Voltage and Source/Drain Resistance", *Elsevier's Journal of Material Science in Semiconductor Processing*, Elsevier, Vol. 16(4), pp. 1131–1137, 2013.

96. Maisagalla Gopal, Siva Sankar D Prasad, Balwinder Raj, "8T SRAM Cell Design for Dynamic and Leakage Power Reduction", *International Journal of Computer Applications*, Foundation of Computer Science, New York, USA, Vol. 71(9), pp. 43–48, Jun 2013.

97. Manisha Pattanaik, B Raj, Shashikant Sharma, Anjan Kumar, "Diode Based Trimode Multi-Threshold CMOS Technique for Ground Bounce Noise Reduction in Static CMOS Adders", *Advanced Materials Research*, Trans Tech Publications, Switzerland, Vol. 548, pp. 885–889, 2012.

98. Balwinder Raj, AK Saxena, S Dasgupta, "Nanoscale FinFET Based SRAM Cell Design: Analysis of Performance metric, Process Variation, Underlapped FinFET and Temperature Effect", *IEEE Circuits and System Magazine*, Vol. 11(2), pp. 38–50, 2011.

99. VK Sharma, M Pattanaik, Balwinder Raj, "Leakage Current ONOFIC Approach for Deep Submicron VLSI Circuit Design", *International Journal of Electrical*, Computer, Electronics and Communication Engineering, World Academy of Sciences, Engineering and Technology, Vol. 7(4), pp. 239–244, 2013.

100. Tulika Chawla, Mamta Khosla, B Raj, "Design and simulation of Triple metal Double-gate Germanium on Insulator Vertical Tunnel Field Effect Transistor', *Microelectronics Journal*, Elsevier; Vol. 114, p. 105125, Aug 2021.

101. Parminder Kaur, Sandeep Singh Gill, B Raj, "Comparative Analysis of OFETs Materials and Devices for Sensor Applications", *Journal of Silicon*, Springer, Vol. 14, pp. 4463–4471, 2022.

102. Sanjeev Kumar Sharma, Parveen Kumar, Balwant Raj, B Raj, "$In_{1-x}Ga_xAs$ Double Metal Gate-Stacking Cylindrical Nanowire MOSFET for Highly Sensitive Photo Detector", *Journal of Silicon,Springer Vol.* 14, pp. 3535–3541, 2022.

103. B Raj, AK Saxena, S Dasgupta, "Analytical Modeling of Quasi Planar Nanoscale Double Gate FinFET with Source/Drain Resistance and Field Dependent Carrier Mobility: A Quantum Mechanical Study", *Journal of Computer (JCP)*, Academy Publisher, Finland, Vol. 4(9), pp. 1–8, 2009.

104. S Bhushan, S Khandelwal, B Raj, "Analyzing Different Mode FinFET based Memory Cell at Different Power Supply for Leakage Reduction", *Seventh International Conference on Bio-Inspired Computing: Theories and Application, (BIC-TA 2012) Advances in Intelligent Systems and Computing*, Vol. 202, pp. 89–100, 2013.

105. Jeetendra Singh, B Raj, "Temperature Dependent Analytical Modeling and Simulations of Nanoscale Memristor", *Journal: Engineering Science and Technology, an International Journal*, Elsevier, Vol. 21, pp. 862–868, Oct 2018.

106. Shradhya Singh, Shashi Bala, Balwant Raj, B Raj, "Improved Sensitivity of Dielectric Modulated Junctionless Transistor for Nanoscale Biosensor Design", *Sensor Letter*, ASP, Vol. 18, pp. 328–333, Apr 2020.

107. Vivek Kumar, Santosh Kumar Vishvakarma, B Raj, "Design and Performance Analysis of ASIC for IoT Applications", *Sensor Letter*, ASP, Vol. 18, pp. 31–38, Jan 2020.

108. Akanksha Jaiswal, RK Sarin, B Raj, Shikha Sukhija, "A Novel Circular Slotted Microstrip-fed Patch Antenna with Three Triangle Shape Defected Ground Structure for Multiband Applications", *Advanced Electromagnetic (AEM)*, Vol. 7(3), pp. 56–63, Aug 2018.

109. Girish Wadhwa, B Raj, "Label Free Detection of Biomolecules using Charge-Plasma-Based Gate Underlap Dielectric Modulated Junctionless TFET", *Journal of Electronic Materials (JEMS)*, Springer, Vol. 47(8), pp. 4683–4693, Aug 2018.

110. Gurmohan Singh, RK Sarin, B Raj, **"Design and Performance Analysis of a New Efficient Coplanar Quantum-Dot Cellular Automata Adder"**, *Indian Journal of Pure & Applied Physics (IJPAP)*, Vol. 55, pp. 97–103, Feb 2017.

111. Amandeep Singh, Mamta Khosla, Balwinder Raj, "Design and Analysis of Electrostatic Doped Schottky Barrier CNTFET Based Low Power SRAM", *International Journal of Electronics and Communications, (AEÜ)*, Elsevier, Vol. 80, pp. 67–72, 2017.

112. Parminder Kaur, Vikas Pandey, B Raj, "Comparative Study of Efficient Design, Control and Monitoring of Solar Power using IoT", *Sensor Letter*, ASP, Vol. 18, pp. 419–426, May 2020.

113. Anil Kumar Bhardwaj, Sumeet Gupta, B Raj, "Development & Analysis of Compact Model for Double Gate Schottky Barrier CNTFET", *Journal of Nanoelectronics and Optoelectronics*, ASP, Vol. 15, pp. 1199–1208, Aug 2020.

114. Girish Wadhwa, Priyanka Kamboj, Jeetendra Singh, B Raj, "Design and Investigation of Junctionless DGTFET for Biological Molecule Recognition", *Transactions on Electrical and Electronic Materials*, Springer, Vol. 22, pp. 282–289, 2021.

115. Tulika Chawla, Mamta Khosla, B Raj, "Optimization of Double-gate Dual material GeOI-Vertical TFET for VLSI Circuit Design", *IEEE VLSI Circuits and Systems Letter*, Vol. 6(2), pp. 13–25, Aug 2020.

116. Sachin Kumar Verma, Shailendra Singh, Girish Wadhwa, B Raj, "Detection of Biomolecules using Charge-Plasma Based Gate Underlap Dielectric Modulated Dopingless TFET", *Transactions on Electrical and Electronic Materials (TEEM)*, Springer, Vol. 21, pp. 528–535, Jun 2020.

117. Neeraj Jain, B Raj, "Impact of Underlap Spacer Region Variation on Electrostatic and analog/RF Performance of Symmetrical High-k SOI FinFET at 20 nm Channel Length", *Journal of Semiconductors (JoS)*, IOP Science, Vol. 38(12), pp. 122002, Dec 2017.

118. Shailendra Singh, B Raj, "Analytical Modeling and Simulation analysis of T-shaped III-V heterojunction Vertical T-FET", *Superlattices and Microstructures*, Elsevier, Vol. 147, pp. 106717, Nov 2020.

119. Gurmohan Singh, RK Sarin, B Raj, "Design and Analysis of Area Efficient QCA Based Reversible Logic Gates", *Journal of Microprocessors and Microsystems*, Elsevier, Vol. 52, pp. 59–68, May 2017.

120. Amandeep Singh, Mamta Khosla, B Raj, "Compact Model for Ballistic Single Wall CNTFET under Quantum Capacitance Limit", *Journal of Semiconductors (JoS)*, IOP Science, Vol. 37, pp. 104001-8, Oct 2016.

121. Sonal Singh, Mamta Khosla, Girish Wadhwa, B Raj, "Design and Analysis of Double-Gate Junctionless Vertical TFET for Gas Sensing Applications", *Applied Physics A*, Springer, Vol. 127(16), 2 Jan 2021.

122. Inderjit Singh, B Raj, Mamta Khosla, B Rajesh Kumar Kaushik, "Potential MRAM Technologies for Low Power SoCs", *SPIN World Scientific Publisher*, SCIE; Vol. 10(4), pp. 2050027, Dec 2020.

123. Shailendra Singh, B Raj, "Parametric Variation Analysis on Hetero-junction Vertical t-shape TFET for Supressing Ambipolar Conduction", *Indian Journal of Pure and Applied Physics*, Vol. 58, pp. 478–485, Jun 2020.

124. Shailendra Singh, Girish Wadhwa, Balwinder Raj, "Design and Analysis of Dual Source Vertical Tunnel Field Effect Transistor for high performance", *Transactions on Electrical and Electronics Materials*, Springer, Vol. 21, pp. 74–82, Oct 2019.

125. Manjit Kaur, Neena Gupta, Sanjeev Kumar, B Raj, Arun Kumar Singh, "RF Performance Analysis of Intercalated Graphene Nanoribbon Based Global Level Interconnects", *Journal of Computational Electronics*, Springer, Vol. 19, pp. 1002–1013, Jun 2020.

126. Girish Wadhwa, B Raj, "Design and Performance Analysis of Junctionless TFET Biosensor for high sensitivity", *IEEE Nanotechnology*, Vol. 18, pp. 567–574, 2019.
127. Jeetendra Singh, B Raj, "Enhanced Nonlinear Memristor Model Encapsulating Stochastic Dopant Drift", *JNO, ASP*, Vol. 14, pp. 958–963, 2019.

Chapter 3

A perspective on smart environments in Indian smart cities

Charul Sharma, Sanjay Kumar Sharma, and Dharmendra Gill

3.1 INTRODUCTION

Cities are the bedrock of civilization. Reference [1] predicts that by 2050, 66% percent of the world's population will be residing in urban areas. At present, cities accommodate over 50% of the world's population in 2% of the geographic space available. The constant tussle between limited resources and urbanized living degrades the city environment, produces 80% of greenhouse gases (GHGs) and consumes 80% of the world's resources [2, 3]. India is expected to surpass China as the most populous country in the world by 2025, and Indian cities will experience a similar pattern of environmental degradation [4]. Against this backdrop, the ambitious plan of the Indian government to develop 100 smart cities across the nation is a welcome initiative that attempts to solve the problems associated with urban infrastructure in India [5].

Reference [6] defines a smart city in India as 'A city that provides core infrastructure and gives a decent quality of life to its citizens, a clean and sustainable environment and application of "Smart" solutions'. This underscores the central role of citizens in building smart cities and highlights that citizen participation forms an integral part of developing Indian smart cities. Reference [7] explores the Indian perspective of smart cities and identifies smart city components in Indian smart cities based on vision statements and citizen preferences. Smart city components further contain dimensions and indicators that are the building blocks of smart city performance assessment schemes [8]. After conducting a detailed analysis of 34 smart city performance assessment schemes, reference [9] points out that more than 80% of smart city assessment schemes around the globe select indicators based on literature review or expert opinion. References [10, 11] have developed performance assessment schemes for Indian smart cities; however, the indicators used for ranking Indian cities ignore the perspective of citizens and city authorities [12]. Clearly, the past studies overlook the two most crucial stakeholders in smart city development: citizens and city

DOI: 10.1201/9781003487692-3

officials; consequently, there exists a significant gap in incorporating their opinions in developing performance assessment schemes for smart cities.

We pioneer the process of selecting smart city indicators on the basis of stakeholder consultation to bridge this gap. This study is unique because it integrates online citizen survey data for selecting indicators for various smart city components for Indian smart cities. The survey data combines the viewpoints of both city officials, who set the development priorities in smart cities, and citizens, who vote for the most crucial issues among these set priorities. We collect the data for 64 smart cities from the smart city forum website, synthesize this data and justify the selection of smart environment indicators for Indian smart cities. The process of selecting indicators can be replicated for other smart city components, but for the purpose of this study, we focus on the smart environment, which is a critical smart city component with 90% coverage in various smart city performance assessment studies [13].

3.2 BACKGROUND

3.2.1 Smart City Mission

The Smart City Mission is a first-of-its-kind urban regeneration program in India, as it involves citizen consultation at an extensive level. The concept of smart cities was marketed to the masses as a promise of 100 new cities in Bharatiya Janata Party's pre-election manifesto [14]. The political party kept its promise when states were approached in August 2014 by the Ministry of Urban Development (MoUD) to submit names of existing cities to turn them into smart cities.

Initially, an equitable criterion of population weighing and capacity benchmarks was used for distributing 100 smart cities among the union territories (UTs) and states. Thereafter, the UTs and states participated in the mission through a smart cities challenge that comprised a two-stage selection process. For the first stage, prospective smart cities were shortlisted from UTs/states based on certain preconditions. These mandated the selection of smart cities in the various UTs/states on the basis of equitable criteria which gives equal weightage (50:50) to the number of statutory towns in each UT/state and its urban population. Therefore, it caps smart cities in a state at a certain number, and each UT/state consists of at least one smart city [15]. The second stage of the competition required each of the prospective 100 smart cities to compete for central government funding by preparing their smart city proposals through extensive citizen engagement. Smart city proposals comprised a bouquet of projects representing the people's hopes and aspirations. Aggregated at the national level, the 100 smart cities proposed 5,151 projects worth 205,018 crores in five years from their respective dates of selection [16].

3.2.2 Citizen participation in the Smart City Mission

The office memorandum on citizen consultation for preparing the smart city proposal stresses citizen involvement as a crucial criterion for assessing the smart city proposals of cities [17]. The memorandum directs smart cities to engage citizens in dialogue through various methods. These methods include information and communication technology (ICT) tools such as social media platforms (e.g., Twitter, Facebook, local media outlets (newspapers, radio, TV), bulk SMS, visual aids and the Government of India's MyGov.in platform [18]. MyGov.in acts as the primary facilitator of citizen consultation for the Smart Cities Mission. It offers diverse consultation methodologies such as discussion forums, tasks, online polls, public talks and blogs. Municipal authorities of shortlisted cities utilized MyGov.in to gather citizen suggestions for their vision of a smart city and incorporated them into smart city proposals. In addition, the discussion forum of MyGov.in was used for shortlisting ideas regarding smart solutions for improving infrastructure services in the city through an online poll. The online poll involved a list of potential smart solutions and priority areas for development that were put up in the public domain for a limited period of time so that citizens could deliberate and vote on the most appropriate solutions and the most crucial issues in their respective smart city. The most popular suggestions identified through the poll could be included in the city's proposal. MyGov.in platform garnered a total of 263,413 comments from 64 cities [19].

Thus, the online citizen survey for developing smart cities gives a two-fold perspective of Indian smart cities from the viewpoint of the two most important stakeholders in smart city development, namely, city officials/government [20] and citizens [10]. It reflects the priority areas for development that were provided by the city officials and respective votes secured in each area, as per the choice of citizens.

3.2.3 Smart city components and indicators

Reference [21] suggests a hierarchic structure for developing a smart city index that describes a smart city and its components for ranking purposes, as shown in Figure 3.1. In this structure, each level is described by the results of the level below. Each component is therefore defined by several dimensions. Furthermore, each dimension is described by a number of indicators. This is identical to the practice of constructing a composite indicator where several indicators are incorporated into a single index as per an underlying model; a similar procedure is used to create a composite indicator [22]. Composite indicators are the preferred choice when it comes to monitoring multifaceted concepts, such as multifaceted concepts like smart city performance, which are difficult to evaluate with a single indicator. The indicators that make up a composite indicator are referred to as components. A

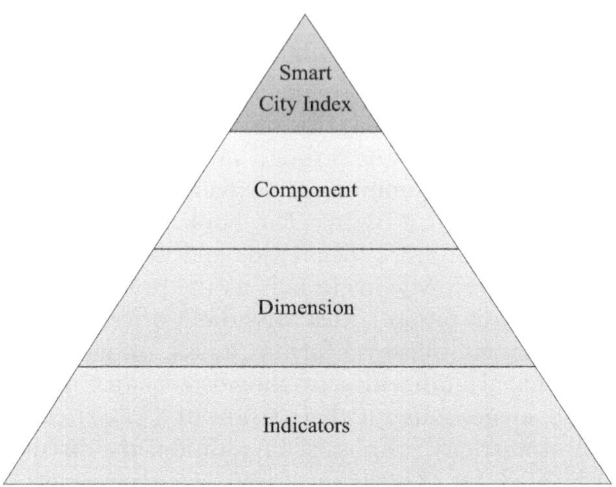

Figure 3.1 Hierarchal structure of a smart city index.

component may have several dimensions, each of which represents a distinct domain or facet of the phenomenon of interest. Each dimension comprises specific indicators that play a crucial role in quantifying, simplifying and communicating complex information, such as environmental impact, into easily understandable and usable forms [23]. They are essential for cities to establish targets, monitor progress and track performance [24]. Reference [13] identifies nine categories of smart city components. Out of these, people, environment, economy, living, mobility and governance are considered inextricable from the holistic outlook of smart cities.

3.2.4 Smart environment

'Smart environment' is one of the main predefined categories with the highest level of inclusion in smart city performance assessment studies [25]. This component focuses on sustainable urban development that provides desirous environmental outcomes for the city and its residents, both people and nonhumans. The strategy of the smart environment component reduces the city's ecological impact and enhances its capacity to endure climate shifts. The use of clean, renewable energy sources, waste management and resource optimization are all fostered by several aspects and indicators of the smart environment [26–30].

3.3 METHODOLOGY

A citizen-participatory approach is key for selecting indicators for assessing the performance of Indian smart cities. The methodology followed for identifying the dimensions and indicators is depicted in Figure 3.2. To integrate citizens' perspectives in the assessment of smart city performance, we started with a content analysis of an online citizen survey [7] available on the MyGov.in forum [18].

We collected data specific to each city from a total of 64 Indian smart cities. This data comprised the priority areas for development set by city

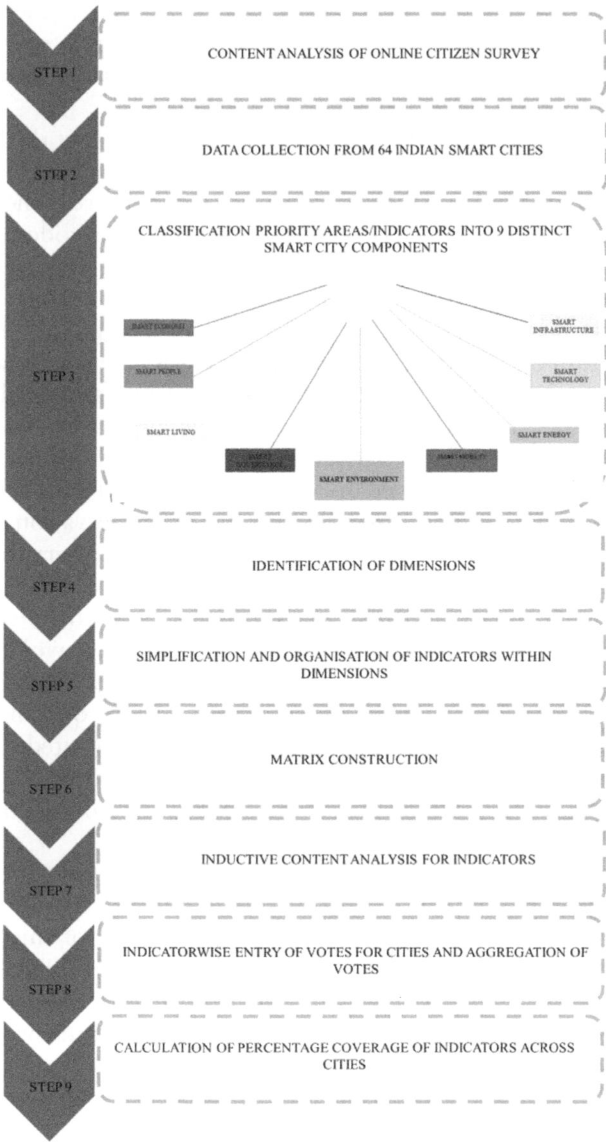

Figure 3.2 Stepwise methodology for identifying dimensions and indicators.

officials, along with the corresponding votes obtained by each priority area through citizen consultations, thus highlighting the most important issues for development according to the citizens in the respective smart city. This information was recorded on a Microsoft Excel sheet. Subsequently, the priority areas across all smart cities were classified into nine distinct categories of smart city components: 'smart environment', 'smart economy', 'smart people', 'smart mobility', 'smart living', 'smart governance', 'smart energy', 'smart infrastructure' and 'smart technology' [31–38]. These components were identified based on an extensive literature review of performance assessment schemes [39–43], which also encompass various dimensions and indicators within the smart city components [44–52].

For the purpose of this chapter, we focus exclusively on the 'smart environment' component, and the priority areas are considered as potential indicators in our study. For classifying the priority areas/indicators into individual categories, each indicator was manually color-coded for all cities, with each color representing a specific component of smart cities. Assigning indicators to their respective smart city components was primarily based on keywords associated with a particular smart city component derived from an extensive literature review of smart city assessment schemes. In some instances, we encountered indicators that could be associated with more than one category. For example, the indicator 'environment control – pollution (air/noise/water)' clearly falls under the 'smart environment' category. However, 'conservation of parks, playgrounds, lakes and greenery' belongs to 'smart environment' and 'smart living', as the conservation of lakes and greenery pertains to 'smart environment', while the conservation of parks and playgrounds falls under 'smart living'. In such cases, the indicator is assigned to both components, and the votes received by the indicator are distributed equally between them. In our study, we avoided considering 'energy', 'mobility' and 'infrastructure' indicators such as renewable energy initiatives, water conservation measures and eco-friendly mobility options in the 'smart environment' category since we wanted to investigate the extent of energy efficiency and infrastructure measures incorporated in Indian smart cities as a distinct topic from purely environmental indicators [53–61].

The allocation of indicators to their respective categories involves a lot of personal interpretation of indicators based on the author's judgment, which makes it a subjective process. The reduced objectivity of this process has also been noted in earlier works [9, 31]. Therefore, to introduce a degree of consistency and accuracy in counting, the first and the third authors independently performed the manual counting and the assignment of indicators to respective categories twice, with minor corrections [7, 9]. This iterative process was conducted for all cities. It ensured that every indicator was properly categorized, and this resulted in a comprehensive list of indicators.

In the next step, the collected indicators were simplified in language for consistency across various cities and organized systematically within the

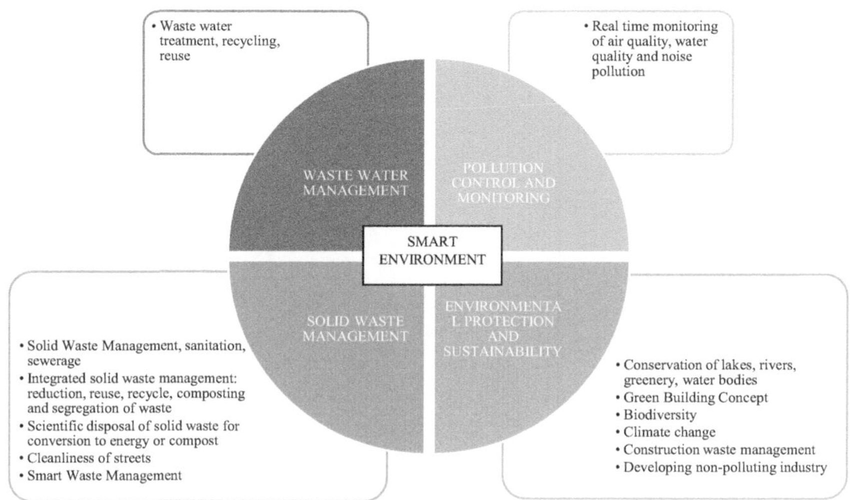

Figure 3.3 Smart city dimensions and indicators for the Indian context.

dimensions. These indicators were then analyzed by creating individual sheets for each smart city component using Microsoft Excel. However, in this context, we will focus on the 'smart environment' category [62–68].

For each component, we constructed matrices as shown in Table 3.1(a) and Table 3.1(b), where the rows represent dimensions and their corresponding indicators, while the columns represent the 64 smart cities. For organizing the indicators based on the cities in which they occurred, we conducted an inductive content analysis of the color-coded data [32, 33]. The process of inductive content analysis began alphabetically with the city of Agartala, whereby its relevant indicators were added to the list. While analyzing the indicators for the subsequent city (Agra), new indicators were added to the matrix only if they were absent in the sheet. Since different cities used different descriptions for similar indicators, we prevented redundancy in the database by avoiding duplicity of indicators in any form. For instance, some cities used the term 'scientific disposal of solid waste' while others used terms such as 'waste to energy' or 'waste to compost'. Despite the varying terminology, these indicators shared the same underlying meaning and were consolidated into a single indicator: 'scientific disposal of solid waste for conversion to energy or compost'.

After completing the initial organization of indicators, the votes obtained by the different indicators in various smart cities were entered into the sheet. This step served a dual purpose. First, it allowed us to record the relevant data, and second, it refined the process of indicator identification in the respective cities and ensured accuracy in our results with minor corrections made in the second round. Once the indicators with common features

Table 3.1(a) Occurrence of 'smart environment' indicators across Indian smart cities

DIMENSION	INDICATOR	AGARTALA	AGRA	AHMEDABAD	AIZAWL	ALIGARH	ALLAHABAD	AMRAVATI	AURANGABAD	BAREILLY	BHAGALPUR	BHUBHANESHWAR	BIHARSHARIF	BILASPUR	CHANDIGARH	CHENNAI	COIMBATORE
Wastewater management	Wastewater treatment, recycling, reuse	Y	Y							Y	Y	Y	Y	Y			
Pollution control and monitoring	Real-time monitoring of air quality, water quality and noise pollution	Y	Y	Y	Y					Y	Y			Y	Y		
Solid waste management	Solid waste management, sanitation, sewerage		Y	Y	Y	Y	Y	Y	Y	Y		Y	Y	Y	Y		Y
	Integrated solid waste management: reduction, reuse, recycle, composting and segregation of waste														Y		
	Scientific disposal of solid waste for conversion to energy or compost	Y			Y			Y	Y		Y						
	Cleanliness of streets		Y			Y											
	Smart waste management										Y		Y	Y	Y		
Environmental protection and sustainability	Conservation of lakes, rivers, greenery, water bodies				Y	Y			Y								
	Green building concept	Y			Y						Y						
	Biodiversity																
	Climate change																
	Construction waste management																
	Developing non-polluting industry							Y						Y			

DIMENSION	INDICATOR	DAHOD	DHARAMSHALA	ERODE	FARIDABAD	GANDHINAGAR	GUWAHATI	GWALIOR	ITANAGAR	JALANDHAR	JAMMU	JHANSI	KAKINADA	KANPUR	KARIMNAGAR	KALYAN DOMBIVILI	KARNAL
Wastewater management	Wastewater treatment, recycling, reuse							Y		Y							Y
Pollution control and monitoring	Real-time monitoring of air quality, water quality and noise pollution				Y					Y							
Solid waste management	Solid waste management, sanitation, sewerage	Y	Y	Y	Y			Y	Y		Y	Y	Y	Y	Y	Y	Y
	Integrated solid waste management: reduction, reuse, recycle, composting and segregation of waste											Y					Y
	Scientific disposal of solid waste for conversion to energy or compost							Y								Y	Y
	Cleanliness of streets																
	Smart waste management		Y			Y											
Environmental protection and sustainability	Conservation of lakes, rivers, greenery, water bodies		Y									Y	Y				
	Green building concept							Y									
	Biodiversity																
	Climate change																
	Construction waste management									Y							
	Developing non-polluting industry																

Table 3.1(b)   Occurrence of 'smart environment' indicators across Indian smart cities

DIMENSION	INDICATOR	KOHIMA	LUCKNOW	LUDHIANA	MANGALORE	MORADABAD	NAMCHI	NDMC	NEW RAIPUR	NEW TOWN KOLKATA	PANAJI	PASIGHAT
Wastewater management	Wastewater treatment, recycling, reuse		y		y						y	
Pollution control and monitoring	Real-time monitoring of air quality, water quality and noise pollution	y	y	y	y		y		y	y	y	y
Solid waste management	Solid waste management, sanitation, sewerage		y	y		y	y		y	y	y	y
	Integrated solid waste management: reduction, reuse, recycle, composting and segregation of waste		y					y				
	Scientific disposal of solid waste for conversion to energy or compost		y									
	Cleanliness of streets											
	Smart waste management											
Environmental protection and sustainability	Conservation of lakes, rivers, greenery, water bodies											
	Green building concept		y	y								
	Biodiversity								y			
	Climate change								y			
	Construction waste management											
	Developing non-polluting industry		y									

DIMENSION	INDICATOR	PATNA	PUDUCHERRY	RAJKOT	RANCHI	ROURKELA	SAGAR	SAHARANPUR	SHILLONG	SHIMLA	SHIVAMOGGA	SOLAPUR
Wastewater management	Wastewater treatment, recycling, reuse			y	y							
Pollution control and monitoring	Real-time monitoring of air quality, water quality and noise pollution			y							y	
Solid waste management	Solid waste management, sanitation, sewerage	y	y	y	y	y	y	y	y	y	y	y
	Integrated solid waste management: reduction, reuse, recycle, composting and segregation of waste			y								
	Scientific disposal of solid waste for conversion to energy or compost											
	Cleanliness of streets											
	Smart waste management											y
Environmental protection and sustainability	Conservation of lakes, rivers, greenery, water bodies							y				
	Green building concept											y
	Biodiversity											
	Climate change											
	Construction waste management											
	Developing non-polluting industry											

DIMENSION	INDICATOR	SURAT	THOOTHUKUDI	TIRUCHIRAPALLI	TIRUPATI	TIRUPPUR	UDAIPUR	VARANASI	VELLORE	VIZAG	WARANGAL
Wastewater management	Wastewater treatment, recycling, reuse	y			y			y			y
Pollution control and monitoring	Real-time monitoring of air quality, water quality and noise pollution	y								y	y
Solid waste management	Solid waste management, sanitation, sewerage	y	y		y	y	y			y	y
	Integrated solid waste management: reduction, reuse, recycle, composting and segregation of waste		y	y							
	Scientific disposal of solid waste for conversion to energy or compost							y			
	Cleanliness of streets										
	Smart waste management								y		
Environmental protection and sustainability	Conservation of lakes, rivers, greenery, water bodies	y	y		y	y	y				y
	Green building concept							y			
	Biodiversity										
	Climate change										
	Construction waste management							y			
	Developing non-polluting industry										

were grouped under the same dimension, the matrices were finally complete [69–76].

We aggregated the total number of votes obtained by different indicators across the various smart cities for the purpose of analysis. Lastly, we calculated the percentage coverage of each indicator across 64 smart cities by manually counting the number of cities that included a specific indicator and dividing it by the total number of smart cities.

3.4 RESULTS AND DISCUSSIONS

The online citizen survey for smart city development provides the number and relative percentage of votes secured by each priority area. Thus, a priority area with a higher percentage meant that it was more relevant for the citizens [78–86]. Despite the emphasis laid on a participatory approach for Indian smart city development, data on poll responses is only available for 64 out of the total 100 cities selected for the smart city challenge. The lack of data casts a shadow on the transparency of the process of selecting cities in the smart cities challenge. Data accessibility plays a crucial role in developing smart cities globally, as it extrapolates the complex urban dynamics and the interlinkages between different indicators [8, 34]. Therefore, the unavailability of poll data for 36 cities hampers the pro-citizen-centric conceptualization of Indian smart cities. Nevertheless, the indicators extracted from the poll data of 64 cities offer valuable insights into the priority areas identified by both city authorities and citizens [87–93].

3.4.1 City officials' perspectives

Table 3.1(a) and Table 3.1(b) depict the outcome of the content analysis conducted on the online poll data, illustrating the dimensions and indicators that constitute a smart environment in Indian smart cities. From the poll data, we extract four main dimensions of a smart environment: 'wastewater management', 'pollution control and monitoring', 'solid waste management' and 'environmental protection and sustainability'. Each dimension comprises specific indicators. The number of indicators within each dimension varies across the 64 cities. Some dimensions, such as 'wastewater management' and 'pollution control and monitoring', consist of a single indicator, while others, such as 'solid waste management' and 'environmental protection and sustainability', encompass five to six indicators, respectively. This discrepancy in the number of indicators reflects the diverse priorities and preferences of city authorities in the development of Indian smart cities [94–102]. Dimensions with a greater number of indicators indicate a higher level of relevance among the majority cities, as evident through the significant number of votes obtained in the online poll as shown in Figure 3.5. This corroborates the observations of [35] regarding the significance of indicators in performance assessment frameworks. Moreover, it suggests that there are additional challenges and a broader scope for integrating smart solutions into those dimensions.

Figure 3.4 illustrates the percentage coverage of each indicator across 64 smart cities. City officials perceive indicators with higher coverage across cities as more impactful and significant, leading them to prioritize the development of those areas in their smart cities. It highlights the indicators' importance in enhancing the overall well-being of the city. Among the individual indicators, 'solid waste management, sanitation and sewerage'

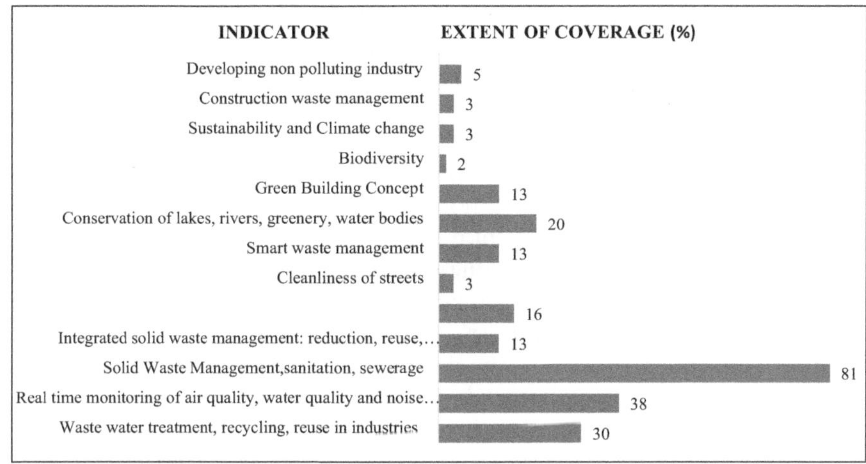

Figure 3.4 Percentage coverage of 'smart environment' indicators across 64 smart cities.

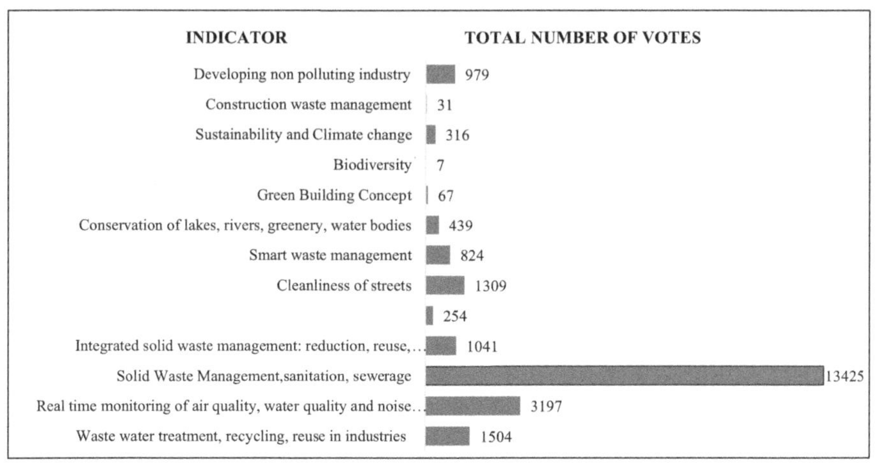

Figure 3.5 Number of votes secured by different indicators across the 64 smart cities.

exhibits the highest coverage, as 52 cities included it in the online poll. Consequently, it emerges as a crucial indicator for assessing 'smart environment' in smart city development [103–107]. Conversely, indicators such as 'cleanliness of streets', 'biodiversity', 'sustainability and climate change', 'construction waste management' and 'developing non-polluting industry' appear in fewer than five cities and have the lowest coverage. The limited coverage of these indicators may be attributed to their contextual specificity. These indicators hold importance for only a few cities that face distinct environmental challenges. For instance, 'developing non-polluting industries' is an indicator with low coverage, but it garners significant support in the cities of Bareilly and Ludhiana. Both cities are textile manufacturing hubs in North India, experiencing issues with air and water quality [36, 37]. Similarly, Panaji is the sole city that prioritizes biodiversity and climate change, as it is a coastal area that suffers from biodiversity erosion as a result of estuary pollution [38]. Therefore, these indicators become endemic to specific regions due to their contextual relevance. These findings highlight the role of regional distinctiveness in selecting indicators in smart city development and corroborate the notion that each city has its own idiosyncrasies in terms of its environmental challenges [7, 39].

On average, 12 cities cover a single indicator. However, indicators such as 'integrated solid waste management', 'scientific disposal of solid waste for conversion to energy and compost', 'smart waste management' and 'green building concept' are covered by eight to ten cities. The relatively lower coverage of indicators may be attributed to the technological aspects associated with these indicators and their novelty [108–115]. Nevertheless, infusing technological smart solutions into smart cities is a significant aspect of developing smart cities. The below-average inclusion of these indicators reflects the hesitance of Indian smart cities to prioritize smart solutions and technology in development.

On the other hand, indicators such as 'conservation of lakes, rivers, greenery, water bodies', wastewater treatment, recycling and reuse in industries' and 'real-time monitoring of air, water quality and noise pollution' appear in more than 15 cities. This indicates the persistent environmental issues that Indian smart city development focuses on. Among the 64 cities surveyed, Jammu and Coimbatore stand out as the only two cities that conducted online surveys but did not include any indicators classified under the 'smart environment' category. The lack of emphasis on smart environment indicators stresses the limited awareness regarding holistic smart city development in these cities [116–121].

3.4.2 Citizens' perspectives

The online survey for developing smart cities gathers citizens' input, fosters citizen engagement and aligns development priorities with people's needs and aspirations [40]. Figure 3.5 shows that 'solid waste management, sanitation and sewerage' received the highest number of votes across 64 cities. It underscores the importance of solid waste management, sanitation and sewerage for developing environmentally sustainable smart cities in India. It also indicates that the city officials have successfully captured the aspirations of the citizens, as the high coverage and maximum votes for this indicator show that there is a consensus between city authorities and citizens for improving this aspect of the environment [122–139].

In contrast, indicators such as 'construction waste management', 'sustainability and climate change', 'biodiversity', 'green building concept', 'conservation of lakes and water bodies' and 'scientific disposal of solid waste for conversion to energy or compost' received fewer than 500 votes. This suggests that citizens do not prioritize these environmental issues for developing smart cities.

The voting trend for different indicators indicates the existing priorities among citizens of a developing nation as opposed to those in developed nations of Europe and North America where such indicators are prioritized [41, 42]. This observation aligns with the comments of [7], who propose the theory that economies with comparable development problems to India could forward an alternative approach for developing smart cities., thereby challenging the Euro-American-centric discourse on smart cities, which recalibrates the authoritative knowledge by considering the diverse perspectives and priorities of developing nations [42].

'Real-time monitoring of air quality, water quality and noise pollution' received the second highest number of votes, highlighting the importance of a non-polluted environment for health-conscious citizens. There is a connection between 'real-time monitoring of air quality, water quality and noise pollution' and 'developing non-polluting industries', as the latter is the only indicator in the 'environment protection and sustainability' dimension that received a significant number of votes [130–137]. This reflects the awareness level of citizens as they focus not only on achieving the main indicator but also recognize the importance of addressing all the governing aspects of a particular environmental issue. Alternatively, indicators like 'pollution control and monitoring' receive no votes in the cities of Agartala, Mangalore and Panaji, reinforcing the impact of context specificity of smart city indicators [138–143]. This verifies the notion that citizens in different cities have varied expectations based on the environmental challenges endemic to their region. Hence, the development path for each smart city, even within the same country, is a pluralistic practice that takes cognizance of the specificities of the local context and can vary on account of the strengths and weaknesses of different cities [7, 43].

Contrarily, the low number of votes and inadequate representation of indicators such as 'biodiversity', 'construction waste management' and 'green building concept', all falling under the 'environment protection and sustainability dimension', indicates that the conservation of biodiversity is not a high priority in India, considering its status as a developing nation.

Given the emphasis on infrastructure development in a growing economy, 'construction waste management' also receives less attention [144–148]. The 'green building concept', which promotes the construction of energy-efficient buildings, is prominently prioritized in only seven cities across India, mostly Tier 1 cities with populations ranging from 100,000 to 500,000 [7]. This highlights that, despite the increasing popularity of the 'green building concept' in developed nations, most major cities in India, with the exception of Lucknow, exclude the 'green building concept' in their smart city development plans.

3.5 CONCLUSION

Selecting the right indicator framework for smart city performance assessment is a crucial but challenging process. We analyze the citizen survey data of 64 cities for integrating stakeholder perspectives in smart environment indicators for Indian smart cities. The study identifies four key dimensions: 'wastewater management', 'pollution control and monitoring', 'solid waste management' and 'environmental protection and sustainability'. This research emphasizes the importance of stakeholder engagement in the selection of smart city indicators. By considering the perspectives of both city authorities and citizens, decision-makers can align development goals with public aspirations, ensuring sustainable and inclusive urban development. City officials and citizens unanimously prioritize indicators such as 'solid waste management, sanitation and sewerage'. However, context-specific challenges and the limited coverage of indicators such as 'biodiversity', 'climate change' and 'construction waste management' add to the complexity of selecting indicators. But context-specific indicators account for the distinctiveness of smart cities, and thereby the findings of our study endorse adding context-specific recent urban issues as smart city indicators. This practice enhances the overall perception of smart cities, adds precision and diversity to the assessment process, and maintains the relevance of smart cities in urban development.

Moving forward, the insights gained from this study upgrade the decision-making processes and guide the smart city initiative in India. The future scope of study involves defining the various indicators for quantifying them into numerical values and assigning weights to the different indicators to account for their relative importance. By incorporating a diverse range of stakeholder perspectives and considering indicators that cover a broad spectrum of environmental concerns, cities strive toward a holistic and more effective approach to urban development. Ultimately, this study contributes to understanding smart city indicators and lays the foundation for more inclusive, sustainable and resilient cities in India and beyond.

ACKNOWLEDGMENT

The author acknowledges the contribution of Parul Sharma, Principal, ASHA School for Special Children, Jhansi, India, and Nitin Ankur, PhD Scholar, Dr B R Ambedkar National Institute of Technology, Jalandhar, India, for proofreading the manuscript and giving their valuable feedback.

REFERENCES

1. United Nations Department of Economic and Social Affairs, "*World Urbanization Prospects: The 2018 Revision*", Population Division. Accessed: Jun 30, 2023. [Online]. Available: https://population.un.org/wup/Country-Profiles/

2. R. Arbolino, F. Carlucci, A. Cira, G. Ioppolo, T. Yigitcanlar, "Efficiency of the EU regulation on greenhouse gas emissions in Italy: The hierarchical cluster analysis approach", *Ecological Indicators*, vol. 81, pp. 115–123, 2017, doi: 10.1016/j.ecolind.2017.05.053.

3. G. Ioppolo, S. Cucurachi, R. Salomone, G. Saija, L. Shi, "Sustainable local development and environmental governance: A strategic planning experience", *Sustainability (Switzerland)*, vol. 8, no. 2, 2016, doi: 10.3390/su8020180.

4. W. Jr. Charles, "*China and India, 2025: A Comparative Assessment*", Santa Monica, CA, 2011. doi: 10.1080/07448481.1987.9939004.

5. A. Sarkar, "*Shaping Indian Cities: Planning and Design with Smart City Technologies*", Msc Thesis, Delft University of Technology, BL Delft. [Online]. Available: http://repository.tudelft.nl/

6. Government of India, "*Smart Cities: Mission Statement & Guidelines*", New Delhi, 2015. doi: 10.1016/B978-0-08-097086-8.74017-7.

7. K. Gupta and R. P. Hall, "The Indian perspective of smart cities", in *Smart Cities Symposium Prague, SCSP 2017*, Prague: IEEE PROCEEDINGS, 2017. doi: 10.1109/SCSP.2017.7973837.

8. A. Sharifi, "A critical review of selected smart city assessment tools and indicator sets", *Journal of Cleaner Production*, vol. 233, 2019, doi: 10.1016/j.jclepro.2019.06.172.

9. A. Sharifi, "A typology of smart city assessment tools and indicator sets", *Sustainable Cities and Society*, vol. 53, no. May 2019, p. 101936, 2020, doi: 10.1016/j.scs.2019.101936.

10. M. Ashish, D. Gaurav, A. Farhan, S. Asis, "*Smart Cities Index: A tool for Evaluating Cities*", Mohali, 2017.

11. R. Ranjan, P. Chatterjee, D. Panchal, D. Pamucar, "Performance Evaluation of Sustainable Smart Cities in India", in *Advances in Environmental Engineering and Green Technologies (AEEGT)*, January, 2019, pp. 14–40. doi: 10.4018/978-1-5225-8579-4.ch002.

12. R. K. R. Kummitha, N. Crutzen, "Smart cities and the citizen-driven internet of things: A qualitative inquiry into an emerging smart city", *Technological Forecasting and Social Change*, vol. 140, no. May 2018, pp. 44–53, 2019, doi: 10.1016/j.techfore.2018.12.001.

13. C. Sharma, S. K. Sharma, D. Gill, "Holistic Smart Cities Viewed Through the Lens of Performance Evaluation Schemes", in *2022 International Conference on Computing, Communication and Intelligent Systems (ICCCIS)*, P. Nand, M. Singh, M. Kaur, V. Jain, Eds., Greater Noida, Uttar Pradesh, India: IEEE, 2022, pp. 983–989. doi: 10.1109/ICCCIS56430.2022.10037622.
14. Bharatiya Janata Party, "*Election Manifesto 2014.*" Accessed: Jun 30, 2023. [Online]. Available: https://www.bjp.org/images/pdf_2014/full_manifesto _english_07.04.2014.pdf
15. G. of I. Ministry of Housing and Urban Affairs, "*Smart Cities Mission.*" Accessed: Jun 01, 2023. [Online]. Available: http://164.100.161.224/content/ innerpage/no-of-smart-cities-in-each-state.php
16. Smart Cities Mission, "*Making a City Smart: Learnings from the Smart Cities Mission*", New Delhi, 2021. [Online]. Available: https://smartnet.niua.org/ content/2dae72ca-e25b-4575-8302-93e8f93b6bf6#:~:text=The purpose of the Smart,that leads to Smart outcomes.
17. Ministry of Urban Development, "*Office Memorandum No.K. 14012/101 (28)/2015-SC-III-A.*" Government of India, New Delhi, 2015.
18. "MyGov.in." Accessed: Jun 30, 2023. [Online]. Available: https://www. mygov.in/
19. G. Dwivedi, "*Smart City Consultations on MyGov.in*", MyGov team.
20. C. Chakhtoura, D. Pojani, "Indicator-based evaluation of sustainable transport plans: A framework for Paris and other large cities", *Transport Policy (Oxford)*, vol. 50, pp. 15–28, Aug 2016, doi: 10.1016/j.tranpol.2016.05.014.
21. R. Giffinger, C. Fertner, H. Kramar, E. Meijers, "City-ranking of European medium-sized cities", 2007.
22. OECD, "*Handbook on Constructing Composite Indicators: Methodology and User Guide*", Paris, France, 2008. doi: 10.1111/jgs.13392.
23. "United Nations Economic Commission for Europe Seminar", United Nations Economic Commission for Europe, 2017, pp. 27–34.
24. C. Colldahl, S. Frey, J. E. Kelemen, "*Smart Cities: Strategic Sustainable Development for an Urban World*", p. 63, 2013.
25. C. Sharma, S. K. Sharma, D. Gill, "Reassessing Smart City Components: An Overview of the Dynamic Nature of Smart City Concept", in *IOP Conference Series: Earth and Environmental Science*, 2023, p. 012017. doi: 10.1088/1755-1315/1186/1/012017.
26. P. Bosch, S. Jongeneel, V. Rovers, H.-M. Neumann, M. Airaksinen, A. Huovila, "CITYkeys: indicators for smart city projects and smart cities", 2017. doi: 10.13140/RG.2.2.17148.23686.
27. B. N. Silva, M. Khan, K. Han, "Towards sustainable smart cities: A review of trends, architectures, components, and open challenges in smart cities", *Sustainable Cities and Society*, vol. 38, no. Aug 2017, pp. 697–713, 2018, doi: 10.1016/j.scs.2018.01.053.
28. T. Yigitcanlar, N. Kankanamge, L. Butler, K. Vella, K. C. Desouza, "*Smart Cites Down Under: Performance of Australian Local Government Areas*", Brisbane Australia, 2020.
29. S. Chatterjee, H. Khas, A. K. Kar, H. Khas, "*Review and Policy Insights*", IEEE, pp. 2335–2340, 2015.
30. B. Cohen, "*What Exactly Is A Smart City?*", Fast Company, Sep 2012.

31. R. Kitchin, T. P. Lauriault, G. McArdle, "Knowing and governing cities through urban indicators, city benchmarking and real-time dashboards", *Regional Studies, Regional Science*, vol. 2, no. 1, pp. 6–28, 2015, doi: 10.1080/21681376.2014.983149.

32. A. Sharifi, "Smart city indicators: Towards exploring potential linkages to disaster resilience abilities", *APN Science Bulletin*, vol. 12, no. 1, pp. 75–89, 2022, doi: 10.30852/sb.2022.1873.

33. P. Mayring, "*Qualitative Content Analysis: Theoretical Foundation, Basic Procedures and Software Solution*", Klagenfurt, Austria, 2014. doi: 10.4135/9781446282243.n12.

34. G. White, A. Zink, L. Codecá, S. Clarke, "A digital twin smart city for citizen feedback", *Cities*, vol. 110, no. 2020, 2021, doi: 10.1016/j.cities.2020.103064.

35. H. Ahvenniemi, A. Huovila, I. Pinto-seppä, M. Airaksinen, "What are the differences between sustainable and smart cities ?", *Cities*, vol. 60, pp. 234–245, 2017, doi: 10.1016/j.cities.2016.09.009.

36. A. Mezzadri, "Globalisation, informalisation and the state in the Indian garment industry", *International Review of Sociology*, vol. 20, no. 3, pp. 491–511, 2010, doi: 10.1080/03906701.2010.511910.

37. A. Mezzadri, "Indian garment clusters and CSR norms: Incompatible agendas at the bottom of the garment commodity chain", *Oxford Development Studies*, vol. 42, no. 2, pp. 238–258, 2014, doi: 10.1080/13600818.2014.885939.

38. S. M. Parvez Al Usmani, Z. A. Ansari, "Status of coastal marine biodiversity of Goa and challenges for sustainable management -an overview", *Journal of Ecophysiology and Occupational Health*, vol. 20, no. 3 & 4, pp. 222–231, 2020, doi: 10.18311/jeoh/2020/25771.

39. A. Sharifi, "A critical review of selected smart city assessment tools and indicator sets", *Journal of Cleaner Production*, vol. 233, pp. 1269–1283, 2019, doi: 10.1016/j.jclepro.2019.06.172.

40. J. Tadili, H. Fasly, "Citizen participation in smart cities: A survey", *PervasiveHealth: Pervasive Computing Technologies for Healthcare*, Oct 2019, 2019, doi: 10.1145/3368756.3368976.

41. M. E. Cortés-Cediel, I. Cantador, M. P. R. Bolívar, "Analyzing Citizen participation and engagement in European smart cities", *Social Science Computer Review*, vol. 39, no. 4, pp. 592–626, 2019, doi: 10.1177/0894439319877478.

42. D. Prasad, T. Alizadeh, "What makes Indian cities smart? A policy analysis of smart cities mission", *Telematics and Informatics*, vol. 55, p. 101466, 2020, doi: 10.1016/j.tele.2020.101466.

43. A. Sharifi, A. Murayama, "Viability of using global standards for neighbourhood sustainability assessment: Insights from a comparative case study", *Journal of Environmental Planning and Management*, vol. 58, no. 1, pp. 1–23, 2015, doi: 10.1080/09640568.2013.866077.

44. Anjana Bhardwaj, Pradeep Kumar, B. Raj, Sunny Anand, "Design and performance optimization of doping-less vertical nanowire TFET using gate stack technique", *Journal of Electronic Materials (JEMS)*, Springer, vol. 41, no. 7, pp. 4005–4013, 2022.

45. Jeetendra Singh, B. Raj, "Tunnel current model of asymmetric MIM structure levying various image forces to analyze the characteristics of filamentary memristor", *Applied Physics A*, Springer, vol. 125, no.3, p. 203.1, Feb 2019.

46. Candy Goyal, Jagpal Singh Ubhi, B. Raj, "Low leakage zero ground noise nanoscale full adder using source biasing technique", *Journal of Nanoelectronics and Optoelectronics*, American Scientific Publishers, vol. 14, pp. 360–370, Mar 2019.

47. Gurmohan Singh, R. K. Sarin, B. Raj, "A novel robust exclusive-OR function implementation in QCA nanotechnology with energy dissipation analysis", *Journal of Computational Electronics*, Springer, vol. 15, no. 2, pp. 455–465, Jun 2016.

49. Girish Wadhwa, Priyanka Kamboj, Balwinder Raj" Design optimisation of junctionless TFET biosensor for high sensitivity", Advances in Natural Sciences: Nanoscience and Nanotechnology, vol. 10, p. 045001, 2019.

50. Priya Bansal, B. Raj, "Memristor modeling and analysis for linear dopant drift kinetics", *Journal of Nanoengineering and Nanomanufacturing*, American Scientific Publishers, vol. 6, pp. 1–7, 2016.

51. Amandeep Singh, Mamta Khosla, B. Raj, "Circuit compatible model for electrostatic doped schottky barrier CNTFET", *Journal of Electronic Materials*, Springer, vol. 45, no. 12, pp. 4825–4835, 2016.

52. Ashima, D Vaithiyanathan, B. Raj", Performance analysis of charge plasma induced graded channel SI nanotube", *Journal of Engineering Research (JER)*, EMSME Special Issue, pp. 146–154, Aug 2021.

53. Abhishek Singh Tomar, Vijay Kumar Magraiya, B. Raj, Scaling of access and data transistor for high performance DRAM cell design", *Quantum Matter*, vol. 2, pp. 412–416, Oct 2013.

54. Neeraj Jain, B. Raj, "Parasitic capacitance and resistance model development and optimization of raised source/drain SOI FinFET structure for analog circuit applications", *Journal of Nanoelectronics and Optoelectronins*, ASP, USA, vol. 13, pp. 531–539, Apl 2018.

55. Shradhya Singh, S. K. Vishvakarma, B. Raj, "Analytical modeling of split-gate junction-less transistor for a biosensor application", *Sensing and Biosensing*, Elsevier, vol. 18, pp. 31–36, Apr 2018.

56. Maisagalla Gopal, Balwinder Raj, "Low power 8T SRAM cell design for high stability video applications", *ITSI Transaction on Electrical and Electronics Engineering*, vol. 1, no. 5, pp. 91–97, 2013.

57. Balwinder Raj, Jatin Mitra, Deepak Kumar Bihani, Rangharajan V, A. K. Saxena, S. Dasgupta, "Analysis of noise margin, power and process variation for 32 nm FinFET based 6T SRAM cell", *Journal of Computer (JCP)*, Academy Publisher, Finland, vol. 5, no. 6, pp. 1–8, 2010.

58. Divya Sharma, Rajesh Mehra, B. Raj, "Comparative analysis of photovoltaic technologies for high efficiency solar cell design", *Superlattices and Microstructures*, Elsevier, vol. 153, p. 106861, May 2021.

59. Pawandeep Kaur, Avtar Singh Buttar, B. Raj, "A comprehensive analysis of nanoscale transistor based biosensor: A review", *Indian Journal of Pure and Applied Physics*, vol. 59, pp. 304–318, Apr 2021.

60. Divya Yadav, Balwant Raj, B. Raj, "Design and simulation of low power microcontroller for IoT applications", *Journal of Sensor Letters*, ASP, vol. 18, pp. 401–409, May 2020.

61. Shailendra Singh, B. Raj, "A 2-D analytical surface potential and drain current modeling of double-gate vertical t-shaped tunnel FET", *Journal of Computational Electronics*, Springer, vol. 19, pp. 1154–1163, Apr 2020.

62. Jeetendra Singh, B. Raj, "An accurate and generic window function for non-linear memristor model", *Journal of Computational Electronics*, Springer, vol. 18 no. 2, pp. 640–647, Jun 2019.

63. Manjit Kaur, Neena Gupta, Sanjeev Kumar, B. Raj, Arun Kumar Singh, "Comparative RF and crosstalk analysis of carbon based nano interconnects", *IET Circuits, Devices & Systems*, vol. 15, no. 6, pp. 493–503, Feb 2021.

64. Nehru Kandasamy, Firdous Ahmad, D. Ajitha, B. Raj, Nagarjuna Telagam "Quantum dot cellular automata based scan flip flop and boundary scan register", *IETE Journal of Research*, vol. 66, pp. 535–548, 2020.

65. S. K. Sharma, B. Raj, M. Khosla, "Enhanced photosensivity of highly spectrum selective cylindrical gate In1-xGaxAs nanowire MOSFET photodetector", *Modern Physics Letter-B*, vol. 33, No. 12, p. 1950144, 2019.

66. Jeetendra Singh, B. Raj, "Design and investigation of 7T2M NVSARM with enhanced stability and temperature impact on store/restore energy", *IEEE Transactions on Very Large Scale Integration Systems*, vol. 27, no. 6, pp. 1322–1328, Jun 2019.

67. Anil Kumar Bhardwaj, Sumeet Gupta, B. Raj, Amandeep Singh, "Impact of double gate geometry on the performance of carbon nanotube field effect transistor structures for low power digital design", *Computational and Theoretical Nanoscience*, ASP, vol. 16, pp. 1813–1820, 2019.

68. Neeraj Jain, B. Raj, "Thermal stability analysis and performance exploration of asymmetrical dual-k underlap spacer (ADKUS) SOI FinFET for security and privacy applications", *Indian Journal of Pure & Applied Physics (IJPAP)*, vol. 57, pp. 352–360, May 2019.

69. Amandeep Singh, Mamta Khosla, B. Raj, "Design and analysis of dynamically configurable electrostatic doped carbon nanotube tunnel FET", *Microelectronics Journal*, Elsevier, vol. 85, pp. 17–24, Mar 2019.

70. Neeraj Jain, Balwinder Raj, "Dual-k spacer region variation at the drain side of asymmetric SOI FinFET structure: Performance analysis towards the Analog/RF design applications", *Journal of Nanoelectronics and Optoelectronics*, American Scientific Publishers, vol. 14, pp. 349–359, Mar 2019.

71. Jeetendra Singh, Sanjeev Sharma, B. Raj, Mamta Khosla, "Analysis of barrier layer thickness on performance of In1-xGaxAs based gate stack cylindrical gate nanowire MOSFET," *JNO*, ASP, vol. 13, pp. 1473–1477, Oct 2018.

72. Neeraj Jain, B. Raj, "Analysis and performance exploration of high-k SOI FinFETs over the conventional low-k SOI FinFET toward analog/RF design", *Journal of Semiconductors (JoS)*, IOP Science, vol. 39, no. 12, p. 124002-1-7, Dec 2018.

73. Candy Goyal, Jagpal Singh Ubhi, B. Raj, "A reliable leakage reduction technique for approximate full adder with reduced ground bounce noise", *Journal of Mathematical Problems in Engineering*, Hindawi, vol. 2018, Article ID 3501041, p. 16, 15 Oct 2018.

74. Anuradha, Jeetendra Singh, B. Raj, Mamta Khosla, "Design and performance analysis of nano-scale memristor-based nonvolatile SRAM", *Journal of Sensor Letter*", American Scientific Publishers, vol. 16, pp. 798–805, Oct 2018.

75. Girish Wadhwa, B. Raj, "Parametric variation analysis of charge-plasma-based dielectric modulated JLTFET for biosensor application", *IEEE Sensor Journal*, vol. 18, no. 15, Aug 1, pp. 6070–6077, 2018.

76. Jeetendra Singh, B. Raj, "Comparative analysis of memristor models for memories design", *JoS*, IoP, vol. 39, no. 7, p. 074006-1-12, Jul 2018.

77. Divya Yadav, Shailesh Singh Chouhan, Santosh Kumar Vishvakarma, B. Raj, "Application specific microcontroller design for IoT based WSN", *Sensor Letter*, ASP, vol. 16, pp. 374–385, May 2018.

78. Gurmohan Singh, R. K. Sarin, B. Raj, "Fault-tolerant design and analysis of quantum-dot cellular automata based circuits", *IEEE/IET Circuits, Devices & Systems*, vol. 12, pp. 638–664, 2018.

79. Jeetendra Singh, B. Raj, "Modeling of mean barrier height levying various image forces of metal insulator metal structure to enhance the performance of conductive filament based memristor model", *IEEE Nanotechnology*, vol. 17, No. 2, pp. 268–267, Mar 2018.

80. Aakash Jain, Sanjeev Sharma, B. Raj, "Analysis of triple metal surrounding gate (TM-SG) III-V nanowire MOSFET for photosensing application", *Opto-Electronics Journal*, Elsevier, vol. 26, no. 2, pp. 141–148, May 2018.

81. Aakash Jain, Sanjeev Sharma, B. Raj, "Design and analysis of high sensitivity photosensor using cylindrical surrounding gate MOSFET for low power sensor applications", *Engineering Science and Technology, an International Journal*, Elsevier, vol. 19, no. 4, pp. 1864–1870, Dec 2016.

82. Amandeep Singh, Mamta Khosla, B. Raj, "Analysis of electrostatic doped schottky barrier carbon nanotube FET for low power applications", *Journal of Materials Science: Materials in Electronics*, Springer, vol. 28, pp. 1762–1768, 2017.

83. G. Saiphani Kumar, Amandeep Singh, B. Raj, "Design and analysis of gate all around CNTFET based SRAM cell design", *Journal of Computational Electronics*, Springer, vol. 17, no. 1, pp. 138–145, Mar 2018.

84. Gurinder pal Singh, B. S. Sohi, Balwinder Raj, "Material properties analysis of graphene base transistor (GBT) for VLSI analog circuits", *Indian Journal of Pure & Applied Physics (IJPAP)*, vol. 55, pp. 896–902, Dec 2017.

85. Amandeep Singh, Mamta Khosla, B. Raj, "Comparative analysis of carbon nanotube field effect transistor and nanowire transistor for low power circuit design", *Journal of Nanoelectronics and Optoelectronics*, American Scientific Publishers, USA, vol. 11, pp. 388–393, Jun 2016.

86. Sunil Kumar, B. Raj, "Estimation of stability and performance metric for inward access transistor based 6T SRAM cell design using n-type/p-type DMDG-GDOV TFET", *IEEE VLSI Circuits and Systems Letter*, vol. 3, no. 2, pp. 25–39, Jun 2017.

87. Shashikant Sharma, Anjan Kumar, Manisha Pattanaik, B. Raj, "Forward body biased multimode multi-threshold CMOS technique for ground bounce noise reduction in static CMOS adders", *International Journal of Information and Electronics Engineering*, vol. 3, no. 3, pp. 567–572, 2013.

88. Hamendra Singh, Pankaj Kumar, Balwinder Raj, "Performance analysis of majority gate set based 1-bit full adder", *International Journal of Computer and Communication Engineering (IJCCE)*, IACSIT Press Singapore, ISSN: 2010-3743, vol. 2, no. 4, 2013.

89. Anil Kumar Bhardwaj, Sumeet Gupta, B. Raj, "Investigation of parameters for schottky barrier (SB) height for schottky barrier based carbon nanotube field effect transistor device", *Journal of Nanoelectronics and Optoelectronics*, ASP, vol. 15, pp. 783–791, Jul 2020.

90. Priya Bansal, B. Raj, "Memristor: A versatile nonlinear model for dopant drift and boundary issues", *JCTN*, American Scientific Publishers, vol. 14, no. 5, pp. 2319–2325, May 2017.

91. Neeraj Jain, B. Raj, "An analog and digital design perspective comprehensive approach on Fin-FET (Fin-Field Effect transistor) technology: A review", *Reviews in Advanced Sciences and Engineering (RASE)*, ASP, vol. 5, pp. 1–14, 2016.

92. Sanjeev Sharma, B. Raj, Mamta Khosla, "Subthreshold performance of In1-xGaxAs based dual metal with gate stack cylindrical/surrounding gate nanowire MOSFET for low power analog applications", *Journal of Nanoelectronics and Optoelectronics*, American Scientific Publishers, USA, vol. 12, pp. 171–176, 2017.

93. Balwinder Raj, A. K. Saxena, S. Dasgupta, "Analytical modeling for the estimation of leakage current and subthreshold swing factor of nanoscale double gate FinFET device", *Microelectronics International, UK*, vol. 26, pp. 53–63, 2009.

94. Soniya, Shailendra Singh, Girish Wadhwa, Balwinder Raj, "An analytical modeling for dual source vertical tunnel field effect transistor", *International Journal of Recent Technology and Engineering (IJRTE)*, vol. 8, no. 2, pp. 603–608, Jul 2019.

95. Shailendra Singh, B. Raj, "Design and analysis of hetrojunction vertical T-shaped tunnel field effect transistor", *Journal of Electronics Material*, Springer, vol. 48, no. 10, pp. 6253–6260, Oct 2019.

96. Candy Goyal, Jagpal Singh Ubhi, B. Raj, "A low leakage CNTFET based inexact full adder for low power image processing applications", *International Journal of Circuit Theory and Applications*, Wiley, vol. 47, no. 9, pp. 1446–1458, Sep 2019.

97. B. Raj, A. K. Saxena, S. Dasgupta, "A compact drain current and threshold voltage quantum mechanical analytical modeling for FinFETs", *Journal of Nanoelectronics and Optoelectronics (JNO)*, USA, vol. 3, no. 2, pp. 163–170, 2008.

98. Girish Wadhwa, B. Raj, "An analytical modeling of charge plasma based tunnel field effect transistor with impacts of gate underlap region", *Superlattices and Microstructures*, Elsevier, vol. 142, p. 106512, Jun 2020.

99. Shailendra Singh, B. Raj",Modeling and simulation analysis of SiGe hetrojunction double GateVertical t-shaped tunnel FET", *Superlattices and Microstructures*, Elsevier, vol. 142, p. 106496, Jun 2020.

100. Amandeep Singh, Dinesh Kumar Saini, Dinesh Agarwal, Sajal Aggarwal, Mamta Khosla, B. Raj, "Modeling and simulation of carbon nanotube field effect transistor and its circuit application", *Journal of Semiconductors (JoS)*, IOP Science, vol. 37, p. 074001-6, Jul 2016.

101. Neeraj Jain, Balwinder Raj, "Device and circuit co-design perspective comprehensive approach on FinFET technology: A review", *Journal of Electron Devices*, vol. 23, no. 1, pp. 1890–1901, 2016.

102. Sunil Kumar, B. Raj, "Analysis of ION and ambipolar current for dual-material gate-drain overlapped DG-TFET", *Journal of Nanoelectronics and Optoelectronics*, American Scientific Publishers, USA, vol. 11, pp. 323–333, Jun 2016.

103. Naveed Anjum, Tarun Bali, B. Raj, "Design and simulation of handwritten multiscript character recognition", *International Journal of Advanced*

Research in Computer and Communication Engineering, vol. 2, no. 7, pp. 2544–2549, Jul 2013.

104. Sanjeev Sharma, B. Raj, Mamta Khosla, "A gaussian approach for analytical subthreshold current model of cylindrical nanowire FET with quantum mechanical effects", *Microelectronics Journal*, Elsevier, vol. 53, pp. 65–72, Apr 2016.

105. Karmjit Singh, B. Raj, "Performance and analysis of temperature dependent multi-walled carbon nanotubes as global interconnects at different technology nodes", *Journal of Computational Electronics*, Springer, vol. 14 no. 2, pp. 469–476, Jun 2015.

106. Sunil Kumar, B. Raj, "Compact channel potential analytical modeling of DG-TFET based on evanescent–mode approach", *Journal of Computational Electronics*, Springer, vol. 14, no. 2, pp. 820–827, Jul 2015.

107. Karmjit Singh, B. Raj, "Temperature dependent modeling and performance evaluation of multi-walled CNT and single-walled CNT as global interconnects", *Journal of Electronic Materials*, Springer, vol. 44, no. 12, pp. 4825–4835, Dec 2015.

108. V. K. Sharma, M. Pattanaik, B. Raj, "INDEP approach for leakage reduction in nanoscale CMOS circuits", *International Journal of Electronics*, Taylor & Francis, vol. 102, no. 2, pp. 200–215, 2014.

109. Karmjit Singh, B. Raj, "Influence of temperature on MWCNT bundle, SWCNT bundle and copper interconnects for nanoscaled technology nodes", *Journal of Materials Science: Materials in Electronics*, Springer, vol. 26, no. 8, pp. 6134–6142, 2015.

110. Naveed Anjum, Tarun Bali, B. Raj, "Design and simulation of handwritten gurumukhi and devanagri numerical recognition", *International Journal of Computer Applications*, Foundation of Computer Science, New York, USA, vol. 73, no. 12, pp. 16–21, 2013.

111. S. Khandelwal, V. Gupta, B. Raj, R. D. Gupta, "Process variability aware low leakage reliable nano scale DG-FinFET SRAM cell design technique", *Journal of Nanoelectronics and Optoelectronics*, vol. 10, no. 6, pp. 810–817, Dec 2015.

112. V. K. Sharma, M. Pattanaik, B. Raj, "ONOFIC approach: Low power high speed nanoscale VLSI circuits design", *International Journal of Electronics*, Taylor & Francis, vol. 101, no. 1, pp. 61–73, 2014.

113. S. Khandelwal, Balwinder Raj, R. D. Gupta, "FinFET based 6T SRAM cell design: Analysis of performance metric, process variation and temperature effect", *Journal of Computational and Theoretical Nanoscience*, ASP, USA, vol. 12, pp. 2500–2506, 2015.

114. Sumit Singh, Shekhar Yadav, Jagdeep Rahul, Anurag Srivastava, B. Raj, "Impact of HfO2 in graded channel dual insulator double gate MOSFET", *Journal of Computational and Theoretical Nanoscience*, American Scientific Publishers, vol. 12, no. 6, pp. 950–953, Apr 2015.

115. Vijay Kumar Sharma, Manisha Pattanaik, B. Raj, "PVT variations aware low leakage INDEP approach for nanoscale CMOS circuits", *Microelectronics Reliability*, Elsevier, vol. 54, pp. 90–99, 2014.

116. B. Raj, A. K. Saxena, S. Dasgupta, "Quantum mechanical analytical modeling of nanoscale DG FinFET: Evaluation of potential, threshold

voltage and source/drain resistance", *Elsevier's Journal of Material Science in Semiconductor Processing*, Elsevier, vol. 16, no. 4, pp. 1131–1137, 2013.

117. Maisagalla Gopal, Siva Sankar D. Prasad, Balwinder Raj, "8T SRAM cell design for dynamic and leakage power reduction", *International Journal of Computer Applications*, Foundation of Computer Science, New York, USA, vol. 71, no. 9, pp. 43–48, Jun 2013.

118. Manisha Pattanaik, B. Raj, Shashikant Sharma, Anjan Kumar, "Diode based trimode multi-threshold CMOS technique for ground bounce noise reduction in static CMOS adders", *Advanced Materials Research*, Trans Tech Publications, Switzerland, vol. 548, pp. 885–889, 2012.

119. Balwinder Raj, A. K. Saxena, S. Dasgupta, "Nanoscale FinFET based SRAM cell design: Analysis of performance metric, process variation, underlapped FinFET and temperature effect", *IEEE Circuits and System Magazine*, vol. 11, no. 2, pp. 38–50, 2011.

120. V. K. Sharma, M. Pattanaik, Balwinder Raj, "Leakage current ONOFIC approach for deep submicron VLSI circuit design", *International Journal of Electrical, Computer, Electronics and Communication Engineering, World Academy of Sciences, Engineering and Technology*, vol. 7, no. 4, pp. 239–244, 2013.

121. Tulika Chawla, Mamta Khosla, B. Raj, "Design and simulation of triple metal double-gate germanium on insulator vertical tunnel field effect transistor", *Microelectronics Journal*, Elsevier, vol. 114, p. 105125, Aug 2021.

122. Parminder Kaur, Sandeep Singh Gill, B. Raj, "Comparative analysis of OFETs materials and devices for sensor applications", *Journal of Silicon*, Springer, vol. 14, pp. 4463–4471, 2022.

123. Sanjeev Kumar Sharma, Parveen Kumar, Balwant Raj, B. Raj, "In1-xGaxAs double metal gate-stacking cylindrical nanowire MOSFET for highly sensitive photo detector", *Journal of Silicon*, Springer, vol. 14, pp. 3535–3541, 2022.

124. B. Raj, A. K. Saxena, S. Dasgupta, "Analytical modeling of quasi planar nanoscale double gate FinFET with source/drain resistance and field dependent carrier mobility: A quantum mechanical study", *Journal of Computer (JCP)*, Academy Publisher, Finland, vol. 4, no. 9, pp. 1–8, 2009.

125. S. Bhushan, S. Khandelwal, B. Raj, "Analyzing different mode FinFET based memory cell at different power supply for leakage reduction", *Seventh International Conference on Bio-Inspired Computing: Theories and Application, (BIC-TA 2012) Advances in Intelligent Systems and Computing*, vol. 202, pp. 89–100, 2013.

126. Jeetendra Singh, B. Raj, "Temperature dependent analytical modeling and simulations of nanoscale memristor", *Journal: Engineering Science and Technology, an International Journal*, Elsevier, vol. 21, pp. 862–868, Oct 2018.

127. Shradhya Singh, Shashi Bala, Balwant Raj, B. Raj, "Improved sensitivity of dielectric modulated junctionless transistor for nanoscale biosensor design", *Sensor Letter*, ASP, vol. 18, pp. 328–333, Apr 2020.

128. Vivek Kumar, Santosh Kumar Vishvakarma, B. Raj, "Design and performance analysis of ASIC for IoT applications", *Sensor Letter*, ASP, vol. 18, pp. 31–38, Jan 2020.

129. Akanksha Jaiswal, R. K. Sarin, B. Raj, Shikha Sukhija, "A novel circular slotted microstrip-fed patch antenna with three triangle shape defected ground structure for multiband applications", *Advanced Electromagnetic (AEM)*, vol. 7, no. 3, pp. 56–63, Aug 2018.

130. Girish Wadhwa, B. Raj, "Label free detection of biomolecules using charge-plasma-based gate underlap dielectric modulated junctionless TFET", *Journal of Electronic Materials (JEMS)*, Springer, vol. 47, no. 8, pp. 4683–4693, Aug 2018.

131. Gurmohan Singh, R. K. Sarin, B. Raj, "Design and performance analysis of a new efficient coplanar quantum-dot cellular automata adder", *Indian Journal of Pure & Applied Physics (IJPAP)*, vol. 55, pp. 97–103, February 2017.

132. Amandeep Singh, Mamta Khosla, Balwinder Raj, "Design and analysis of electrostatic doped schottky barrier CNTFET based low power SRAM", *International Journal of Electronics and Communications, (AEÜ)*, Elsevier, vol. 80, pp. 67–72, 2017.

133. Parminder Kaur, Vikas Pandey, B. Raj, "Comparative study of efficient design, control and monitoring of solar power using IoT", *Sensor Letter*, ASP vol. 18, pp. 419–426, May 2020.

134. Anil Kumar Bhardwaj, Sumeet Gupta, B. Raj, "Development & analysis of compact model for double gate schottky barrier CNTFET", *Journal of Nanoelectronics and Optoelectronics*, ASP, vol. 15, pp. 1199–1208, Aug 2020.

135. Girish Wadhwa, Priyanka Kamboj, Jeetendra Singh, B. Raj, "Design and investigation of junctionless DGTFET for biological molecule recognition", *Transactions on Electrical and Electronic Materials*, Springer, vol. 22, pp. 282–289, 2021.

136. Tulika Chawla, Mamta Khosla, B. Raj, "Optimization of double-gate dual material GeOI-vertical TFET for VLSI circuit design", *IEEE VLSI Circuits and Systems Letter*, vol. 6, no. 2, pp. 13–25, Aug 2020.

137. Sachin Kumar Verma, Shailendra Singh, Girish Wadhwa, B. Raj, "Detection of biomolecules using charge-plasma based gate underlap dielectric modulated dopingless TFET", *Transactions on Electrical and Electronic Materials (TEEM)*, Springer, vol. 21, pp. 528–535, Jun 2020.

138. Neeraj Jain, B. Raj, "Impact of underlap spacer region variation on electrostatic and analog/RF performance of symmetrical high-k SOI FinFET at 20 nm channel length", *Journal of Semiconductors (JoS)*, IOP Science, vol. 38, no. 12, pp. 122002, Dec 2017.

139. Shailendra Singh, B. Raj", Analytical modeling and simulation analysis of T-shaped III-V heterojunction Vertical T-FET", *Superlattices and Microstructures*, Elsevier, vol. 147, p. 106717, Nov 2020.

140. Gurmohan Singh, R. K. Sarin, B. Raj, "Design and analysis of area efficient QCA based reversible logic gates", *Journal of Microprocessors and Microsystems*, Elsevier, vol. 52, pp. 59–68, May 2017.

141. Amandeep Singh, Mamta Khosla, B. Raj, "Compact model for ballistic single wall CNTFET under quantum capacitance limit", *Journal of Semiconductors (JoS)*, IOP Science, vol. 37, p. 104001-8, Oct 2016.

142. Sonal Singh, Mamta Khosla, Girish Wadhwa, B. Raj, "Design and analysis of double-gate junctionless vertical TFET for gas sensing applications", *Applied Physics A*, Springer, vol. 127, no.16, 2 Jan 2021.

143. Inderjit Singh, B. Raj, Mamta Khosla, B. Rajesh Kumar Kaushik, "Potential MRAM technologies for low power SoCs", *SPIN World Scientific Publisher*, SCIE, vol. 10, no. 4, p. 2050027, Dec 2020.

144. Shailendra Singh, B. Raj, "Parametric variation analysis on hetero-junction vertical t-shape TFET for supressing ambipolar conduction", *Indian Journal of Pure and Applied Physics*, vol. 58, pp. 478–485, Jun 2020.

145. Shailendra Singh, Girish Wadhwa, Balwinder Raj, "Design and analysis of dual source vertical tunnel field effect transistor for high performance", *Transactions on Electrical and Electronics Materials*, Springer, vol. 21, pp. 74–82, Oct 2019.

146. Manjit Kaur, Neena Gupta, Sanjeev Kumar, B. Raj, Arun Kumar Singh, "RF performance analysis of intercalated graphene nanoribbon based global level interconnects", *Journal of Computational Electronics*, Springer, vol. 19, pp. 1002–1013, Jun 2020.

147. Girish Wadhwa, B. Raj, "Design and performance analysis of junctionless TFET biosensor for high sensitivity", *IEEE Nanotechnology*, vol. 18, pp. 567–574, 2019.

148. Jeetendra Singh, B. Raj, "Enhanced nonlinear memristor model encapsulating stochastic dopant drift", *JNO, ASP*, vo. 14, pp. 958–963, 2019.

Chapter 4

Evolutions of semiconductor solar cells

*Balwinder Raj, Mandeep Singh,
Balwant Raj, and Meenakshi Devi*

4.1 INTRODUCTION

Semiconductor solar panels are electrical devices made to transform sunlight into energy. They are also known as photovoltaic panels or solar cells. They are essential for producing clean, renewable energy. The interesting path of technical development and scientific investigation may be seen in the growth of semiconductor solar cells. These devices have completely changed how we harvest solar energy and have been essential in the advancement of alternative sources of energy. Let's look at the major turning points in the development of semiconductor solar cells [1].

Historical discoveries: In the 19th century, researchers Antoine Becquerel and Alexandre-Edmond Becquerel discovered the photovoltaic effect, which showed that some materials produced electric current when under ultraviolet light. These preliminary studies paved the way for more investigation.

i. **Initial solar cell (1880s):** Employing a thin coating of gold-coated selenium, American inventor Charles Fritts built the first real solar cell in 1888. Despite its poor efficiency, it was the first use of solar energy [2].
ii. **Einstein's explanation (1905):** Albert Einstein's seminal research on the photoelectric phenomenon in 1905 gave rise to empirical knowledge of how light and electrons behave in semiconductors, which was crucial for the advancement of solar power cells.
iii. **1950s through 1960s:** The dominance of silicon began when it was used as the main building block for solar cell manufacturing in the 1950s and 1960s. In 1954, scientists Calvin Fuller, Daryl Chapin, and Gerald Pearson of Bell Labs created the first usable silicon solar cell. Such early cells had an efficiency of about 6%.
iv. **Space exploration (1950s–1960s):** Because of its dependability and capacity to produce power in the thin air of space, solar cells became increasingly popular during this time. The initial spacecraft to employ solar cells for electricity was Vanguard 1, which went into orbit in 1958.
v. **Performance improvements: 1970s–1980s:** The performance of solar cells was the primary area of study. Substantial enhancements in

DOI: 10.1201/9781003487692-4

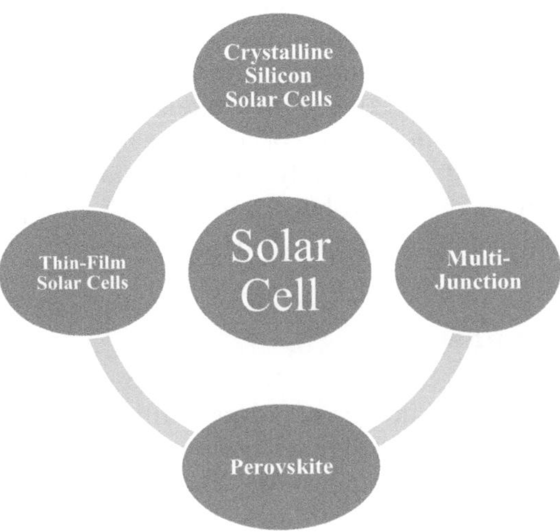

Figure 4.1 Classifications of a solar cell.

efficiency were made possible by discoveries like the creation of the initial solar cells with multiple junctions and the use of anti-reflective coatings.

vi. **Thin-film innovations (1980s–1990s):** As a substitute to conventional silicon cells, thin-film solar cell technologies such as amorphously formed silicon (a-Si), cadmium telluride (CdTe), and copper indium gallium selenide (CIGS) developed. These methods provided benefits in the areas of production simplicity, affordability, and adaptability.

vii. **21st century:** Revolutionary developments in semiconductor solar cell fabrication have occurred in the twenty-first century. Efficiency rates have surpassed 40% thanks to high-efficiency solar cell technologies including tandem cells and perovskite-silicon tandem cells [3–14].

4.2 CLASSIFICATION OF SOLAR CELLS

Photovoltaic cells, commonly referred to as solar cells, are available in a variety of forms, with each having unique properties and uses. The following sections discuss a few typical categories for solar cells, as shown in Figure 4.1.

4.3 FIRST-GENERATION SOLAR CELLS

Most solar cells presently on the market are based on silicon wafers, the so-called first-generation technology. As this technology has matured, costs have become increasingly dominated by material costs, mostly those of the silicon wafer, the strengthened low-iron glass cover sheet, and other encapsulates [15–22]. One of the major factors of energy loss in a solar cell is the gap between the photon energy and the bandgap energy Eg of the photovoltaic material. No absorption would occur if the photon energy was smaller than the bandgap energy, and merely the part equal to the bandgap energy out of the photon energy could be extracted as electric power leaving the other part wasted as heat if larger [23–31].

4.4 IBC SOLAR PANEL

A single more significant subtype of polycrystalline solar panels is the interdigitated back-contact (IBC) solar cell. Among its benefits is the fact that it may assist in creating a customized idea that matches the energy needs of certain local circumstances [32–38]. IBC cells use an additional creativity: connections are positioned on the back of the cell rather than the top. Due to less shadowing on the exterior of the cell, it is able to operate more efficiently, and at the same time, electron-hole pairs produced by the sunlight that is absorbed can still be gathered on the back of the cell. In the 1980s, Stanford University created the initial IBC solar cell, which had a 21.3% performance [4]. For effective light absorption among the silicon wafers, the front side of the IBC solar cell is texturized with random pyramidal during the manufacturing operation. The back side is cleaned. The front

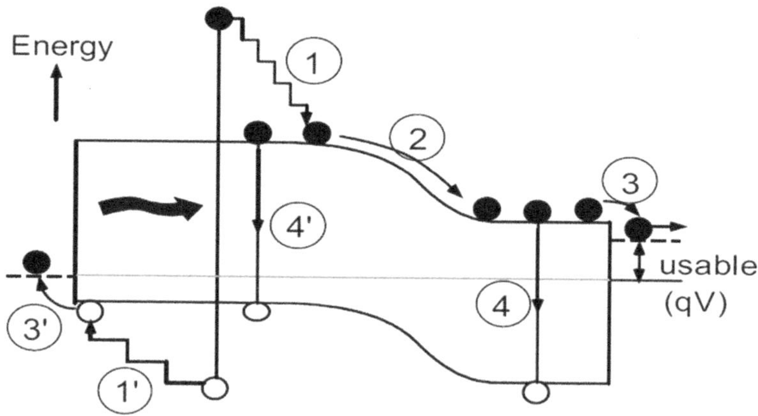

Fig. 1. Loss processes in a standard solar cell: (1) thermalisation loss; (2) and (3) junction and contact voltage loss; (4) recombination loss.

Figure 4.2 First-generation solar cell.

and rear area fields are created by ion implementation or POCl$_3$ diffusion. The process of screen printing or lithography determines the back emitter design. On the reverse end of the cell, there is an interdigitated architecture that contains the dual emitter and the rear-surface field doping levels [39–47]. Figure 4.3 illustrates an IBC solar cell. Optical shadowing effects on the outer end are eliminated by using all of the back connections in the cell's construction. As a consequence, the IBC solar cell's ability to absorb and bypass current density both rises. Additional benefits include (1) more space due to the lack of frontal metal fingers, (2) lower serial resistance on the reverse side, and (3) easier module layouts. The majority of IBC solar power panels are made of n-type silicon panels with scattered boron emitters. A pyramid shape, having a reflective layer, is employed to enhance the capture of light effect. The front and back heat SiO$_2$ passivation layers exhibit a long-wave reaction and have less surface interaction. In order to achieve a direct connection with the silicon base, an electrode of metal is produced. By getting rid of front panels and busbars, the efficiency of the absorbent may be enhanced [48–56]. Higher photo-absorption ability is the result of an expanded functional surface. The cell's structure still has to be improved, though. A number of problems arise when dual lithography is in use. To prevent unequal distribution and elevated temperature impairment, expenses rise. A boron-doped diffusing masking sheet and an interdigitated pattern printed on the back surface can simplify the procedure. Yet the expense of protective measures to prevent uneven dispersion and elevated temperature damage is substantially higher. Additionally, the positioning accuracy and printing reproducibility of the printing by screen method are poor. Doping by ion deployment, which carefully controls the amount of doping and enables the production of homogeneous p- and n-regions as well as a regulated junction dimension, is often used to circumvent these issues. PV companies find it challenging to obtain the extremely high annealing temperatures needed for this doping process. Alternatives include the use of laser doping methods [5]. This provides consistent doping level, depth

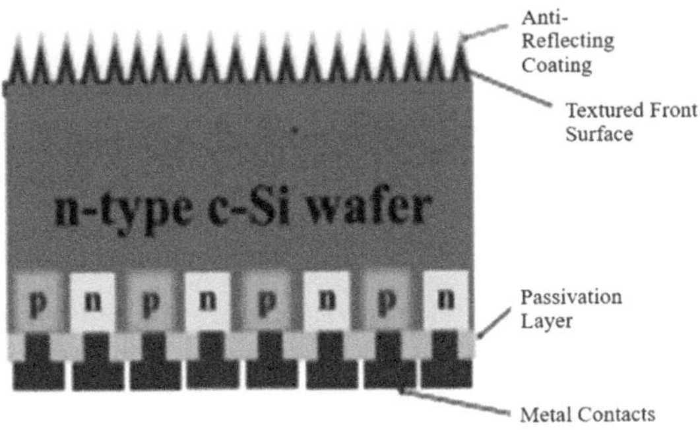

Figure 4.3 IBC solar cell structure [4].

variability, patterning capacity, and doped region uniformity. The selected doping region shields all of the silicon from being damaged by high temperatures. The capability of laser texture, laser treatment for connection launching, and laser-fired connections in solar cells are some further benefits of laser doping. In their initial iteration, IBC solar cells had a yield of 21.3%. In 2010, Sun Energy attained an unparalleled success of over 24% [6]. In 2016, Sun Power achieved the greatest IBC structural performance of 25.2% [7]. In uses wherein greater current ratings are required, such as CPV, solar race cars, and light planes, IBC cells are extremely valuable. IBC cells have an 8% revenue share worldwide, based on the International Technology Roadmap study from 2018, and in the next ten years, that percentage is predicted to grow by at least 5%.

Advantages of IBC:

i. **High efficiency:** IBC solar cells are renowned for having excellent performance. The interdigitated structure improves the extraction of charge carriers and lowers coupling damages, increasing the overall effectiveness in the conversion of solar energy into electricity [57–62].

ii. **Minimizes shading losses:** The layout of IBC cells minimizes shading losses. Because the contact points are located on the exterior of the cell, absorption by the grids and busbars has less of an influence on the outer appearance, which is more vulnerable to sunlight.

iii. **Increased effectiveness:** Since the connections are less visible on the underside of the solar cell, there is less chance of rust along with other environmental impacts with the back-contact layout.

iv. **Improved aesthetics:** IBC solar panels frequently exhibit a smoother, more even aspect since their front face is devoid of grids and busbars. This gives them an overall more visually pleasant look.

v. **Improved energy production:** IBC solar panels can generate greater power for a given region because of their greater effectiveness and lower shading losses, which makes them especially suited for situations with restricted space [63–68].

Disadvantages of IBC:

i. **Complex fabrications:** IBC solar cells require more complicated fabrication procedures than conventional solar cells do. The back-contact layout necessitates extra manufacturing stages, which might raise the end price.

ii. **Cost:** IBC solar cells may be more costly than normal solar systems because of the difficulty of production and the usage of specialized materials, which may be a big drawback for certain clients.

iii. **Sensitivity to temperatures:** IBC solar panels can be highly susceptible to temperature changes, which could have an impact on how well they function in severe weather. Temperature control measures that are suitable can reduce this level of sensitivity.

iv. **Restricted industrial accessibility:** IBC solar cells could be more difficult to locate and might offer fewer alternatives for producers and vendors because they aren't as readily accessible in the market as conventional solar cells.

v. **Installation complexity:** IBC solar cells' distinctive shape presents opportunities for specialized deployment methods, which might increase the price and difficulty of the job generally.

4.5 THIN-FILM SOLAR CELL

In a thin-film solar panel, a layer that absorbs radiation is sandwiched between two contacting sheets [69–77]. The points of contact among the layers may have to be passivity with one or more additional layers if they are not to operate as recombine sites and lower the quantity of produced electrons and, consequently, the process's productivity. Figure 4.4 shows an electron microscope picture of a thin-film solar cell made of the material $Cu(In,Ga)Se_2$ (CIGS). A thin-film $Cu_2ZnSn(S,Se)_4$ (CZTS) solar panel has an analogous design. Electric connecting is additionally required for the absorbers and interface layers, whether for interacting with specific cells or to create solidly connected solar cell modules. Electron-hole pairings are

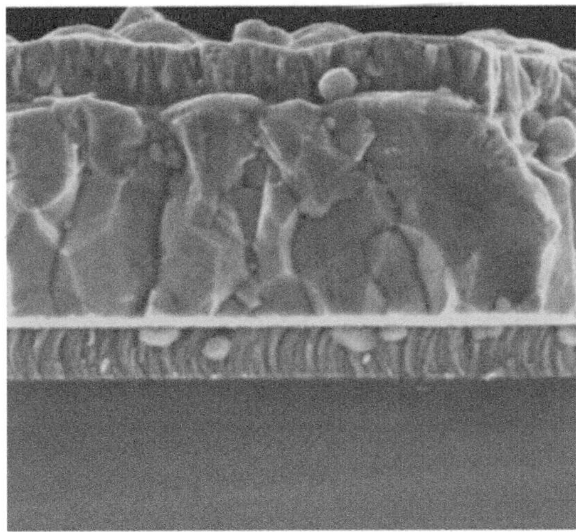

Figure 4.4 An illustration of a $Cu(In,Ga)Se_2$-dependent thin-film solar panel. The window levels, absorbers level, and subsequent interface levels are labeled in the diagram from top to bottom with the various layers. Molybdenum metal, $Cu(In,Ga)Se_2$ as the absorber, two buffer layers, and an additional layer of opaque conductive oxides make up the rear connection [8].

produced by absorbing the light from the sun. The p-n junction, which is created among the n-type buffering and windows layer and the p-type absorbent layer, acts as an electric field that divides such electron-hole pairs. Electrons flow through the p-n junction during the dissociation operation and are captured at the negatively charged solar cell surface near the opposite end of the p-n junction. Therefore, as a result of the electron shortage, the rear of the solar cell acquires a positive charge [78–86]. The variation in electric potential is caused by the different numbers of electric charge carriers across both positive and negative electrodes. Electrons will move from the negative connection to the positive one after making contact with a load from the outside. The electric load may be changed to maximize the solar cell's production at its peak power. The solar cell's performance in converting sunlight into electricity is calculated by dividing the quantity of energy lost by the outside load at its peak output by the intensity of incident light. The two CIGS and CZTS have significant benefits due to their significant solar radiation absorbing rates. In contrast to crystallized silicon, which requires layers that are 50–100 times thicker, this allows the light-absorbing sheets to be between 1 and 2 lm thick. Because electron-hole pairs are created near the charge-separating p-n junction, a small width relaxes a few of the restrictions on the performance of the material [89–93]. The flexibility to adjust the bandgap for ideal alignment with the sun spectra is a further advantage. In CIGS, this is created by varying the indium and gallium concentrations, whereas in CZTS, sulfur and selenium can be combined [8].

Advantages of thin-film solar cells:

    i. They are more lightweight than c-Si PV modules.
   ii. PV panels use less material to manufacture.
  iii. They are excellent for transportable applications.
   iv. They are extremely resistant to deterioration.
    v. They are efficient for the majority of innovations.
   vi. Due to temperatures below freezing coefficient, it experiences some reduction of heat

Disadvantages of thin-film solar cells:

    i. To get identical power production to the c-Si component, an additional mounting area is required.
   ii. It tends to be less widely available in the market.
  iii. Wholesale costs are greater.

4.6 MULTI-JUNCTION SOLAR CELLS

Group III and Group V components are combined to form III-V semiconductors, which are employed as semiconductor connections in MJSCs [94–98].

A collection of III-V layers of semiconductors with optimized width, doped stages, and fundamental structure makes up commercialized triple-junction (3-J) solar cells. The top two cells of 3-J solar cells produced on small bandgap germanium (Ge) substrates have been identified as having the broad govern bandgaps of gallium indium phosphide (GaInP) and gallium indium arsenide (GaInAs) materials. The highest energy level in the band known as valence and the lowest energy level in the band of conduction coincide with regard to velocity in III-V substances because they are directly bandgap semiconductors [99–105]. In contrast, Si, a type of indirect bandgap semiconductor, would have to be 100 m thin to have an identical amount of light absorption. The ability to create bandgaps is yet another significant advantage of III-V semiconductors over traditional Si semiconductors. III-V materials' component proportions may be tuned to narrow the bandgap and increase the conversion of light performance. Every cell is created particularly to create p-n junctions. There are direct band gaps of 1.81 and 1.42 eV for the highest cell emitter and base layers, as well as GaInP for the middle cell, as shown in Figure 4.5. The indirect band gap of the bottom cell's Ge substrate is 0.67 eV. In III-V semiconductors, band gap adjusting is possible by changing the crystal shape by varying the mix of indium (In) and gallium (Ga), which is beneficial for maximizing the absorption of light [106–112]. Ge provides a current that is nearly twice as big as the maximum sub-cell current because it has a tiny bandgap. Additionally, a significant portion of

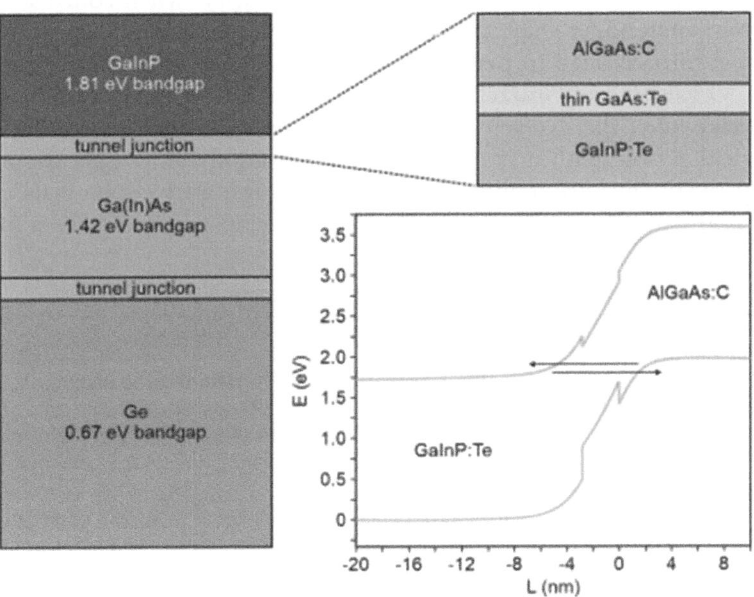

Figure 4.5 Energy bands of the absorbers levels in the 3-J solar cell, as well as a closer look at a tunnel junction and its computed band structure [9].

the production expenses go into the Ge substrate, restricting the possibility of lowering the overall cost of the device [9].

Advantages of multi-junction solar cells:

  i. Sources of renewable power.
  ii. Minimize the use of electricity.
  iii. Variety of uses.
  iv. Not expensive to use.
  v. Technological advancement.

Disadvantages of multi-junction solar cells:

  i. Storing solar energy is costly.
  ii. Reliant on weather conditions.
  iii. Take up a lot of space.
  iv. Linked to pollutants.

4.7 PEROVSKITE SOLAR PANELS

Due to their low processing costs and excellent conversion efficiencies, perovskite solar panels (PSCs) have lately gained popularity in the area of solar cell development. Because of their superiority (when contrasted with different materials), they are thought to have considerable promise and could eventually lead to perovskite becoming the leading cell material. In 1991, O'Regan and Gratzel announced a ground-breaking design of solar cells known as the dye-sensitized solar cell, which can turn sunlight into electrical power with a conversion rate of roughly 7%. This design was motivated by the idea of photosynthesis. These innovative solar cells quickly gained attention when they were developed because they offered a number

Table 4.1 Comparison of single-junction and multi-junction solar cells

Factors	Single-junction solar cells	Multi-junction solar cells
Material	Silicon is primarily utilized as a semiconductor material.	GaInP is one of three distinct semiconductor materials used to create the various layers that make up the device (InGaAs) Deutschium (Ge).
Efficiency	The ability of photovoltaic cells to transform light into power ranges from 22% to 25%.	It was recently demonstrated that these types of cells possess a yield of over 43% throughout the phase of research and development.
Cost	Prices are becoming more reasonable as time goes on and as production techniques improve.	These types of cells have still not been made available for purchase.

of benefits over traditional solar cells, including plentiful raw ingredients, simple production, and low cost. Perovskite was first used to describe a class of ceramic oxides with the basic molecular formula ABY3, which was discovered in 1839 by the German mineralogist Gustav Rose [113–119]. It was given the term "perovskite", and contains calcium titanate (CaTiO3), which is present in calcium titanium ore. In Figure 4.6, a perovskite's crystal arrangement is depicted. Miyasaka and his associates developed the first perovskite-structured solar cell in 2009. Ultimately, they increased their power conversion efficiency (PCE) to 3.13 and 3.81%, respectively [10].

Figure 4.7 shows the normal (non-inverted) architecture of a perovskite photovoltaic cell. As seen, a perovskite solar panel is made up of many layers ranging from top to bottom, including a metal back contacting layer, an electron interaction layer, a hole interface layer, an ITO layer, and glass.

The perovskite solar cell's operation is shown in Figure 4.6. As demonstrated, incoming light forms electron-hole (e- / h+) pairs in the sunlight-absorbing perovskite material. Insufficient binding power causes the particles that are charged to split, and they gradually diffuse via the electrical conduction layers. Following the collection of the charges by the appropriate electrodes, an electric current is created.

Advantages of the perovskite solar cell:

    i. The immediate optical band gap of perovskite material is about 1.5 eV.
    ii. Perovskite materials provide long minority carrier lifespans as well as long diffusing lengths.
    iii. It has an extensive range of absorption ranging from the visible to the nearer infrared radiation (800 nm) and a high coefficient of absorption (105 cm-1).

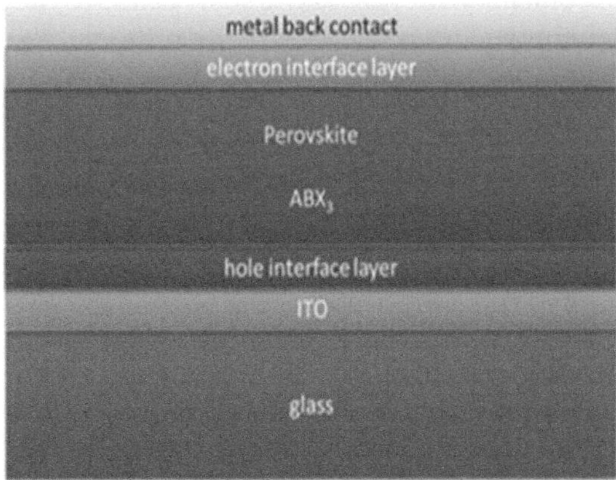

Figure 4.6 Perovskite solar cell [10].

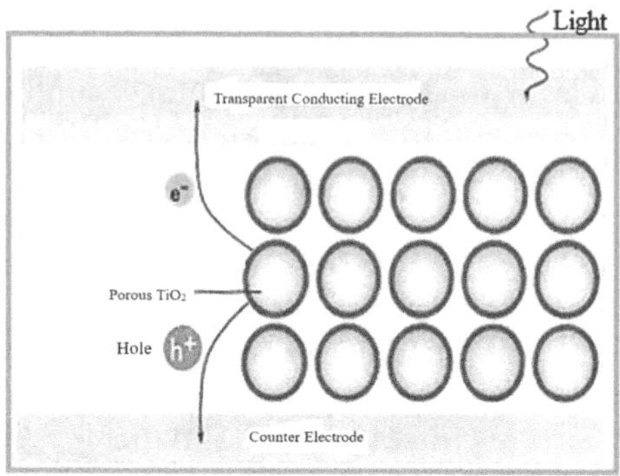

Figure 4.7 Working of perovskite solar cell [10].

   iv. Over 22% efficiency is provided using perovskite cells.

   v. Perovskite materials like methylammonium lead halides are very cheap and straightforward to produce.

   vi. It possesses a high dielectric value, quick splitting of charges, prolonged electron and hole transport distances, and a long separated carrier lifespan.

  vii. This inexpensive material aids in transforming the roofing, walls, and windows of structures to create solar electricity.

 viii. Perovskite requires fewer elements than silicon to absorb the same quantity of light. Thus, it is less expensive than silicon.

Disadvantages of the perovskite solar cell:

   i. Methyl ammonia lead iodide perovskite deterioration has to be researched.

  ii. The main problems with perovskite solar cells are the thickness of the film and reliability.

 iii. The perovskite substance is poisonous by nature and will degrade fast when exposed to humidity, heat, snowy conditions, etc.

4.8 AMORPHOUS SILICON PHOTOVOLTAIC CELL (A-SI:H)

Amorphous silicon thin films are created by chemical vapor deposition (CVD), utilizing gases including silane (SiH4), most often PECVD or hot wire CVD. The multiple layers can be formed on porous substrates like thin metallic sheets and polymers as well as stiff substrates like glass, enabling continuous manufacturing and a variety of uses. The substance utilized in solar cells is really hybridized amorphous silicon, also known as a-Si:H, an

Table 4.2 Comparative table of various solar cells

Property	Crystalline Silicon (c-Si)	Thin-film	Multi-junction	Perovskite
Material composition	Single or polycrystalline silicon	Various materials (e.g., a-Si, CdTe, CIGS)	Multiple layers of semiconductors	Organic-inorganic hybrid
Efficiency	High (15–22% for commercial panels)	Moderate to high (10–22%)	Very high (>40%)	High (20–25%, rapidly improving)
Manufacturing cost	Moderate to high	Moderate to low	High	Moderate to low
Flexibility	Rigid	Flexible	Rigid	Flexible
Durability	Highly durable	Relatively less durable	Highly durable	Less durable
Temperature sensitivity	Sensitive to high temperatures	Less sensitive to high temperatures	Sensitive to high temperatures	Sensitive to high temperatures
Space efficiency	Requires more space	Requires less space	Requires less space	Requires less space
Applications	Widely used in various applications	Diverse applications, including building-integrated PV and portable devices	Specialized applications (e.g., space exploration)	Emerging technology with potential in various applications
Commercial availability	Widely available	Commercially available	Specialized applications	Emerging technology
Environmental impact	Lower manufacturing footprint, but recycling and disposal concerns	Potentially lower environmental impact	Higher due to complex materials and production processes	Environmental concerns related to lead and other chemicals
Light absorption	Efficient at specific wavelengths	Broad absorption spectrum	Multi-bandgap design for broad absorption	Broad absorption spectrum
Thickness	Relatively thick (200–300 micrometers)	Thin (less than 1 micrometer)	Thin (less than 1 micrometer)	Thin (less than 1 micrometer)
Weight	Heavier due to thickness	Lighter	Lighter	Lighter
Cell structure	Single or multi-crystalline wafers	Thin semiconductor layers on a substrate	Multiple semiconductor layers	Thin layers on a substrate

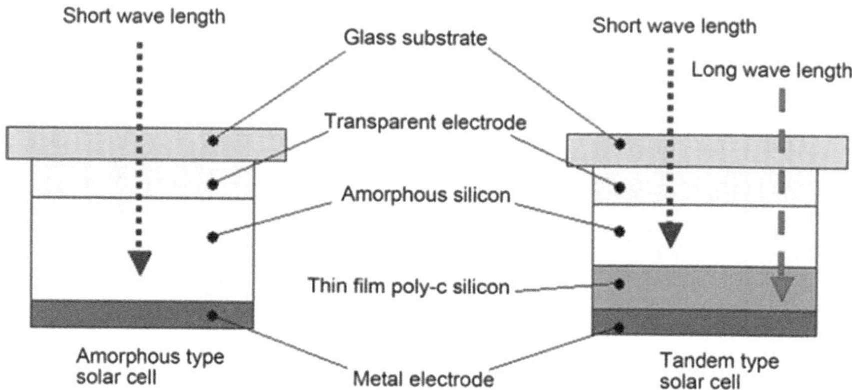

Figure 4.8 Amorphous solar panels made of silicon on the left side and a pair of tandem solar panels on the other side are shown schematically [12].

alloy of silicon and hydrogen (5–20 atomic % hydrogen). Hydrogen performs a crucial role in treating the suspended bonds created by the silicon atoms' haphazard distribution.

The roof of a building with four stories (latitude: 34 41') includes both Kaneka Co. Ltd. kinds of solar cells put at an angle of elevation of 5. The price of construction may be significantly decreased as the elevation angle is considerably lower compared to the latitude. Figure 4.8 shows the schematics of the two varieties of solar cells [11].

4.9 CONCLUSION

Semiconductor solar panels have progressed tremendously over time, moving from their insignificant origins to having an important role in the worldwide quest for renewable and environmentally friendly energy solutions. Although crystalline silicon (c-Si) solar cells have historically dominated the industry, emerging materials, including perovskite solar panels, multijunction cells, and thin-film solar panels, are opening up fresh possibilities for performance advancements and a wide range of uses. We are getting closer to using solar energy to its maximum potential because of increased initiatives to increase productivity, reduce manufacturing expenses, and minimize environmental effects. Future technologies for solar cells are expected to be increasingly more effective and affordable because of ongoing development and research. This will pave the way for an economy with more renewable energy that is more environmentally friendly, which will help fight the effects of climate change and meet our energy demands. Semiconductor solar cells are certainly leading this energy transformation and are set to influence subsequent generations' use of solar energy.

REFERENCES

1. A. Awasthi; A. K. Shukla, S. R. Murali Manohar, C. Dondariya, K. N. Shukla, D. Porwal, G. Richhariya, "Review on sun tracking technology in solar PV system", *Energy Reports*, vol. 6, pp. 392–405, 2020.

2. L. M. Frass, "History of Solar Cell", June, 2014, doi: 10.1007/978-3-319-07530-3_1.

3. Tao Meng, Hamada Hiroki, Druffel Thad, Joon Lee Jae, Rajeshwar Krishan, "Review—research needs for photovoltaics in the 21st century", *ECS Journal of Solid State Science and Technology*, vol. 9, p. 125010, 2020.

4. Y. Lee, C. Park, N. Balaji, Y. J. Lee, V. A. Dao, "High-efficiency silicon solar cells: A review", *Israel Journal of Chemistry*, vol. 55, no. 10, pp. 1050–1063, 2015.

5. G. Masmitja, P. Ortega, I. Martín, G. López, C. Voz, R. Alcubilla, "IBC c-Si (n) solar cells based on laser doping processing for selective emitter and base contact formation", *Energy Procedia*, vol. 92, pp. 956–961, 2016.

6. A. Rehnam, S. H. Lee, "Advancements in n-type base crystalline silicon solar cells and their emergence in the photovoltaic industry", *The Scientific World Journal*, vol. 2013, p. 13, 2013.

7. D. Adachi, J. L. Hernández, K. Yamamoto, "Impact of carrier recombination on fill factor for large area heterojunction crystalline silicon solar cell with 25.1% efficiency", *Applied Physics Letters*, vol. 107, no. 23, p. 233506, 2015.

8. Marika Edoff, "Thin film solar cells: Research in an industrial perspective", *AMBIO*, vol. 41, pp. 112–118, 2012.

9. Baiju Adil, Yarema Maksyam "Status and challenges of multi-junction solar cell technology", *Frontiers Energy Research*, vol. 10, 2022, doi: 10.3389/fenrg.2022.971918.

10. Chowdhury Sanchari, Kumar Mallem, Dutta Subhajit, Park Jinsu, Kim Jaemin, Kim Seyoun, Ju Minkyu, Kim Youngkuk, Cho Younghyun, Cho Eun-Chel, Junsin Y. "High-efficiency crystalline silicon solar cells: A review", *New & Renewable Energy*, vol. 15, no. 3, p. 37, 2019.

11. Ohmukai Mastao, Tsuyoshi Akhira, "Comparison between amorphous and tandem silicon solar cells in practical use", *Journal of Power and Energy Engineering*, vol. 5, no. 4, Apr 2017.

12. A. Mohammad Bagher, "Types of solar cells and application", *American Journal of Optics and Photonics*, vol. 3, pp. 94–113, 2015.

13. Y. Xing, P. Han, S. Wang, P. Liang, S. Lou, Y. Zhang, S. Hu, H. Zhu, C. Zhao, Y. Mi, "A review of concentrator silicon solar cells", *Renewable and Sustainable Energy Reviews*, vol. 51, pp. 1697–1708, 2015.

14. D. K. Shah, D. Kc, M. Muddassir, M. S. Akhtar, C. Y. Kim, O. B. Yang, "A simulation approach for investigating the performances of cadmium telluride solar cells using doping concentrations, carrier lifetimes, thickness of layers, and band gaps", *Solar Energy*, vol. 216, pp. 259–265, 2021.

15. Priya Bansal, B. Raj, "Memristor modeling and analysis for linear dopant drift kinetics," *Journal of Nanoengineering and Nanomanufacturing*, American Scientific Publishers, vol. 6, pp. 1–7, 2016.

16. Amandeep Singh, Mamta Khosla, B. Raj, "Circuit compatible model for electrostatic doped schottky barrier CNTFET, "*Journal of Electronic Materials*, Springer, vol. 45, no. 12, pp. 4825–4835, 2016.

17. D. Vaithiyanathan, B. Raj, "Performance analysis of charge plasma induced graded channel si nanotube", *Journal of Engineering Research (JER)*, EMSME Special Issue, pp. 146–154, Aug 2021.

18. Abhishek Singh Tomar, Vijay Kumar Magraiya, B. Raj, "Scaling of access and data transistor for high performance DRAM cell design", *Quantum Matter*, vol. 2, pp. 412–416, Oct 2013.

19. Neeraj Jain, B. Raj, "Parasitic capacitance and resistance model development and optimization of raised source/drain SOI FinFET structure for analog circuit applications", *Journal of Nanoelectronics and Optoelectronins*, ASP, USA, vol. 13, pp. 531–539, Ap 2018.

20. Shradhya Singh, S. K. Vishvakarma, B. Raj, "Analytical modeling of split-gate junction-less transistor for a biosensor application", *Sensing and Bio-Sensing*, Elsevier, vol. 18, pp. 31–36, Apr 2018.

21. Maisagalla Gopal, Balwinder Raj, "Low power 8T SRAM cell design for high stability video applications", *ITSI Transaction on Electrical and Electronics Engineering*, vol. 1, no. 5, pp. 91–97, 2013.

22. Balwinder Raj, Jatin Mitra, Deepak Kumar Bihani, V. Rangharajan, A. K. Saxena, S. Dasgupta, "Analysis of noise margin, power and process variation for 32 nm FinFET based 6T SRAM cell", *Journal of Computer (JCP)*, Academy Publisher, Finland, vol. 5, no. 6, pp. 1–8, 2010.

23. Divya Sharma, Rajesh Mehra, B. Raj, "Comparative analysis of photovoltaic technologies for high efficiency solar cell design", *Superlattices and Microstructures*, Elsevier, vol. 153, p. 106861, May 2021.

24. Anjana Bhardwaj, Pradeep Kumar, B. Raj, Sunny Anand, "Design and performance optimization of doping-less vertical nanowire TFET using gate stack technique", *Journal of Electronic Materials (JEMS)*, Springer, vol. 41, no. 7, pp. 4005–4013, 2022.

25. Jeetendra Singh, B. Raj, "Tunnel current model of asymmetric MIM structure levying various image forces to analyze the characteristics of filamentary memristor", *Applied Physics A*, Springer, vol. 125, no. 3, p. 203.1, Feb 2019.

26. Candy Goyal, Jagpal Singh Ubhi, B. Raj, "Low leakage zero ground noise nanoscale full adder using source biasing technique", *Journal of Nanoelectronics and Optoelectronics*, American Scientific Publishers, vol. 14, pp. 360–370, Mar 2019.

27. Gurmohan Singh, R. K. Sarin, B. Raj, "A novel robust exclusive-OR function implementation in QCA nanotechnology with energy dissipation analysis", *Journal of Computational Electronics*, Springer, vol. 15, no. 2, pp. 455–465, Jun 2016.

28. Tanu Wadhera, Deepti Kakkar, Girish Wadhwa, B. Raj, "Recent advances and progress in development of the field effect transistor biosensor: A review", *Journal of Electronic Materials*, Springer, vol. 48, no. 12, pp. 7635–7646, Dec 2019.

29. Girish Wadhwa, Priyanka Kamboj, Balwinder Raj, "Design optimisation of junctionless TFET biosensor for high sensitivity", *Advances in Natural Sciences: Nanoscience and Nanotechnology*, vol. 10, p. 045001, 2019.

30. Pawandeep Kaur, Avtar Singh Buttar, B. Raj, "A comprehensive analysis of nanoscale transistor based biosensor: A review", *Indian Journal of Pure and Applied Physics*, vol. 59, pp. 304–318, Apr 2021.

31. Divya Yadav, Balwant Raj, B. Raj, "Design and simulation of low power microcontroller for IoT applications", *Journal of Sensor Letters*, ASP, vol. 18, pp. 401–409, May 2020.

32. Shailendra Singh, B. Raj,"A 2-D analytical surface potential and drain current modeling of double-gate vertical t-shaped tunnel FET", *Journal of Computational Electronics*, Springer, vol. 19, pp. 1154–1163, Apr 2020.

33. Jeetendra Singh, B. Raj, "An accurate and generic window function for nonlinear memristor model", *Journal of Computational Electronics*, Springer, vol. 18, no. 2, pp. 640–647, Jun 2019.

34. Manjit Kaur, Neena Gupta, Sanjeev Kumar, B. Raj, Arun Kumar Singh, "Comparative RF and crosstalk analysis of carbon based nano interconnects", *IET Circuits, Devices & Systems*, vol. 15, no. 6, pp. 493–503, Feb 2021.

35. Nehru Kandasamy, Firdous Ahmad, D. Ajitha, B. Raj, Nagarjuna Telagam, "Quantum dot cellular automata based scan flip flop and boundary scan register", *IETE Journal of Research*, vol. 66, pp. 535–548, 2020.

36. S. K. Sharma, B. Raj, M. Khosla, "Enhanced photosensivity of highly spectrum selective cylindrical gate In1-xGaxAs nanowire MOSFET photodetector", *Modern Physics Letter-B*, vol. 33, no. 12, p. 1950144, 2019.

37. Jeetendra Singh, B. Raj, "Design and investigation of 7T2M NVSARM with enhanced stability and temperature impact on store/restore energy", *IEEE Transactions on Very Large Scale Integration Systems*, vol. 27, no. 6, pp. 1322–1328, Jun 2019.

38. Anil Kumar Bhardwaj, Sumeet Gupta, B. Raj, Amandeep Singh, "Impact of double gate geometry on the performance of carbon nanotube field effect transistor structures for low power digital design", *Computational and Theoretical Nanoscience*, ASP, vol. 16, pp. 1813–1820, 2019.

39. Neeraj Jain, B. Raj, "Thermal stability analysis and performance exploration of asymmetrical dual-k underlap spacer (ADKUS) SOI FinFET for security and privacy applications", *Indian Journal of Pure & Applied Physics (IJPAP)*, vol. 57, pp. 352–360, May 2019.

40. Amandeep Singh, Mamta Khosla, B. Raj, "Design and analysis of dynamically configurable electrostatic doped carbon nanotube tunnel FET", *Microelectronics Journal*, Elesvier, vol. 85, pp. 17–24, Mar 2019.

41. Neeraj Jain, Balwinder Raj, "Dual-k spacer region variation at the drain side of asymmetric SOI FinFET structure: Performance analysis towards the analog/rf design applications", *Journal of Nanoelectronics and Optoelectronics*, American Scientific Publishers, vol. 14, pp. 349–359, Mar 2019.

42. Jeetendra Singh, Sanjeev Sharma, B. Raj, Mamta Khosla, "Analysis of barrier layer thickness on performance of In1-xGaxAs based gate stack cylindrical gate nanowire MOSFET," *JNO*, ASP, vol. 13, pp. 1473–1477, Oct 2018.

43. Neeraj Jain, B. Raj, "Analysis and performance exploration of high-k SOI FinFETs over the conventional low-k SOI FinFET toward analog/RF design", *Journal of Semiconductors (JoS)*, IOP Science, vol. 39, no. 12, p. 124002-1-7, Dec 2018.

44. Candy Goyal, Jagpal Singh Ubhi, B. Raj, "A reliable leakage reduction technique for approximate full adder with reduced ground bounce noise', *Journal of Mathematical Problems in Engineering*, Hindawi, vol. 2018, Article ID 3501041, p. 16, 15 Oct 2018.

45. Jeetendra Singh, B. Raj, Mamta Khosla, "Design and performance analysis of nano-scale memristor-based nonvolatile SRAM", *Journal of Sensor Letter*", American Scientific Publishers, vol. 16, pp. 798–805, Oct 2018.

46. Girish Wadhwa, B. Raj, "Parametric variation analysis of charge-plasma-based dielectric modulated JLTFET for biosensor application", *IEEE Sensor Journal*, vol. 18, no. 15, pp. 6070–6077, 2018.

47. Jeetendra Singh, B. Raj, "Comparative analysis of memristor models for memories design", *JoS*, IoP, vol. 39, no. 7, p. 074006-1-12, Jul 2018.

48. Divya Yadav, Shailesh Singh Chouhan, Santosh Kumar Vishvakarma, B. Raj, "Application specific microcontroller design for IoT based WSN", *Sensor Letter*, ASP, vol. 16, pp. 374–385, May 2018.

49. Gurmohan Singh, R. K. Sarin, B. Raj, "Fault-tolerant design and analysis of quantum-dot cellular automata based circuits", *IEEE/IET Circuits, Devices & Systems*, vol. 12, pp. 638–664, 2018.

50. Jeetendra Singh, B. Raj, "Modeling of mean barrier height levying various image forces of metal insulator metal structure to enhance the performance of conductive filament based memristor model", *IEEE Nanotechnology*, vol. 17, no. 2, pp. 268–267, Mar 2018.

51. Aakash Jain, Sanjeev Sharma, B. Raj, "Analysis of triple metal surrounding gate (TM-SG) III-V nanowire MOSFET for photosensing application", *Optoelectronics Journal*, Elsevier, vol. 26, no. 2, pp. 141–148, May 2018.

52. Aakash Jain, Sanjeev Sharma, B. Raj, "Design and analysis of high sensitivity photosensor using cylindrical surrounding gate MOSFET for low power sensor applications", *Engineering Science and Technology, an International Journal*, Elsevier, vol. 19, no. 4, pp. 1864–1870, Dec 2016.

53. Amandeep Singh, Mamta Khosla, B. Raj, "Analysis of electrostatic doped schottky barrier carbon nanotube FET for low power applications," *Journal of Materials Science: Materials in Electronics*, Springer, vol. 28, pp. 1762–1768, 2017.

54. G. Saiphani Kumar, Amandeep Singh, B. Raj, "Design and analysis of gate all around CNTFET based SRAM cell design", *Journal of Computational Electronics*, Springer, vol. 17, no. 1, pp. 138–145, Mar 2018.

55. Gurinder pal Singh, B. S. Sohi, Balwinder Raj, "Material properties analysis of graphene base transistor (GBT) for VLSI analog circuits", *Indian Journal of Pure & Applied Physics (IJPAP)*, vol. 55, pp. 896–902, Dec 2017.

56. Amandeep Singh, Mamta Khosla, B. Raj, "Comparative analysis of carbon nanotube field effect transistor and nanowire transistor for low power circuit design," *Journal of Nanoelectronics and Optoelectronics*, American Scientific Publishers, USA, vol. 11, pp. 388–393, Jun 2016.

57. Sunil Kumar, B. Raj, "Estimation of stability and performance metric for inward access transistor based 6T SRAM cell design using n-type/p-type DMDG-GDOV TFET", *IEEE VLSI Circuits and Systems Letter*, vol. 3, no. 2, pp. 25–39, Jun 2017.

58. Shashikant Sharma, Anjan Kumar, Manisha Pattanaik, B. Raj, "Forward body biased multimode multi-threshold CMOS technique for ground bounce noise reduction in static CMOS adders", *International Journal of Information and Electronics Engineering*, pp. 567–572, vol. 3, no. 3, 2013.

59. Hamendra Singh, Pankaj Kumar, Balwinder Raj, "Performance Analysis of Majority Gate SET Based 1-bit Full Adder", *International Journal*

of Computer and Communication Engineering (IJCCE), IACSIT Press Singapore, ISSN: 2010-3743, Vol. 2, no. 4, 2013.

60. Anil Kumar Bhardwaj, Sumeet Gupta, B. Raj, "Investigation of parameters for schottky barrier (SB) height for schottky barrier based carbon nanotube field effect transistor device", *Journal of Nanoelectronics and Optoelectronics*, ASP, vol. 15, pp. 783–791, Jul 2020.

61. Priya Bansal, B. Raj, "Memristor: A versatile nonlinear model for dopant drift and boundary issues," *JCTN*, American Scientific Publishers, vol. 14, no. 5, pp. 2319–2325, May 2017.

62. Neeraj Jain, B. Raj, "An analog and digital design perspective comprehensive approach on Fin-FET (Fin-Field Effect transistor) technology: A review", *Reviews in Advanced Sciences and Engineering (RASE)*, ASP, vol. 5, pp. 1–14, 2016.

63. Sanjeev Sharma, B. Raj, Mamta Khosla, "Subthreshold performance of In1-xGaxAs based dual metal with gate stack cylindrical/surrounding gate nanowire MOSFET for low power analog applications", *Journal of Nanoelectronics and Optoelectronics*, American Scientific Publishers, USA, vol. 12, pp. 171–176, 2017.

64. Balwinder Raj, A. K. Saxena, S. Dasgupta, "Analytical modeling for the estimation of leakage current and subthreshold swing factor of nanoscale double gate FinFET device", *Microelectronics International, UK*, vol. 26, pp. 53–63, 2009.

65. Shailendra Singh, Girish Wadhwa, Balwinder Raj, "An analytical modeling for dual source vertical tunnel field effect transistor", *International Journal of Recent Technology and Engineering (IJRTE)*, vol. 8, no. 2, pp. 603–608, Jul 2019.

66. Shailendra Singh, B. Raj, "Design and analysis of hetrojunction vertical T-shaped tunnel field effect transistor", *Journal of Electronics Material*, Springer, vol. 48, no. 10, pp. 6253–6260, Oct 2019.

67. Candy Goyal, Jagpal Singh Ubhi, B. Raj, "A low leakage CNTFET based inexact full adder for low power image processing applications", *International Journal of Circuit Theory and Applications*, Wiley, vol. 47, no. 9, pp. 1446–1458, Sept 2019.

68. B. Raj, A. K. Saxena, S. Dasgupta, "A compact drain current and threshold voltage quantum mechanical analytical modeling for FinFETs", *Journal of Nanoelectronics and Optoelectronics (JNO)*, USA, vol. 3, no. 2, pp. 163–170, 2008.

69. Girish Wadhwa, B. Raj, "An analytical modeling of charge plasma based tunnel field effect transistor with impacts of gate underlap region", *Superlattices and Microstructures*, Elsevier, vol. 142, p. 106512, Jun 2020.

70. Shailendra Singh, B. Raj, "Modeling and simulation analysis of SiGe hetrojunction double GateVertical t-shaped tunnel FET", *Superlattices and Microstructures*, Elsevier vol. 142, p. 106496, Jun 2020.

71. Amandeep Singh, Dinesh Kumar Saini, Dinesh Agarwal, Sajal Aggarwal, Mamta Khosla, B. Raj, "Modeling and simulation of carbon nanotube field effect transistor and its circuit application," *Journal of Semiconductors (JoS)*, IOP Science, vol. 37, p. 074001-6, Jul 2016.

72. Neeraj Jain, Balwinder Raj, "Device and circuit co-design perspective comprehensive approach on FinFET technology: A review", *Journal of Electron Devices*, vol. 23, no. 1, pp. 1890–1901, 2016.

73. Sunil Kumar, B. Raj, "Analysis of I_{ON} and ambipolar current for dual-material gate-drain overlapped DG-TFET," *Journal of Nanoelectronics and Optoelectronics*, American Scientific Publishers, USA, vol. 11, pp. 323–333, Jun 2016.

74. Naveed Anjum, Tarun Bali, B. Raj, "Design and simulation of handwritten multiscript character recognition", *International Journal of Advanced Research in Computer and Communication Engineering*, vol. 2, no. 7, pp. 2544–2549, Jul 2013.

75. Sanjeev Sharma, B. Raj, Mamta Khosla, "A gaussian approach for analytical subthreshold current model of cylindrical nanowire FET with quantum mechanical effects", *Microelectronics Journal*, Elsevier, vol. 53, pp. 65–72, Apr 2016.

76. Karmjit Singh, B. Raj, "Performance and analysis of temperature dependent multi-walled carbon nanotubes as global interconnects at different technology nodes," *Journal of Computational Electronics*, Springer, vol. 14, no. 2, pp. 469–476, Jun 2015.

77. Sunil Kumar, B. Raj, "Compact channel potential analytical modeling of DG-TFET based on evanescent–mode approach," *Journal of Computational Electronics*, Springer, vol. 14, no. 2, pp. 820–827, Jul 2015.

78. Karmjit Singh, B. Raj, "Temperature dependent modeling and performance evaluation of multi-walled CNT and single-walled CNT as global interconnects," *Journal of Electronic Materials*, Springer, vol. 44, no. 12, pp. 4825–4835, Dec 2015.

79. V. K. Sharma, M. Pattanaik, B.Raj, "INDEP approach for leakage reduction in nanoscale CMOS circuits", *International Journal of Electronics*, Taylor & Francis, vol. 102, no. 2, pp. 200–215, 2014.

80. Karmjit Singh, B. Raj, "Influence of temperature on MWCNT bundle, SWCNT bundle and copper interconnects for nanoscaled technology nodes," *Journal of Materials Science: Materials in Electronics*, Springer, vol. 26, no. 8, pp. 6134–6142, 2015.

81. Naveed Anjum, Tarun Bali, B. Raj, "Design and simulation of handwritten gurumukhi and devanagri numerical recognition", *International Journal of Computer Applications*, Foundation of Computer Science, New York, USA, vol. 73, no. 12, pp. 16–21, 2013.

82. S. Khandelwal, V. Gupta, B. Raj, R. D. "Gupta, process variability aware low leakage reliable nano scale DG-FinFET SRAM cell design technique", *Journal of Nanoelectronics and Optoelectronics*, vol. 10, no. 6, pp. 810–817, Dec 2015.

83. V. K. Sharma, M. Pattanaik, B. Raj, "ONOFIC approach: Low power high speed nanoscale VLSI circuits design", *International Journal of Electronics*, Taylor & Francis, vol. 101, no. 1, pp. 61–73, 2014.

84. S. Khandelwal, Balwinder Raj, R. D. Gupta, "FinFET based 6T SRAM cell design: Analysis of performance metric, process variation and temperature effect", *Journal of Computational and Theoretical Nanoscience*, ASP, USA, vol. 12, pp. 2500–2506, 2015.

85. Sumit Singh, Shekhar Yadav, Jagdeep Rahul, Anurag Srivastava; B. Raj, "Impact of HfO_2 in graded channel dual insulator double gate MOSFET", *Journal of Computational and Theoretical Nanoscience*, American Scientific Publishers, vol. 12, no. 6, pp. 950–953, Apr 2015.

86. Vijay Kumar Sharma, Manisha Pattanaik, B. Raj, "PVT variations aware low leakage INDEP approach for nanoscale CMOS circuits", *Microelectronics Reliability*, Elsevier, vol. 54, pp. 90–99, 2014.

87. B. Raj, A. K. Saxena, S. Dasgupta, "Quantum mechanical analytical modeling of nanoscale DG FinFET: Evaluation of potential, threshold voltage and source/drain resistance", *Elsevier's Journal of Material Science in Semiconductor Processing*, Elsevier, vol. 16, no. 4, pp. 1131–1137, 2013.

88. Maisagalla Gopal, Siva Sankar D Prasad, Balwinder Raj, "8T SRAM cell design for dynamic and leakage power reduction", *International Journal of Computer Applications*, Foundation of Computer Science, New York, USA, vol. 71, no. 9, pp. 43–48, Jun 2013.

89. Manisha Pattanaik, B. Raj, Shashikant Sharma, Anjan Kumar, "Diode based trimode multi-threshold CMOS technique for ground bounce noise reduction in static CMOS adders", *Advanced Materials Research*, Trans Tech Publications, Switzerland, vol. 548, pp. 885–889, 2012.

90. Balwinder Raj, A. K. Saxena, S. Dasgupta, "Nanoscale FinFET based SRAM cell design: Analysis of performance metric, process variation, underlapped FinFET and temperature effect", *IEEE Circuits and System Magazine*, vol. 11, no. 2, pp. 38–50, 2011.

91. V. K. Sharma, M. Pattanaik, Balwinder Raj, "Leakage current ONOFIC approach for deep submicron VLSI circuit design", *International Journal of Electrical, Computer, Electronics and Communication Engineering, World Academy of Sciences, Engineering and Technology*, vol. 7, no. 4, pp. 239–244, 2013.

92. Tulika Chawla, Mamta Khosla, B. Raj, "Design and simulation of triple metal double-gate germanium on insulator vertical tunnel field effect transistor", *Microelectronics Journal*, Elsevier, vol. 114, p. 105125, Aug 2021.

93. Parminder Kaur, Sandeep Singh Gill, B. Raj, "Comparative analysis of OFETs materials and devices for sensor applications", *Journal of Silicon*, Springer, vol. 14, pp. 4463–4471, 2022.

94. Sanjeev Kumar Sharma, Parveen Kumar, Balwant Raj, B. Raj, "$In_{1-x}Ga_xAs$ double metal gate-stacking cylindrical nanowire MOSFET for highly sensitive photo detector", *Journal of Silicon*, Springer, vol. 14, pp. 3535–3541, 2022.

95. B. Raj, A. K. Saxena, S. Dasgupta, "Analytical modeling of quasi planar nanoscale double gate FinFET with source/drain resistance and field dependent carrier mobility: A quantum mechanical study", *Journal of Computer (JCP)*, Academy Publisher, Finland, vol. 4, no. 9, pp. 1–8, 2009.

96. S. Bhushan, S. Khandelwal, B. Raj, "Analyzing different mode FinFET based memory cell at different power supply for leakage reduction", *Seventh International Conference on Bio-Inspired Computing: Theories and Application, (BIC-TA 2012) Advances in Intelligent Systems and Computing*, vol. 202, pp. 89–100, 2013.

97. Sonal Singh, Mamta Khosla, Girish Wadhwa, B. Raj, "Design and analysis of double-gate junctionless vertical TFET for gas sensing applications", *Applied Physics A*, Springer, vol. 127, no. 16, 2 Jan 2021.

98. Inderjit Singh, B. Raj, Mamta Khosla, B. Rajesh Kumar Kaushik, "Potential MRAM technologies for low power SoCs", *SPIN World Scientific Publisher, SCIE*; vol. 10, no. 4, p. 2050027, Dec 2020.

99. Shailendra Singh, B. Raj, "Parametric variation analysis on hetero-junction Vertical t-shape TFET for supressing ambipolar conduction", *Indian Journal of Pure and Applied Physics*, vol. 58, pp. 478–485, Jun 2020.

100. Shailendra Singh, Girish Wadhwa, Balwinder Raj, "Design and analysis of dual source vertical tunnel field effect transistor for high performance", *Transactions on Electrical and Electronics Materials*, Springer, vol. 21, pp. 74–82, Oct 2019.

101. Manjit Kaur, Neena Gupta, Sanjeev Kumar, B. Raj, Arun Kumar Singh, "RF performance analysis of intercalated graphene nanoribbon based global level interconnects", *Journal of Computational Electronics*, Springer, vol. 19, pp. 1002–1013, Jun 2020.

102. Girish Wadhwa, B. Raj, "Design and performance analysis of junctionless TFET biosensor for high sensitivity", *IEEE Nanotechnology*, vol. 18, pp. 567–574, 2019.

103. Jeetendra Singh, B. Raj, "Enhanced nonlinear memristor model encapsulating stochastic dopant drift", *JNO*, ASP, vol. 14, pp. 958–963, 2019.

104. Jeetendra Singh, B. Raj, "Temperature dependent analytical modeling and simulations of nanoscale memristor", *Journal: Engineering Science and Technology, an International Journal*, Elsevier, vol. 21, pp. 862–868, Oct 2018.

105. Shradhya Singh, Shashi Bala, Balwant Raj, B. Raj, "Improved sensitivity of dielectric modulated junctionless transistor for nanoscale biosensor design", *Sensor Letter*, ASP, vol. 18, pp. 328–333, Apr 2020.

106. Vivek Kumar, Santosh Kumar Vishvakarma, B. Raj, "Design and performance analysis of ASIC for IoT applications", *Sensor Letter*, ASP, vol. 18, pp. 31–38, Jan 2020.

107. Akanksha Jaiswal, R. K. Sarin, B. Raj, Shikha Sukhija, "A novel circular slotted microstrip-fed patch antenna with three triangle shape defected ground structure for multiband applications", *Advanced Electromagnetic (AEM)*, vol. 7, no. 3, pp. 56–63, Aug 2018.

108. Girish Wadhwa, B. Raj, "Label free detection of biomolecules using charge-plasma-based gate underlap dielectric modulated junctionless TFET", *Journal of Electronic Materials (JEMS)*, Springer, vol. 47, no. 8, pp. 4683–4693, Aug 2018.

109. Gurmohan Singh, R. K. Sarin, B. Raj, **"Design and performance analysis of a new efficient coplanar quantum-dot cellular automata adder"**, *Indian Journal of Pure & Applied Physics (IJPAP)*, vol. 55, pp. 97–103, Feb 2017.

110. Amandeep Singh, Mamta Khosla, Balwinder Raj, "Design and analysis of electrostatic doped schottky barrier CNTFET based low power SRAM," *International Journal of Electronics and Communications, (AEÜ)*, Elsevier, vol. 80, pp. 67–72, 2017.

111. Parminder Kaur, Vikas Pandey, B. Raj, "Comparative study of efficient design, control and monitoring of solar power using IoT", *Sensor Letter*, ASP vol. 18, pp. 419–426, May 2020.

112. Anil Kumar Bhardwaj, Sumeet Gupta, B. Raj, "Development & analysis of compact model for double gate schottky barrier CNTFET", *Journal of Nanoelectronics and Optoelectronics*, ASP, vol. 15, pp. 1199–1208, Aug 2020.

113. Girish Wadhwa, Priyanka Kamboj, Jeetendra Singh, B. Raj, "Design and investigation of junctionless DGTFET for biological molecule recognition", *Transactions on Electrical and Electronic Materials*, Springer, vol. 22, pp. 282–289, 2021.
114. Tulika Chawla, Mamta Khosla, B. Raj, "Optimization of double-gate dual material GeOI-vertical TFET for VLSI circuit design", *IEEE VLSI Circuits and Systems Letter*, vol. 6, no. 2, pp. 13–25, Aug 2020.
115. Sachin Kumar Verma, Shailendra Singh, Girish Wadhwa, B. Raj, "Detection of biomolecules using charge-plasma based gate underlap dielectric modulated dopingless TFET", *Transactions on Electrical and Electronic Materials (TEEM)*, Springer, vol. 21, pp. 528–535, Jun 2020.
116. Neeraj Jain, B. Raj, "Impact of underlap spacer region variation on electrostatic and analog/RF performance of symmetrical high-k SOI FinFET at 20 nm channel length", *Journal of Semiconductors (JoS)*, IOP Science, vol. 38, no. 12, p. 122002, Dec 2017.
117. Shailendra Singh, B. Raj, "Analytical modeling and simulation analysis of T-shaped III-V heterojunction Vertical T-FET", *Superlattices and Microstructures*, Elsevier, vol. 147, p. 106717, Nov 2020.
118. Gurmohan Singh, R. K. Sarin, B. Raj, "Design and analysis of area efficient QCA based reversible logic gates", *Journal of Microprocessors and Microsystems*, Elsevier, vol. 52, pp. 59–68, May 2017.
119. Amandeep Singh, Mamta Khosla, B. Raj, "Compact model for ballistic single wall CNTFET under quantum capacitance limit," *Journal of Semiconductors (JoS)*, IOP Science, vol. 37, p. 104001-8, Oct 2016.

Chapter 5

Multi-junction solar cell

Balwinder Raj, Akam, and Piush Verma

5.1 INTRODUCTION

There is a growing global demand for energy, which is becoming critical as supply falls short of demand. Fossil fuels like petrol, gasoline, etc. emit carbon dioxide when they are used [1–5]. Researchers and engineers worldwide are working tirelessly to identify a suitable solution for the issue. Furthermore, work has been done by implementing the newest smart technologies to reduce the consumption of electrical energy. Power generation has become more challenging due to the scarcity of natural resources such as biofuels, natural gas, and coal. This is because it is becoming more expensive and less profitable to extract certain minerals. However, the burning of these minerals to produce power pollutes the environment severely. The negative impact of this pollution has caused many to rethink how the planet produces power for its citizens. The significance of renewable energy therefore grows. Because of its limitless availability and affordability, renewable energy is becoming more and more popular. Renewable energy supplies power that is nontoxic and pollution-free, whereas all other energy sources produce significant environmental pollutants [6–11].

Solar, wind, tidal, hydroelectric, geothermal heat, bio-diesel, and fuel are examples of renewable energy sources. Energy from the sun is one of these and is becoming more popular as a source of electricity. Energy from the sun is being employed as a source of energy for automobiles and in power generation. In many countries, solar automobiles are operating extremely effectively. Solar energy is transformed into usable electrical energy using solar cells, also known as photovoltaic (PV) cells. Distribution and installation are both relatively straightforward. The best and only viable option for places where it is difficult to supply the generated power is a solar cell. There are differences in solar radiation around the earth. Additionally, air mass affects it [12–17].

Figure 5.1 depicts the varied sun radiation for various air masses. The requirements for solar cells change along with their efficiency as the solar spectrum changes [2]. The effort to boost solar cell efficiency is still ongoing. Every researcher is working to increase the efficiency of solar cells. The

DOI: 10.1201/9781003487692-5

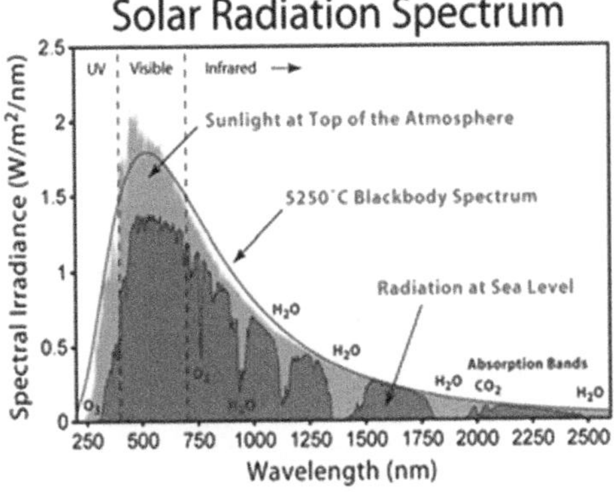

Figure 5.1 Solar spectrum for different air mass conditions.

dual-junction solar cell based on AlAs/GaAs in this research is another effort to increase efficiency while capturing solar energy from the sun [18–26]. Selecting the right material is crucial for producing high-efficiency solar cells. In the following sections, all of the difficulties and results of our study are discussed with appropriate data analysis and citations. To ensure that the world is not dependent on fossil fuels, research into the most effective solar cells has expanded in the last few decades. Being numerous and substantial is one of the benefits of PV technology. Solar energy is a renewable resource that is generally universally accessible. Consequently, PV will always be able to create energy (as long as the sun shines). Single-junction solar cells cannot absorb the whole photon of light [2].

In this study, the optimization of AlAs/GaAs dual-junction solar cells has been carried out. This uses GaAs as a tunnel diode between the two top and bottom solar cells. The simulation was done using the Silvaco ATLAS TCAD tool which is calibrated or tuned with recent experimental results of AlAs/GaAs, InGaP/GaAs-based dual-junction solar cells [27–34].

5.2 MULTI-JUNCTION SOLAR CELL

Single-junction solar cells have only one layer, where as more than one layer of solar cell known as a multi-junction solar cell. The absorbed wavelength range depends on the energy band gap, which varies with the material and the mole fraction. The condition for light to be absorbed is when an incident photon coming from the solar cell has energy greater than or equal to the band gap energy of the material [35–46]. Similarly, if the photon has less energy than the material's energy band gap, it will not be absorbed. The multi-junction solar cell has stacked p-n junctions with different materials or mole fractions of different band gaps. Figure 5.2 shows a schematic

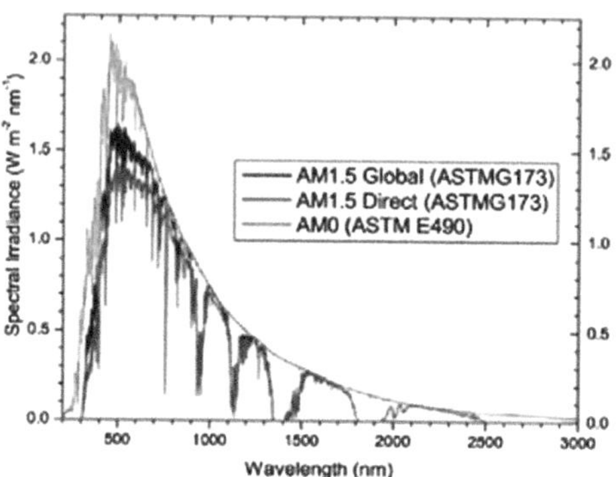

Figure 5.2 AM 1.5 G solar spectrum.

representation of the AlAs/GaAs-based dual-junction solar cell. In between the top and bottom cells, a tunnel diode is there to carry out the tunneling phenomenon. Absorbance wavelength with a wider range is permitted by multi-junction solar cells. Multi-junction solar cells with III-V material semiconductors are found to have the highest reported efficiency of 46% in the commercial market under concentrated sunlight [6]. AlAs with a high energy gap is becoming an important material for high short-circuit current and efficient solar cells. GaAs absorbs the near-infrared spectrum (NIR). The single-junction solar cell is limited by an unabsorbed spectrum. However, multi-junction solar cells have different band gap materials with different layers. Decreasing the order of band gaps is arranged to allow photons to move under the bottom cells. As the solar spectrum is incident on it each layer depending on the band gap of the material absorbs the photon [47–53]. Solar cell material, having a correspondingly lower wavelength than that of the photon wavelength, is well absorbed and utilized in the conversion of electrical energy. From the literature, we have learned that to achieve maximum efficiency, the band gap of solar cell material should be equal to the incident solar energy packet (hʋ) [7]. For this purpose, a layer should be designed for a specific band gap to achieve maximum energy conversion over the spectrum. Therefore, for multi-junction solar cells, as observed from the results in the literature, the absorption of solar radiation in a multi-junction solar cell is higher than that of a single-junction solar cell. Figures 5.3 (a) and (b) show the absorption of a photon in single-junction and multi-junction solar cells [3, 4].

5.3 PRINCIPLES OF MULTI-JUNCTION SOLAR CELLS

Examples of III-V solar cells with a single junction or several junctions can be found in the literature. Tandem devices are the most cutting-edge

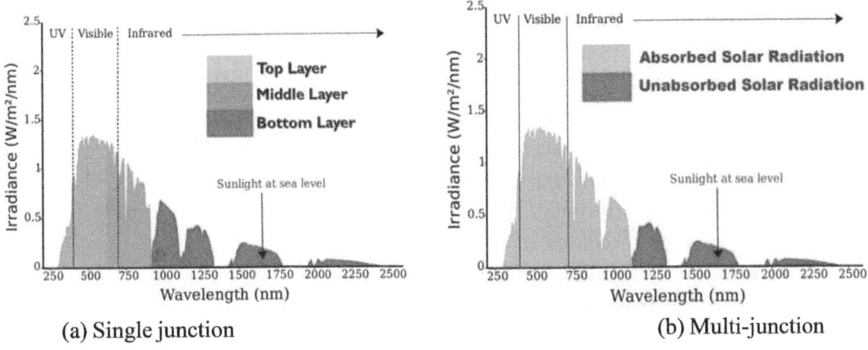

(a) Single junction   (b) Multi-junction

Figure 5.3 (a) Single-junction, (b) multi-junction.

multi-junction technology currently accessible, and third-generation devices have occasionally been created by combining first and second-generation devices [54–61]. However, the most effective devices manufactured are expensive III-Vi semiconductors that are only suitable for space applications or concentrators. Due to the complex fabrication, the cost of the solar cell becomes increasingly high. However, to more effectively collect a wider range of complete photon energy, multi-junction solar cells utilize a variety of semiconductor materials. Modern multi-junction solar cells may produce almost twice as much electricity under the same conditions as conventional silicon solar cells, depending on the specific technology. The primary constraint on the production of multi-junction solar cells is the availability of materials with the most advantageous band gaps that concurrently provide high-efficiency and low-defect probabilities. Such multi-junction cells are thought to be suitable contenders for III-V semiconducting compounds [9]. Since the majority of these substances have direct electronic structures and broad spectral band gaps, they likely have high absorption coefficients [62–68].

5.4 III-V MULTI-JUNCTION SOLAR CELL

The multi-junction solar cell in particular and high-efficiency solar cells in general benefit greatly from the properties of III-V semiconductor materials. Elements from groups III (Al, Ga, and In) and V (N, P, As, and Sb) [9] of the periodic table are grouped in a zinc-blend (or wurtzite) crystal structure to form III-V semiconductors. Because its band gap matches the solar spectrum and high ERE, the GaAs solar cell has continuously maintained the greatest single-junction efficiency. When band gaps are greater, semiconductors can have little interface recombination, relatively long lives, and high mobilities. For hetero-junction passivation (i.e., window and back-surface field (BSF) layers) that produces diffusion lengths, band-gap alloys are utilized, longer than the necessary thickness for single-pass absorption [69–76]. As a result, III-V materials have not necessarily needed to adopt the light trapping methods frequently used in silicon photovoltaics. The main drawback of III-V solar cells is their susceptibility to defects that manifest in deep non-radiative recombination sites, such as phase barriers, impurities, and dislocations. If there are inherent native flaws, they also have a relatively low density. III-Vs may be created with extremely flawless crystals, avoiding the issues of imperfections [10].

The efficiency of the III-V compound solar cell has been recorded as the highest. Multiple-junction solar cells provide a high-efficiency solar cell with different band gaps. In III-V compound solar cells, the material to be chosen has to be suitable for the decreasing band gap. The material with the highest band gap will be placed on the top and the lowest on the lower side. Changing the substrate material, such as using Ge instead of silicon, also helps in efficient multi-junction solar cells. Silicon is an excellent material with many benefits. Silicon can be highly purified [77–83]. A further advantage is that the price of silicon has fallen. Arguably, silicon is a perfect solar cell material, but limitations are always there. The efficiency limit of silicon can be overcome by multi-junction devices. The best examples of SI-based tandem multi-junction solar cells are GaAsP/Si tandem cells that have reached an AM 1.5 g conversion efficiency of 23.4% (two junctions) and GaInP/GaInAsP/Si up to 35.9% (two and four junctions). Growing III-V materials on Si is also a challenge. Some are grown with lattice constant and some junctions [11]. With further research, this can also be solved.

5.5 PHOTON ABSORPTION AND CHARGE CARRIER ABSORPTION

Photon absorption shows the excitation of photon energy and electrons present in the valence band that has shifted to the conduction band after absorption of photon energy. Photon energy can be absorbed in the presence of electron energy levels [84–92]. In electron-hole pairs, recombination occurs when energy is released in the form of the recombination rate (photon emission) or non-radiative combination when the electron returns to its initial energy state. The existence of a semi-permeable membrane on both sides of the absorbent permits energy stored in electron-hole pairs to be used in an external circuit. To restore the non-equilibrium EHP to the equilibrium value, the recombination process occurs. The high recombination rate in the vicinity of a region depletes the region of minority carriers. A parameter known as "surface recombination velocity" in units of cm/sec is used to specify the recombination at the surface (Figure 5.4). Both bulk and surface recombination are necessary for the material to have a minority-carrier lifetime. Recombination surfacing restrictions can slow down the

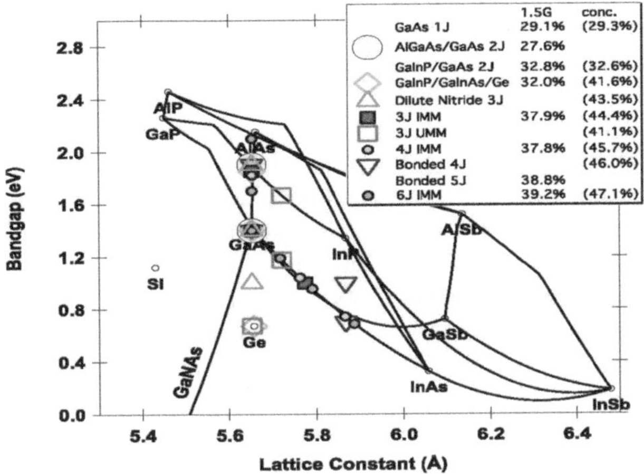

Figure 5.4 Principle of wide photo response of MJ solar cell.

loss of minority carriers. The material's observed minority-carrier lifespan can be increased if the depletion rate of minority carriers can be slowed down. The disruption of the crystal lattice's periodicity, which results in bonds dangling at the surface of the semiconductor, is what generates the defects at semiconductor surfaces [12]. The growth of the layer at the semiconductor surface that ties up some of these dangling bonds allows for the decrease of dangling bonds and, consequently, surface recombination. Surface passivation is the term used to describe this decrease in dangling bonds [93–101].

In Figure 5.4, we can see the electrons becoming excited and leaving their state in the multi-junction solar cell [14]. The solar cell is constructed in such a manner that the electron-hole pair must reach the membrane before recombination can occur. As a result, the time it takes for such charge carriers to reach their respective membrane must be less than their lifetime, limiting the absorber thickness. The energy is used in the external circuit by extracting the photo-generated charge carrier owing to light absorption utilizing externally linked electrical contacts. Light energy is turned into electrical energy at this point. After traveling via the external circuit, electrons form clusters just at the back contact and absorber layers.

5.6 PROPER OPTIMIZATION OF INGAP/GAAS SOLAR CELL WITH THE HELP OF SILVACO TCAD

In this research work, the performance of the InGaP\GaAs-based dual-junction (DJ) solar cell design has been investigated by altering different device parameters such as thickness and doping concentration of BSF and top and bottom window. A developed, simulated, and improved InGaP/GaAs DJ solar cell is shown. Since the window layer plays a key structural role in the creation of the DJ cell, it is necessary to choose the material and design parameters that would result in the best efficiency. Thus, by employing AlGaAs and AlGaInP in the top cell's window layer, the device structure is optimized. Varying the thickness of the top window layer improves efficiency. The next step is to optimize the top base thickness, bottom BSF thickness, and bottom BSF doping concentration. I-V characteristics, P-V characteristics, and external quantum efficiency (EQE) are taken from the simulation results and shown. The optimal top window layer material is determined in this study by comparing the two materials, AlGaAs and AlGaInP, using a thorough simulation analysis. It is discovered that AlGaInP, which is lattice-matched with InGaP and GaAs, is the ideal material for top windows [102–107]. Under the irradiation of one sun with the AM 1.5G spectrum, the optimized DJ cell structure yields VOC = 1.49 V and JSC = 28.65 mA/cm2, with a notable improvement in conversion efficiency up to 37.27%. This efficiency is around 1% greater than that of a simulated study that was done using AlGaAs as the top window material. In our current study, the increased efficiency is a result of optimizing a few parameters in addition to the primary window layer characteristics. Last but not least, the study presented here sheds light on the optimization of the InGaP/GaAs DJ solar cell and demonstrates the significance of using precise numerical TCAD simulations when designing cells of this type that have numerous parameters that cannot be optimized experimentally [108–115].

This study's primary contribution is to enhance the performance of InGaP/GaAs DJ solar cells for two different top window layers, AlGaAs and AlGaInP. To demonstrate the standards for selecting the optimum top window layer material, we provide a thorough comparison of the two materials. Consequently, the top window layer is ideal since it provides the most efficiency. The remaining sections of this chapter are organized as follows. First, an illustration of the design of our InGaP/GaAs DJ solar cell is provided. Additionally, calibration vs. measurements and simulation models are included. The performance of the InGaP/GaAs DJ solar cell is then statistically optimized while taking into account the two different top window layer materials, AlGaAs and AlGaInP. This is then optimized using the composition's optimal value. To get the highest permitted conversion efficiency, the bottom BSF optimal doping and thickness are also examined [116–121]. Then, information on the two optimized InGaP/GaAs DJ solar cells' I-V characteristics, P-V curve, and EQE is given. Also provided are the rates of photo generation and the electrostatic potential. The best material for the top window layer is chosen after comparing the performances of the two optimized InGaP/GaAs DJ solar cell designs. Additionally, the performance of the ideal InGaP/GaAs DJ solar cell structure is evaluated against the results of current earlier

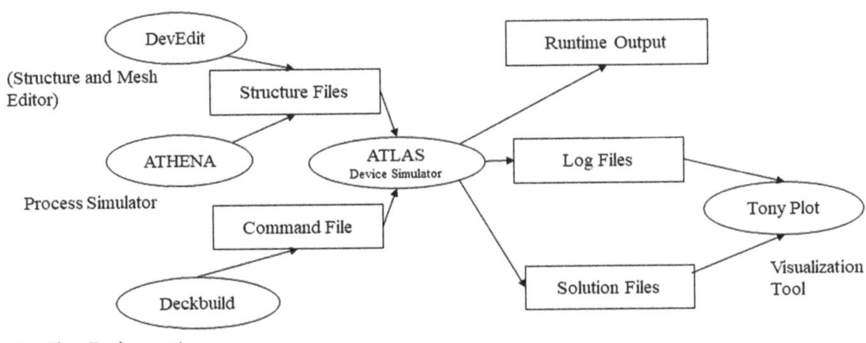

Figure 5.5 Silvaco TCAD simulation flow.

investigations. The Silvaco TCAD simulation tool is used for simulations (Figure 5.5) [15].

5.6.1 Parameters and efficiency of AlAs/ GaAs new DJ junction solar cell

AlGaAs/GaAs-based multi-junction solar cell theoretical efficiency and cell characteristics are presented in this chapter. In this research, a new and distinctive multi-junction solar model is presented. The parameters of the solar cell are calculated using a modified version of the spectral p-n junction solar cell model and shown in Figure 5.6. Silvaco ATLAS TCAD simulation is used to conduct all of the calculations for this chapter using various acceptable equations from various studies [122–128]. The multi-junction solar cell that is presented in this research is thought to be entirely optically transparent and has very high current conductivity from the top layer to the bottom layer. Despite the existence of the tunnel junction, the loss it produces is minimal. Plots of the J-V characteristics are extracted for various solar radiation levels. Calculations for various air mass (AM) conditions, including AM 1.5D, AM 1.5G, and AM 0 solar spectrum, are analyzed in this work. The air mass conditions have an impact on the solar cell's efficiency. Calculations are made for solar cell characteristics such as power, fill factor, short-circuit current density, and open circuit voltage. The efficiency for the AM 0 condition is 50.7484%, while the efficiency for the AM 1.5G condition is 44.5156%. The efficiency is 48.2183% for AM 1.5D. The first section of this chapter introduces the concept of the work that needs to be done, the second section provides the literature review that must be used to conduct this research, the third section discusses the multi-junction solar cell, the fourth section provides information about the proposed solar cell model, the fifth section provides the parameters of simulation and assumption, the sixth and final sections analyze the results obtained by MATLAB

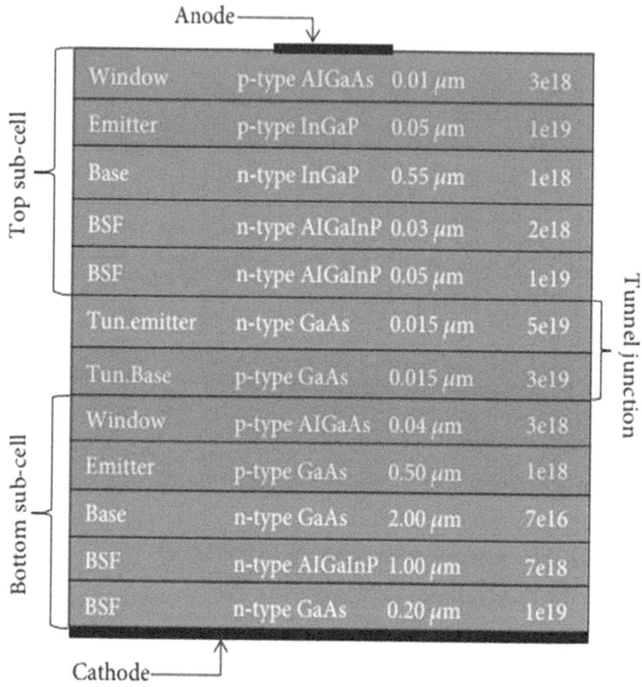

Figure 5.6 Layers that show material thickness and doping level.

simulation, and the work is concluded. In this chapter, for different and all AM conditions such as AM 1.5G, AM 0, and AM 1.5D, the efficiency variations of AlAs/GaAs-based multi-junction solar cells have been studied. The first and second highest efficiency in this work is noted to be found at AM 0, at 50.7484%, and at AM 1.5D, at 48.2183% [16].

5.6.2 Optimization of GaInP/GaAs solar cell using AlGaAs as tunnel junction

The tunnel junction of AlGaAs was grown over GaAs using the molecular beam epitaxy technique, where the GaAs substrate is p-type. In this study, an AM 1.5G spectral of sunlight was used, where G stands for global radiation for calculating current and voltage characteristics [129–135]. In this chapter, comparisons are done between a GaInP/GaAs junction without the AlGaAs tunnel junction and with GaInP/GaAs junction with the AlGaAs tunnel junction. The authors found a 48% improvement with AlGaAs as tunnel junctions. In this chapter, the significant dependency was simulated on the atomic homogeneity of performance of the solar cell structures in the grown layers and interface characteristics. Layer-over-layer growth of the solar cell structure

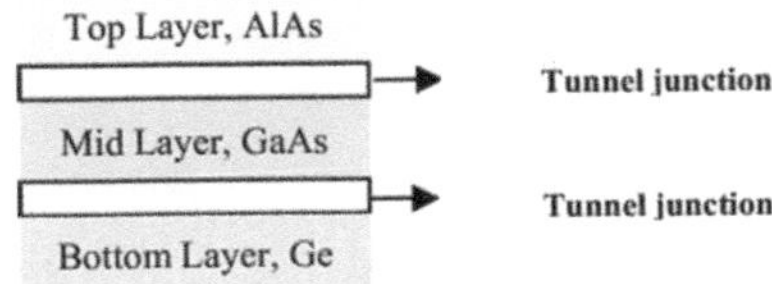

Figure 5.7 Structure of solar cell.

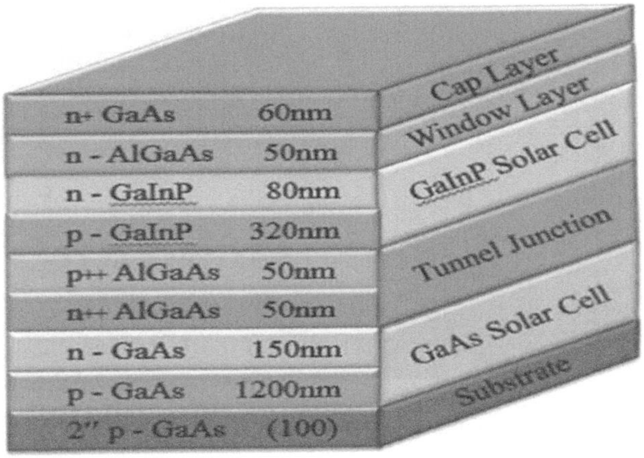

Figure 5.8 Multi-junction solar cell.

of GaInP/GaAs was done using SIMS analysis, and performance was investigated. The structure of a solar cell is given in Figure 5.7.

In this chapter, the authors have mentioned SIMS, which is widely used for dopant profiling within contact and also over the patterned junction. SIMS is an analytical technique that is used for semiconductor solar cells. The authors utilized atomic force microscopy and secondary ion mass spectrometry for the evaluation of the structural and physical properties of the solar cell structure [17]. The multi-junction solar cell is presented in Figure 5.8.

5.6.3 Enhancement of InGaP solar cells grown by epitaxy method

Anti-phase boundary (APB) formation and reduction of anomalous band gap happen because of the atomic ordering of growth of ternary InGaP alloys. In this study, atomic ordering is emphasized. Solid-source molecular beam epitaxy used for GaAs (001) substrates miscut 2° toward (111) B is also studied. Further, lattice-matched In0.52Ga0.48P solar cell performance and the effect on growth temperature and growth rate are dealt

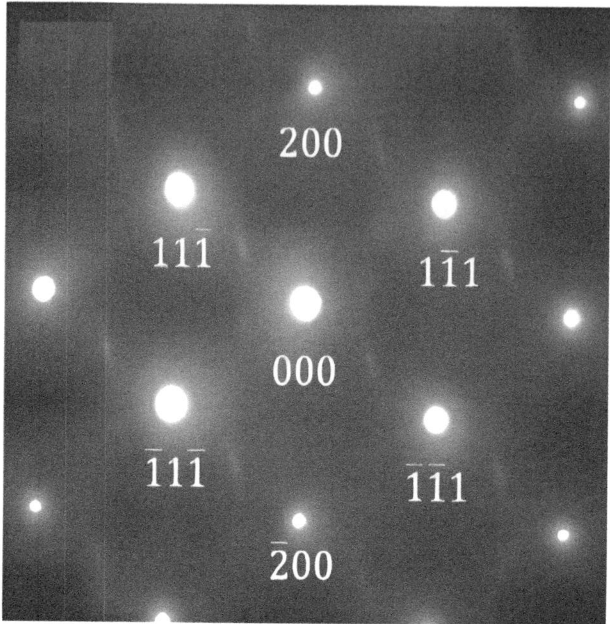

Figure 5.9 At 480 °C and 1.0 μm=h InGaP grown.

with. The highest efficiency for InGaP/GaAs solar cells is found to be 25.57%, and the single-junction one is 14.43% at 510 °C. Open circuit voltage was found to vary from 1.243 to 1.312 V using growth temperature and growth rate. In this paper, the effect is also observed over the band gap, which improves the optical and electrical measurements. Film widening is also observed at 480 °C and film grows at 1.855eV. Similarly, at 510 °C, film grows at 1.880eV. This chapter mostly deals with the growth of InGaP and its alloys and their impact on performance. Figure 5.9 illustrates InGaP layers on miscut substrates [18].

5.6.4 Efficiency enhancement of InGaP/ GaAs dual-junction solar cells with various tunneling diodes

The Silvaco ATLAS tool is used in this study to optimize the top and bottom cells' intermediate short-circuit current while designing InGaP/GaAs (DJ) solar cells. The dual-junction solar cell is shown in Figure 5.10. The base layer of the top cell is significantly thicker than the base layer of the bottom cell and displays a greater short-circuit current density (Jsc). The effects of the tunnel diode architectures of GaAs/GaAs, InGaP/InGaP, and AlGaAs/AlGaAs on the characteristics of solar cells have been investigated

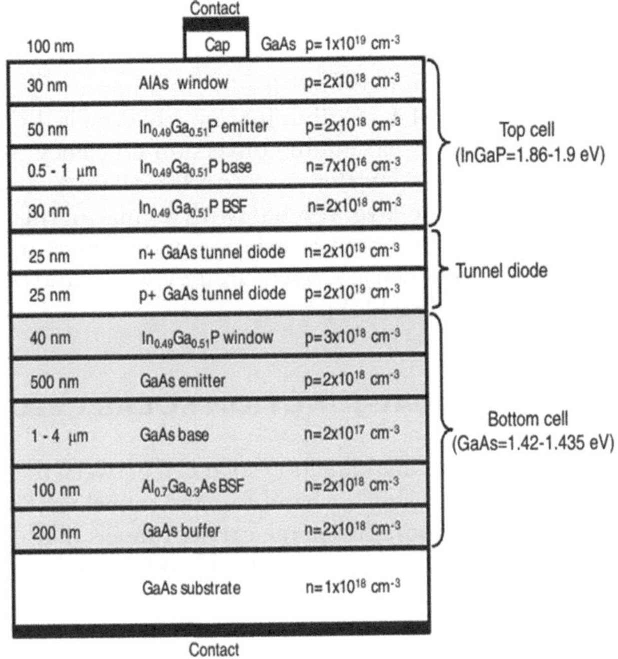

Figure 5.10 Dual-junction InGaP/GaAs solar cell.

in this work. Additionally, the photo production rate in the DJ solar cell structure is assessed using incoming light of various wavelengths. For different incidence wavelengths within the AM 1.5G solar spectrum, the light productivity in the optimal solar cell structure is determined, and the EQE and conversion efficiency are thoroughly investigated. By optimizing the Jsc matching criteria between the top and bottom cells in numerical simulations using ATLAS, InGaP/GaAs DJ solar cells with varied tunnel diodes are constructed [136–140]. The base thicknesses of the top and bottom cells in this study are tuned for greater effectiveness. Based on the aforementioned result, it is found that the tunnel diode construction affects the current matching condition. The use of InGaP/InGaP tunnel diodes demonstrates greater Jsc when compared to GaAs/GaAs tunnel diodes. The base width of the top cell is raised to meet the current under AM 1.5G lighting. In solar cells, the greatest conversion is enhanced.

An efficiency of up to 26.04% was attained using an InGaP/InGaP tunnel diode. These results might result in a better understanding of the physical behaviors of the devices by adding experimental characteristics further into the development of III-V multi-junction solar cells. This work examines the GaInP/GaAs DJ top cell's utilization of dual-layered BSF in solar cells of various thicknesses, using Silvaco ATLAS TCAD tool, a computerized numerical modeling system. The rates of creating detailed photos have been determined. The main modeling stages are provided, and the simulation findings are confirmed using existing experimental data to characterize the accuracy of our results. The maximal Jsc for this improved cell structure is 17.33 mA/cm2 under AM 1.5G illumination, the Voc is 2.66 V, and the FF is 88.67%, displaying a maximum conversion efficiency of 34.52% (1 sun) and 39.15% (1000 suns). In this study, an InGaP/GaAs DJ solar cell is designed employing a double-layer top BSF with a range of thicknesses, and the results are compared to those of earlier experiments. Conversion efficiency, short-circuit current density, open circuit voltage, and other popular solar cell models are calculated [19].

5.7 TUNNELING IN DUAL-JUNCTION SOLAR CELL

The tunneling operation in a dual-junction solar cell between the top and bottom tunneling diode is done using the band-to-band tunneling (BTBT) mechanism. In the BTBT mechanism, the charge carrier tunnels (both electron and hole) across the valence band to the conduction band through the band gap.

5.7.1 Basics of tunneling

In essence, tunneling is a quantum mechanical phenomenon in which particles overcome the energy barrier even though they don't have sufficient energy. It can't be explained using classical mechanics because it is not observed in bigger particles; it will not be accurate. Therefore, we will not focus on classical mechanics in which we mostly consider the particle nature of atoms. Here, the wave nature is prominent for discussing quantum particles and forming a chain between the quantum mechanical behavior and the wave nature. When an electromagnetic wave is an incident on a semiconductor, the energy from the EM wave will lead to the movement of electrons freely inside the semiconductor; if they have sufficient length, the wave found outside the semiconductor will be negligible (as shown in Figure 5.11 (a)). Similarly, if the semiconductor is thin, the electron wave will have sufficient energy to move to the edges of the semiconductor with some decay (as shown in Figure 5.11 (b)).

5.7.2 Quantum mechanical tunneling

This is a quantum phenomenon in which the particle has a non-zero probability of overcoming the potential barrier, yet the particle will be found on the other side due to its wave nature. This phenomenon is known as quantum mechanical tunneling [29]. If a particle is on the other side, then we say it tunnels across the barrier.

Quantum mechanical tunneling is presented in Figure 5.12.

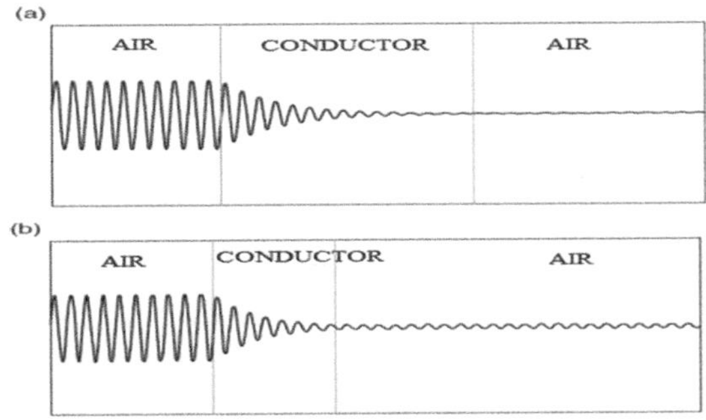

Figure 5.11 (a) and (b) emerging electromagnetic wave.

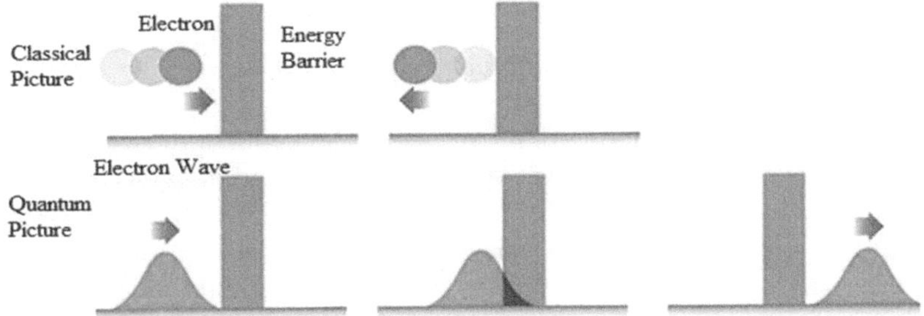

Figure 5.12 (a) Quantum mechanical tunneling.

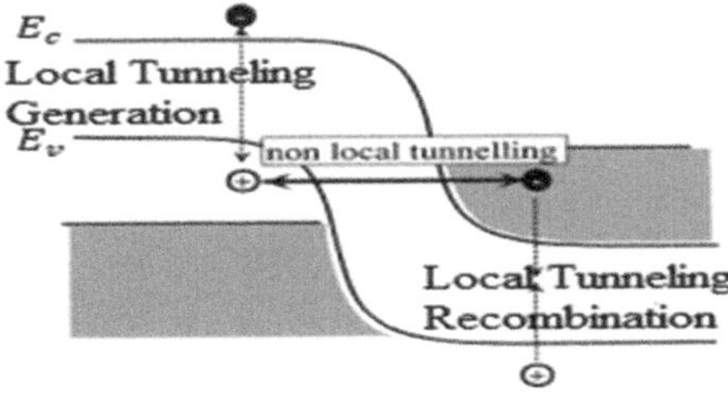

Figure 5.12 (b) Local and non-local tunneling.

5.7.3 Advanced tunneling models

Mostly, in semiconductors, there are two types of models of tunneling observed for calculating the current: local model and non-local models. In short, local models refer to tunneling from one band to another band, and non-local tunneling occurs in specific spatial coordinates, where electrons tunnel from one place to another.

5.7.3.1 Non-local BTBT tunneling

In non-local BTBT models, tunneling, as shown in Figure 5.13 (b), occurs from a spatial perspective, so for spatial parameters, we need to obtain the tunneling probability using Schrodinger's equation. Without external bias, we don't need to solve for the V(x) potential barrier because it is already zero. But with external bias, it becomes very difficult to solve for probability using Schrodinger's equation. Probability also depends on the band structure of the semiconductor, so we need to add the potential V(x) in Schrodinger's equation. These kinds of dependencies make it a little difficult to solve for much more complicated and complex spatial dependencies [30]. This is a reason that we prefer local tunneling models for analysis.

5.7.3.2 Local BTBT tunneling

Band perspective causes tunneling effects in local models. The interplay of the various fields produced by atoms in the semiconductor's lattice results in the different energy bands that are present in semiconductors. Using an E-k diagram, these energy levels are simply illustrated. Indirect or direct energy band gaps are also possible. A potential barrier that may be easily identified using an E-k diagram separates the conduction and valence bands. Even though an intrinsic material may have a very small conduction band of electrons, when there is an externally applied electric field, just those few electrons will flow and produce very little current. However, electrons can tunnel from the valence to the conduction band if the electric field is sufficiently strong. For the electrons within the valence band to tunnel through into the potential barrier and enter the conduction band, the band in the semiconductor has to be pushed down by the electric field. A significant current may now flow through the material because tunneling has produced enough electrons in the conduction band. The direct and indirect band gap of materials is shown in Figure 5.14.

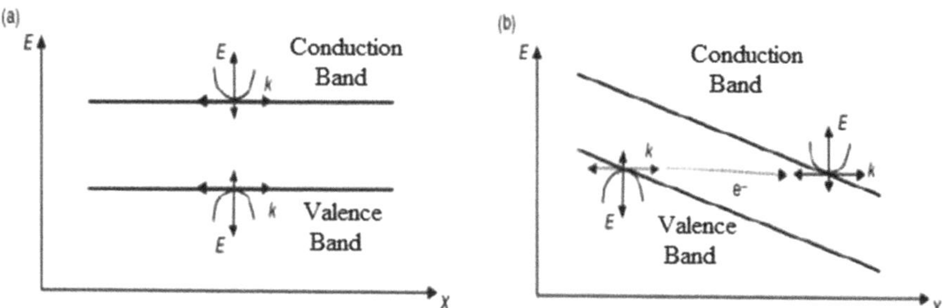

Figure 5.13 Direct and indirect band gap.

5.8 SIMULATION AND DISCUSSION OF RESULTS

Figure 5.14 depicts the AlAs/GaAs DJ solar cell's fundamental architecture. Using the Silvaco TCAD tool, (b) exhibits the simulated cell structure while Figure 5.14 (a) depicts the various levels of the structure. The structure is made up of an AlAs top cell and a GaAs bottom cell. The top cells consist of four layers, namely window, p-type emitter, n-type base, and BSF layer,

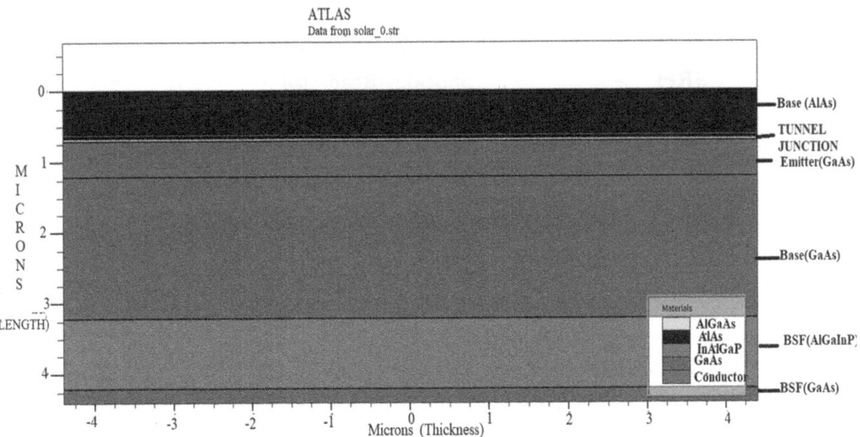

Figure 5.14 (a) Different layers of the structure.

Figure 5.14 (b) Simulated dual-junction solar cell structure.

while the bottom cell consist of five layers, namely, window, emitter, base, and two BSF layers. The top and bottom cell systems are interconnected with a GaAs/GaAs tunnel diode, which must be transparent to optimize the performance of the structure. Any solar radiation entering it should be absorbed.

The physical model used for designing the dual-junction solar cell can be included in the simulation, and it should be carefully chosen in order to optimize the solar cell in the most efficient way possible. Concentration-dependent low field mobility (CONMOB), optical recombination (OPTR), Shockley-Read-Hall (SRH), Auger recombination, and Fermi statistics are the primary physical models utilized in the construction of every solar cell. Replicating the non-local band-to-band tunneling for the DJ cell can be enabled through the tunneling junction between both the bottom and top cells. Table 5.1 lists the physical models that the Silvaco ATLAS simulator used to create the DJ solar cell, and Table 5.2 provides data for the main parameters with their standard value for designing the AlAs/GaAs DJ solar cell.

5.9 CONCLUSION

Simulation and optimization of the AlAs/GaAs solar cell have been assessed in this chapter. As all the layers, including the window layer, are essential to the solar cell configuration, the selection of all supported layers and their

Table 5.1 Physical models used by the Silvaco tool [34]

Physical Models Used	Physical Models Used
OR	Optical recombination
CONMOB	Concentration-dependent mobility
AUGER	AUGER recombination
SRH	Shockley-Read-Hall
FERMI	Fermi distribution
BBT.NONLOCAL	Non-local band-to-band tunneling

Table 5.2 Main parameters with their standard value for designing the AlAs/GaAs DJ solar cell [35]

Parameters	GaAs	GaAs	AlGaAs
The energy band gap (eV)	1.42	2.16	2.07
Lattice constant (Å)	5.65	5.65	5.65
Electron mobility (cm2/V-s)	212.2	8800	1945
Permittivity	13.2	10.3	10.9
Hole mobility (cm2/V-s)	67.6	400	141

parameters is essential for the better efficiency of a solar cell. AlGaAs are thus used as the top cell absorber layer to enhance the device. BSF layer thickness optimizes efficiency. The mole fraction of BSF doping in the window layer of the top cell is also optimized. Simulation parameters, such as I-V characteristics and EQE vs wavelength, have also been plotted. Significant improvements are obtained in the proposed solar cell structure and results. Our research was very useful, as we have seen that solar energy can be used to power electronic appliances in a non-polluting environment. Our goal is to learn how solar energy works and in which situation it can be used. The solar research has provided us with the necessary details. In this research, we have seen that the efficiency of a dual-junction solar cell is increased with the help of optimization of a solar cell in various ways. By changing the BSF thickness and doping concentration of the solar cell, there is a change in the efficiency.

REFERENCES

1. Marwa S. Salem, Omar M. Saif Ahmed Shaker Mohamed Abouelatta, Abdullah J. Alzahrani, Adwan Alanazi, M. K. Elsaid, Rabie A, "Performance optimization of the InGaP/GaAs dual-junction solar cell using SILVACO TCAD", *Ramadan International Journal of Photoenergy*, vol. 2021, Article ID 8842975, p. 12, Feb 2021.
2. Y. Özen, N. Akın, B. Kınacı, S. Özçelik, "Performance evaluation of a GaInP/GaAs solar cell structure with the integration of AlGaAs tunnel junction," *Solar Energy Materials and Solar Cells*, vol. 137, pp. 1–5, 2015.
3. R. Oshima, Y. Nagato, Y. Okano, T. Sugaya, "Enhancement of open-circuit voltage in InGaP solar cells grown by solid source molecular beam epitaxy," *Japanese Journal of Applied Physics*, vol. 57, no. 8S3, Article 08RD07, 2018.
4. J. Leem, Y. Lee, J. Yu, "Optimum design of InGaP/GaAs dual-junction solar cells with different tunnel diodes," *Optical and Quantum Electronics*, vol. 41, no. 8, pp. 605–612, 2009.
5. P. Nayak, J. Dutta, G. Mishra, "Efficient InGaP/GaAs DJ solar cell with double back surface field layer," *Engineering Science and Technology, an International Journal*, vol. 18, no. 3, pp. 325–335, 2015.
6. M. M. Rana, S. M. F. Khan, M. S. Shafayat, "Design and optimization of AlGaAs/InP multi- junction solar cell," *2019 1st International Conference on Advances in Science, Engineering and RoboticsTechnology (ICASERT)*, IEEE, vol. 2019, pp. 1–5, 2019, doi: 10.1109/ICASERT.
7. Y. Guo, Q. Liang, B. Shu, J. Wang, Q. Yang, "The III–V triple-junction solar cell characteristics andoptimization with a Fresnel lens concentrator," *International Journal of Photoenergy*, vol. 2018, p. 10, 2018.
8. K. Attari, L. Amhaimar, A. Asselman, M. Bassou, "The design and optimization of GaAs single solarcells us- ing the genetic algorithm and Silvaco ATLAS," *International Journal of Photoenergy*, vol. 2017, pp. 1–7, 2017.
9. Gang Lu, Zeyulin Zhang, Fengqin He, Hailong You, Dazheng Chen, Weidong Zhu, Jincheng Zhang, Chunfu Zhang, "Performance improvement of all-inorganic, hole transport-layer- free perovskite solarcells through

dipoles- adjustion by polyethyleneimine incorporating", *IEEE Electron Deviceletters*, vol. 42, no. 4, Apr 2021.

10. Junichi Nakamura, Naoki Asano, Takeshi Hieda, Chikao Okamoto, Hiroyuki Katayama, Kyotaro Nakamur, "Development of heterojunction back contact si solar cells", *IEEE Journal Ofphotovoltaics*, vol. 4, no. 6, Nov 2014.

11. H. Tang, S. He, and C. Peng, "A short progress report on high-efficiency perovskite solar cells. *Nanoscale Research Letters*, vol. 12, p. 410, 2017. doi: 10.1186/s11671-017-2187-5.

12. Gavin Conibeer, "*ARC Photovoltaics Centre of Excel- lence, School of Photovoltaic and Renewable EnergyEngineering*", University of New South Wales, Sydney NSW 2052, Australia Third-Generation Photovoltaics, 2009.

13. Gang Lu, Zeyulin Zhang, Fengqin He, Hailong You, Dazheng Chen, Weidong Zhu, Jincheng Zhang, Chunfu Zhang, "Performance improvement of all-inorganic, hole- transport- layer-free perovskite solarcells through dipoles- adjustion by polyethyleneimine incorporating", *IEEE Electron Device Letters*, vol. 42, no. 4, Apr 2021.

14. Jing Ma, Jingjing Chang, Jie Su, Zhenhua Lin, "*Optimize the Oxide/ Perovskite Heterojunction Contact for- Low Temperature High Efficiency and Stable All-inorganic CsPbI2Br Perovskite Solar Cells*", 2020 47th IEEE Photo- voltaic Specialists Conference (PVSC), Jan 2021.

15. Xingtao Wang, Yong Wang, Yuetian Chen, Xiaomin Liu, Yixin Zhao, "Efficient and stable CsPbI3 inorganic perovskite photovoltaics enabled by crystal secondary growth", *Advanced Materials*, vol. 33, no. 44, p. 2103688, 2 Nov 2021.

16. R. RafieiRad, B. Azizollah Ganji, "Efficiency im- provement of perovskite solar cells by UtilizingCuInS2 thin layer: Modeling and numerical study", *IEEE Transaction Son Electron Devices*, vol. 68, no. 10, Oct 2021.

17. Jakapan Chantana, Kanta Tai, Haruki Hayashi, Takahito Nishimura, Yu Kawano, Takashi Minemoto, "Investigation of carrier recombination of Na-doped Cu2SnS3 solar cell for its improved conversionefficiency of 5.1%", *Solar Energy Materials and Solar Cells*, vol. 206, p. 110261, Mar 2020.

18. Qiufeng Ye, Yang Zhao, Shaiqiang Mu, Fei Ma, Feng Gao, Zema Chu, Zhigang Yin, Pingqi Gao, XingwangZhang, Jingbi You, "Cesium lead inorganic solar cell with efficiency beyond 18% via reducedcharge recombination", *Advanced Materials*, vol. 31, no. 49, p. 1905143, 6 Dec, 2019 .

19. P. Nayak, J. Dutta, G. Mishra, "Efficient In- GaP/GaAs DJ solar cell with double back surface fieldlayer," *Engineering Science and Technology, an International Journal*, vol. 18, no. 3, pp. 325–335, 2015.

20. J. Plá, M. Barrera, F. Rubinelli, "The influence of the InGaP window layer on the optical and electrical performance of GaAs solar cells," *Semiconductor Science and Technology*, vol. 22, no. 10, pp. 1122–1130, 2007.

21. M. Lueck, C. Andre, A. Pitera, M. Lee, E. Fitzgerald, S. Ringel, "Dual junction GaInP/GaAs solar cells grown on metamorphic SiGe/Si substrates with high open circuit voltage," *IEEE Electron Device Letters*, vol. 27, no. 3, pp. 142–144, 2006.

22. J. Leem, Y. Lee, J. Yu, "Optimum design of InGaP/GaAs dual-junction solar cells with different tunnel-2017.iodes," *Optical and Quantum Electronics*, vol. 41, no. 8, pp. 605–612, 2009.

23. K. J. Singh, S. K. Sarkar, "An effective modeling approach for high efficient solar cell using virtual wafer fabrication tools," *Journal of Nano-and Electronic Physics*, vol. 3, no. 1, p. 792, 2011.

24. A. Kirk, "High efficacy thinned four-junction solar cell," *Semiconductor Science and Technology*, vol. 26, no. 12, Article 125013, 2011.

25. P. Nayak, J. Dutta, G. Mishra, "Efficient InGaP/GaAs DJ solar cell with double back surface field layer," *Engineering Science and Technology, an International Journal*, vol. 18, no. 3, pp. 325–335, 2015.

26. M. H. Tsutagawa, S. Michael, "*Triple junction InGaP/- GaAS/Ge solar cell optimization: The design parameters for a 36.2% efficient space cell using Silvaco ATLAS modeling & simulation*," 2009 34th IEEE Photovoltaic Specialists Conference (PVSC), pp. 001954–001957, Philadelphia, PA, USA, 2009.

27. G. Sahoo, P. Nayak, G. Mishra, "An ARC less InGaP/GaAs DJ solar cell with hetero tunnel junction," *Superlattices and Microstructures*, vol. 95, pp. 115–127, 2016.

28. H. Arzbin, A. Ghadimi, "Efficiency improvement of ARC less InGaP/GaAs DJ solar cell with InGaP tunnel junction and optimized two BSF layer in top and bottom cells," *Optik*, vol. 148, pp. 358–367, 2017.

29. M. Green, A. Ho-Baillie, H. Snaith, "The emergence of perovskite solar cells", *Nature Photon*, vol. 8, pp. 506–514, 2014.

30. J. Adams, V. Elarde, T. Major, H. Miyamoto, M. Os owski, N. Pan, R. Tatavarti, F. Tuminello, A. Wibowo, C. Youtsey, G. Ragunathan, "Flexible and lightweight epitaxial lift-off GaAs multi-junction solar cells for portable power and UAV applications", *2015 IEEE 42nd Photovoltaic Specialist Conference*, PVSC, pp. 8–11, 2015. doi: 10.1109/PVSC.2015.7356137.

31. X. Li, P. C. Li, L. Ji, C. Stender, S. R. Tatavarti, K. Sablon, E. T. Yu, "Integration of sub- wavelength optical nano structures for improved anti reflection performance of mechanically flexible GaAs solar cells fabricated by epitaxial lift-off", *Solar Energy Materials and Solar Cells*, vol. 143, pp. 567–572, 2015. doi: 10.1016/j.solmat.(2015).

32. M. A. Green, Y. Hishikawa, W. Warta, E. D. Dunlop, D. H. Levi, J. Hohl-Ebinger, A. W. Ho-Baillie, "Solar cell efficiency tables (version 50)", *Progress in Photovoltaics: Research and Applications*, vol. 25, pp. 668–676, 2017.

33. S. Essig, M. A. Steiner, C. Allebé, J. F. Geisz, B. Paviet-Salomon, S. Ward, A. Descoeudres, V. LaSalvia, L. Barraud, N. Badel, A. Faes, A. "Realization of GaInP/Si dual-junction solar cells with 29.8% 1-sun efficiency," *IEEE Journal of Photovoltaics*, vol. 6, no. 4, pp. 1012–1019, 2016.

34. Yiming Liu, Mehrad Ahmadpour, Jost Adam, Jakob Kjelstrup-Hansen, Horst-Günter Rubahn, Morten Madsen, "Modeling multijunction solar cells by non local tunneling and subcell analysis", *IEEE Journal of Photovoltaics*, vol. 8, no. 5, pp. 1363–1369, 2018.

35. M. A. Mintairov, N. A. Kalyuzhnyy, V. V. Evstropov, V. M. Lantratov, S. A. Mintairov, M. Z. Shvarts, V. M. Andreev, A. Luque, "The segmental approximation in multijunction solar cells", *IEEE Journal of Photovoltaics*, vol. 5, no. 4, pp. 1229–1236, 2015.

36. Anjana Bhardwaj, Pradeep Kumar, B. Raj, Sunny Anand, "Design and performance optimization of doping-less vertical nanowire TFET using gate stack technique", *Journal of Electronic Materials (JEMS)*, Springer, vol. 41, no. 7, pp. 4005–4013, 2022.

37. Jeetendra Singh, B. Raj, "Tunnel current model of asymmetric MIM structure levying various image forces to analyze the characteristics of filamentary memristor", *Applied Physics A*, Springer, vol. 125, no. 3, p. 203.1, Feb 2019.

38. Candy Goyal, Jagpal Singh Ubhi, B. Raj, "Low leakage zero ground noise nanoscale full adder using source biasing technique", *Journal of Nanoelectronics and Optoelectronics*, American Scientific Publishers, vol. 14, pp. 360–370, Mar 2019.

39. Gurmohan Singh, R. K. Sarin, B. Raj, "A novel robust exclusive-OR function implementation in QCA nanotechnology with energy dissipation analysis", *Journal of Computational Electronics*, Springer, vol. 15, no. 2, pp. 455–465, Jun 2016.

40. Tanu Wadhera, Deepti Kakkar, Girish Wadhwa, B. Raj, "Recent advances and progress in development of the field effect transistor biosensor: A review", *Journal of Electronic Materials*, Springer, vol. 48, no. 12, pp. 7635–7646, Dec 2019.

41. Girish Wadhwa, Priyanka Kamboj, Balwinder Raj, "Design optimisation of junctionless TFET biosensor for high sensitivity", *Advances in Natural Sciences: Nanoscience and Nanotechnology*, vol. 10, p. 045001, 2019.

42. Priya Bansal, B. Raj, "Memristor modeling and analysis for linear dopant drift kinetics," *Journal of Nanoengineering and Nanomanufacturing*, American Scientific Publishers, vol. 6, pp. 1–7, 2016.

43. Amandeep Singh, Mamta Khosla, B. Raj, "Circuit compatible model for electrostatic doped schottky barrier CNTFET, "*Journal of Electronic Materials*, Springer, vol. 45, no. 12, pp. 4825–4835, 2016.

44. D. Vaithiyanathan, B. Raj, "Performance analysis of charge plasma induced graded channel si nanotube", *Journal of Engineering Research (JER)*, EMSME Special Issue, pp. 146–154, Aug 2021.

45. Abhishek Singh Tomar, Vijay Kumar Magraiya, B. Raj, "Scaling of access and data transistor for high performance DRAM cell design", *Quantum Matter*, vol. 2, pp. 412–416, Oct 2013.

46. Neeraj Jain, B. Raj, "Parasitic capacitance and resistance model development and optimization of raised source/drain SOI FinFET structure for analog circuit applications", *Journal of Nanoelectronics and Optoelectronins*, ASP, USA, vol. 13, pp. 531–539, Ap 2018.

47. Shradhya Singh, S. K. Vishvakarma, B. Raj, "Analytical modeling of split-gate junction-less transistor for a biosensor application", *Sensing and Bio-Sensing*, Elsevier, vol. 18, pp. 31–36, Apr 2018.

48. Maisagalla Gopal, Balwinder Raj, "Low power 8T SRAM cell design for high stability video applications", *ITSI Transaction on Electrical and Electronics Engineering*, vol. 1, no. 5, pp. 91–97, 2013.

49. Balwinder Raj, Jatin Mitra, Deepak Kumar Bihani, V. Rangharajan, A. K. Saxena, S. Dasgupta, "Analysis of noise margin, power and process variation for 32 nm FinFET based 6T SRAM cell", *Journal of Computer (JCP)*, Academy Publisher, Finland, vol. 5, no. 6, pp. 1–8, 2010.

50. Divya Sharma, Rajesh Mehra, B. Raj, "Comparative analysis of photovoltaic technologies for high efficiency solar cell design", *Superlattices and Microstructures*, Elsevier, vol. 153, p. 106861, May 2021.

51. Pawandeep Kaur, Avtar Singh Buttar, B. Raj, "A comprehensive analysis of nanoscale transistor based Biosensor: A review", *Indian Journal of Pure and Applied Physics*, vol. 59, pp. 304–318, Apr 2021.

52. Divya Yadav, Balwant Raj, B. Raj, "Design and simulation of low power microcontroller for IoT applications", *Journal of Sensor Letters*, ASP, vol. 18, pp. 401–409, May 2020.

53. Shailendra Singh, B. Raj,"A 2-D analytical surface potential and drain current modeling of double-gate vertical t-shaped tunnel FET", *Journal of Computational Electronics*, Springer, vol. 19, pp. 1154–1163, Apr 2020.

54. Jeetendra Singh, B. Raj, "An accurate and generic window function for nonlinear memristor model", *Journal of Computational Electronics*, Springer, vol. 18, no. 2, pp. 640–647, Jun 2019.

55. Manjit Kaur, Neena Gupta, Sanjeev Kumar, B. Raj, Arun Kumar Singh, "Comparative RF and crosstalk analysis of carbon based nano interconnects", *IET Circuits, Devices & Systems*, vol. 15, no. 6, pp. 493–503, Feb 2021.

56. Nehru Kandasamy, Firdous Ahmad, D. Ajitha, B. Raj, Nagarjuna Telagam, "Quantum dot cellular automata based scan flip flop and boundary scan register", *IETE Journal of Research*, vol. 66, pp. 535–548, 2020.

57. S. K. Sharma, B. Raj, M. Khosla, "Enhanced photosensivity of highly spectrum selective cylindrical gate In1-xGaxAs nanowire MOSFET photodetector", *Modern Physics Letter-B*, vol. 33, no. 12, p. 1950144, 2019.

58. Jeetendra Singh, B. Raj, "Design and investigation of 7T2M NVSARM with enhanced stability and temperature impact on store/restore energy", *IEEE Transactions on Very Large Scale Integration Systems*, vol. 27, no. 6, pp. 1322–1328, Jun 2019.

59. Anil Kumar Bhardwaj, Sumeet Gupta, B. Raj, Amandeep Singh, "Impact of double gate geometry on the performance of carbon nanotube field effect transistor structures for low power digital design", *Computational and Theoretical Nanoscience*, ASP, vol. 16, pp. 1813–1820, 2019.

60. Neeraj Jain, B. Raj, "Thermal stability analysis and performance exploration of asymmetrical dual-k underlap spacer (ADKUS) SOI FinFET for security and privacy applications", *Indian Journal of Pure & Applied Physics (IJPAP)*, vol. 57, pp. 352–360, May 2019.

61. Amandeep Singh, Mamta Khosla, B. Raj, "Design and analysis of dynamically configurable electrostatic doped carbon nanotube tunnel FET", *Microelectronics Journal*, Elesvier, vol. 85, pp. 17–24, Mar 2019.

62. Neeraj Jain, Balwinder Raj, "Dual-k spacer region variation at the drain side of asymmetric SOI FinFET structure: Performance analysis towards the analog/rf design applications", *Journal of Nanoelectronics and Optoelectronics*, American Scientific Publishers, vol. 14, pp. 349–359, Mar 2019.

63. Jeetendra Singh, Sanjeev Sharma, B. Raj, Mamta Khosla, "Analysis of barrier layer thickness on performance of In1-xGaxAs based gate stack cylindrical gate nanowire MOSFET," *JNO*, ASP, vol. 13, pp. 1473–1477, Oct 2018.

64. Neeraj Jain, B. Raj, "Analysis and performance exploration of high-k SOI FinFETs over the conventional low-k SOI FinFET toward analog/RF design", *Journal of Semiconductors (JoS)*, IOP Science, vol. 39, no. 12, p. 124002-1-7, Dec 2018.

65. Candy Goyal, Jagpal Singh Ubhi, B. Raj, "A reliable leakage reduction technique for approximate full adder with reduced ground bounce noise', *Journal of Mathematical Problems in Engineering*, Hindawi, vol. 2018, Article ID 3501041, p. 16, 15 Oct 2018.

66. Jeetendra Singh, B. Raj, Mamta Khosla, "Design and performance analysis of nano-scale memristor-based nonvolatile SRAM", *Journal of Sensor Letter*, American Scientific Publishers, vol. 16, pp. 798–805, Oct 2018.

67. Girish Wadhwa, B. Raj, "Parametric variation analysis of charge-plasma-based dielectric modulated JLTFET for biosensor application", *IEEE Sensor Journal*, vol. 18, no. 15, pp. 6070–6077, 2018.

68. Jeetendra Singh, B. Raj, "Comparative analysis of memristor models for memories design", *JoS*, IoP, vol. 39, no. 7, p. 074006-1-12, Jul 2018.

69. Divya Yadav, Shailesh Singh Chouhan, Santosh Kumar Vishvakarma, B. Raj, "Application specific microcontroller design for IoT based WSN", *Sensor Letter*, ASP, vol. 16, pp. 374–385, May 2018.

70. Gurmohan Singh, R. K. Sarin, B. Raj, "Fault-tolerant design and analysis of quantum-dot cellular automata based circuits", *IEEE/IET Circuits, Devices & Systems*, vol. 12, pp. 638–664, 2018.

71. Jeetendra Singh, B. Raj, "Modeling of mean barrier height levying various image forces of metal insulator metal structure to enhance the performance of conductive filament based memristor model", *IEEE Nanotechnology*, vol. 17, no. 2, pp. 268–267, Mar 2018.

72. Aakash Jain, Sanjeev Sharma, B. Raj, "Analysis of triple metal surrounding gate (TM-SG) III-V nanowire MOSFET for photosensing application", *Opto-electronics Journal*, Elsevier, vol. 26, no. 2, pp. 141–148, May 2018.

73. Aakash Jain, Sanjeev Sharma, B. Raj, "Design and analysis of high sensitivity photosensor using cylindrical surrounding gate MOSFET for low power sensor applications", *Engineering Science and Technology, an International Journal*, Elsevier, vol. 19, no. 4, pp. 1864–1870, Dec 2016.

74. Amandeep Singh, Mamta Khosla, B. Raj, "Analysis of electrostatic doped schottky barrier carbon nanotube FET for low power applications," *Journal of Materials Science: Materials in Electronics*, Springer, vol. 28, pp. 1762–1768, 2017.

75. G. Saiphani Kumar, Amandeep Singh, B. Raj, "Design and analysis of gate all around CNTFET based SRAM cell design", *Journal of Computational Electronics*, Springer, vol. 17, no. 1, pp. 138–145, Mar 2018.

76. Gurinder pal Singh, B. S. Sohi, Balwinder Raj, "Material properties analysis of graphene base transistor (GBT) for VLSI analog circuits", *Indian Journal of Pure & Applied Physics (IJPAP)*, vol. 55, pp. 896–902, Dec 2017.

77. Amandeep Singh, Mamta Khosla, B. Raj, "Comparative analysis of carbon nanotube field effect transistor and nanowire transistor for low power circuit design," *Journal of Nanoelectronics and Optoelectronics*, American Scientific Publishers, USA, vol. 11, pp. 388–393, Jun 2016.

78. Sunil Kumar, B. Raj, "Estimation of stability and performance metric for inward access transistor based 6T SRAM cell design using n-type/p-type DMDG-GDOV TFET", *IEEE VLSI Circuits and Systems Letter*, vol. 3, no. 2, pp. 25–39, Jun 2017.

79. Shashikant Sharma, Anjan Kumar, Manisha Pattanaik, B. Raj, "Forward body biased multimode multi-threshold CMOS technique for ground bounce noise reduction in static CMOS adders", *International Journal of Information and Electronics Engineering*, pp. 567–572, vol. 3, no. 3, 2013.

80. Hamendra Singh, Pankaj Kumar, Balwinder Raj, "Performance Analysis of Majority Gate SET Based 1-bit Full Adder", *International Journal*

of Computer and Communication Engineering (IJCCE), IACSIT Press Singapore, ISSN: 2010-3743, Vol. 2, no. 4, 2013.

81. Anil Kumar Bhardwaj, Sumeet Gupta, B. Raj, "Investigation of parameters for schottky barrier (SB) height for schottky barrier based carbon nanotube field effect transistor device", *Journal of Nanoelectronics and Optoelectronics*, ASP, vol. 15, pp. 783–791, Jul 2020.

82. Priya Bansal, B. Raj, "Memristor: A versatile nonlinear model for dopant drift and boundary issues," *JCTN*, American Scientific Publishers, vol. 14, no. 5, pp. 2319–2325, May 2017.

83. Neeraj Jain, B. Raj, "An analog and digital design perspective comprehensive approach on Fin-FET (Fin-Field Effect transistor) technology: A review", *Reviews in Advanced Sciences and Engineering (RASE)*, ASP, vol. 5, pp. 1–14, 2016.

84. Sanjeev Sharma, B. Raj, Mamta Khosla, "Subthreshold performance of In1-xGaxAs based dual metal with gate stack cylindrical/surrounding gate nanowire MOSFET for low power analog applications", *Journal of Nanoelectronics and Optoelectronics*, American Scientific Publishers, USA, vol. 12, pp. 171–176, 2017.

85. Balwinder Raj, A. K. Saxena, S. Dasgupta, "Analytical modeling for the estimation of leakage current and subthreshold swing factor of nanoscale double gate FinFET device", *Microelectronics International, UK*, vol. 26, pp. 53–63, 2009.

86. Shailendra Singh, Girish Wadhwa, Balwinder Raj, "An analytical modeling for dual source vertical tunnel field effect transistor", *International Journal of Recent Technology and Engineering (IJRTE)*, vol. 8, no. 2, pp. 603–608, Jul 2019.

87. Shailendra Singh, B. Raj, "Design and analysis of hetrojunction vertical T-shaped tunnel field effect transistor", *Journal of Electronics Material*, Springer, vol. 48, no. 10, pp. 6253–6260, Oct 2019.

88. Candy Goyal, Jagpal Singh Ubhi, B. Raj, "A low leakage CNTFET based inexact full adder for low power image processing applications", *International Journal of Circuit Theory and Applications*, Wiley, vol. 47, no. 9, pp. 1446–1458, Sept 2019.

89. B. Raj, A. K. Saxena, S. Dasgupta, "A compact drain current and threshold voltage quantum mechanical analytical modeling for FinFETs", *Journal of Nanoelectronics and Optoelectronics (JNO)*, USA, vol. 3, no. 2, pp. 163–170, 2008.

90. Girish Wadhwa, B. Raj, "An analytical modeling of charge plasma based tunnel field effect transistor with impacts of gate underlap region", *Superlattices and Microstructures*, Elsevier, vol. 142, p. 106512, Jun 2020.

91. Shailendra Singh, B. Raj, "Modeling and simulation analysis of SiGe hetrojunction double GateVertical t-shaped tunnel FET", *Superlattices and Microstructures*, Elsevier vol. 142, p. 106496, Jun 2020.

92. Amandeep Singh, Dinesh Kumar Saini, Dinesh Agarwal, Sajal Aggarwal, Mamta Khosla, B. Raj, "Modeling and simulation of carbon nanotube field effect transistor and its circuit application," *Journal of Semiconductors (JoS)*, IOP Science, vol. 37, p. 074001-6, Jul 2016.

93. Neeraj Jain, Balwinder Raj, "Device and circuit co-design perspective comprehensive approach on FinFET technology: A review", *Journal of Electron Devices*, vol. 23, no. 1, pp. 1890–1901, 2016.

94. Sunil Kumar, B. Raj, "Analysis of I_{ON} and ambipolar current for dual-material gate-drain overlapped DG-TFET," *Journal of Nanoelectronics and Optoelectronics*, American Scientific Publishers, USA, vol. 11, pp. 323–333, Jun 2016.

95. Naveed Anjum, Tarun Bali, B. Raj, "Design and simulation of handwritten multiscript character recognition", *International Journal of Advanced Research in Computer and Communication Engineering*, vol. 2, no. 7, pp. 2544–2549, Jul 2013.

96. Sanjeev Sharma, B. Raj, Mamta Khosla, "A gaussian approach for analytical subthreshold current model of cylindrical nanowire FET with quantum mechanical effects", *Microelectronics Journal*, Elsevier, vol. 53, pp. 65–72, Apr 2016.

97. Karmjit Singh, B. Raj, "Performance and analysis of temperature dependent multi-walled carbon nanotubes as global interconnects at different technology nodes," *Journal of Computational Electronics*, Springer, vol. 14, no. 2, pp. 469–476, Jun 2015.

98. Sunil Kumar, B. Raj, "Compact channel potential analytical modeling of DG-TFET based on evanescent–mode approach," *Journal of Computational Electronics*, Springer, vol. 14, no. 2, pp. 820–827, Jul 2015.

99. Karmjit Singh, B. Raj, "Temperature dependent modeling and performance evaluation of multi-walled CNT and single-walled CNT as global interconnects," *Journal of Electronic Materials*, Springer, vol. 44, no. 12, pp. 4825–4835, Dec 2015.

100. V. K. Sharma, M. Pattanaik, B.Raj, "INDEP approach for leakage reduction in nanoscale CMOS circuits", *International Journal of Electronics*, Taylor & Francis, vol. 102, no. 2, pp. 200–215, 2014.

101. Karmjit Singh, B. Raj, "Influence of temperature on MWCNT bundle, SWCNT bundle and copper interconnects for nanoscaled technology nodes," *Journal of Materials Science: Materials in Electronics*, Springer, vol. 26, no. 8, pp. 6134–6142, 2015.

102. Naveed Anjum, Tarun Bali, B. Raj, "Design and simulation of handwritten gurumukhi and devanagri numerical recognition", *International Journal of Computer Applications*, Foundation of Computer Science, New York, USA, vol. 73, no. 12, pp. 16–21, 2013.

103. S. Khandelwal, V. Gupta, B. Raj, R. D. "Gupta, process variability aware low leakage reliable nano scale DG-FinFET SRAM cell design technique", *Journal of Nanoelectronics and Optoelectronics*, vol. 10, no. 6, pp. 810–817, Dec 2015.

104. V. K. Sharma, M. Pattanaik, B. Raj, "ONOFIC approach: Low power high speed nanoscale VLSI circuits design", *International Journal of Electronics*, Taylor & Francis, vol. 101, no. 1, pp. 61–73, 2014.

105. S. Khandelwal, Balwinder Raj, R. D. Gupta, "FinFET based 6T SRAM cell design: Analysis of performance metric, process variation and temperature effect", *Journal of Computational and Theoretical Nanoscience*, ASP, USA, vol. 12, pp. 2500–2506, 2015.

106. Sumit Singh, Shekhar Yadav, Jagdeep Rahul, Anurag Srivastava; B. Raj, "Impact of HfO_2 in graded channel dual insulator double gate MOSFET", *Journal of Computational and Theoretical Nanoscience*, American Scientific Publishers, vol. 12, no. 6, pp. 950–953, Apr 2015.

107. Vijay Kumar Sharma, Manisha Pattanaik, B. Raj, "PVT variations aware low leakage INDEP approach for nanoscale CMOS circuits", *Microelectronics Reliability*, Elsevier, vol. 54, pp. 90–99, 2014.

108. B. Raj, A. K. Saxena, S. Dasgupta, "Quantum mechanical analytical modeling of nanoscale DG FinFET: Evaluation of potential, threshold voltage and source/drain resistance", *Elsevier's Journal of Material Science in Semiconductor Processing*, Elsevier, vol. 16, no. 4, pp. 1131–1137, 2013.

109. Maisagalla Gopal, Siva Sankar D Prasad, Balwinder Raj, "8T SRAM cell design for dynamic and leakage power reduction", *International Journal of Computer Applications*, Foundation of Computer Science, New York, USA, vol. 71, no. 9, pp. 43–48, Jun 2013.

110. Manisha Pattanaik, B. Raj, Shashikant Sharma, Anjan Kumar, "Diode based trimode multi-threshold CMOS technique for ground bounce noise reduction in static CMOS adders", *Advanced Materials Research*, Trans Tech Publications, Switzerland, vol. 548, pp. 885–889, 2012.

111. Balwinder Raj, A. K. Saxena, S. Dasgupta, "Nanoscale FinFET based SRAM cell design: Analysis of performance metric, process variation, underlapped FinFET and temperature effect", *IEEE Circuits and System Magazine*, vol. 11, no. 2, pp. 38–50, 2011.

112. V. K. Sharma, M. Pattanaik, Balwinder Raj, "Leakage current ONOFIC approach for deep submicron VLSI circuit design", *International Journal of Electrical, Computer, Electronics and Communication Engineering, World Academy of Sciences, Engineering and Technology*, vol. 7, no. 4, pp. 239–244, 2013.

113. Tulika Chawla, Mamta Khosla, B. Raj, "Design and simulation of triple metal double-gate germanium on insulator vertical tunnel field effect transistor", *Microelectronics Journal*, Elsevier, vol. 114, p. 105125, Aug 2021.

114. Parminder Kaur, Sandeep Singh Gill, B. Raj, "Comparative analysis of OFETs materials and devices for sensor applications", *Journal of Silicon*, Springer, vol. 14, pp. 4463–4471, 2022.

115. Sanjeev Kumar Sharma, Parveen Kumar, Balwant Raj, B. Raj, "$In_{1-x}Ga_xAs$ double metal gate-stacking cylindrical nanowire MOSFET for highly sensitive photo detector", *Journal of Silicon*, Springer, vol. 14, pp. 3535–3541, 2022.

116. Tulika Chawla, Mamta Khosla, B. Raj, "Optimization of double-gate dual material GeOI-vertical TFET for VLSI circuit design", *IEEE VLSI Circuits and Systems Letter*, vol. 6, no. 2, pp. 13–25, Aug 2020.

117. Sachin Kumar Verma, Shailendra Singh, Girish Wadhwa, B. Raj, "Detection of biomolecules using charge-plasma based gate underlap dielectric modulated dopingless TFET", *Transactions on Electrical and Electronic Materials (TEEM)*, Springer, vol. 21, pp. 528–535, Jun 2020.

118. Neeraj Jain, B. Raj, "Impact of underlap spacer region variation on electrostatic and analog/RF performance of symmetrical high-k SOI FinFET at 20 nm channel length", *Journal of Semiconductors (JoS)*, IOP Science, vol. 38, no. 12, p. 122002, Dec 2017.

119. Shailendra Singh, B. Raj, "Analytical modeling and simulation analysis of T-shaped III-V heterojunction Vertical T-FET", *Superlattices and Microstructures*, Elsevier, vol. 147, p. 106717, Nov 2020.

120. Gurmohan Singh, R. K. Sarin, B. Raj, "Design and analysis of area efficient QCA based reversible logic gates", *Journal of Microprocessors and Microsystems*, Elsevier, vol. 52, pp. 59–68, May 2017.

121. Amandeep Singh, Mamta Khosla, B. Raj, "Compact model for ballistic single wall CNTFET under quantum capacitance limit," *Journal of Semiconductors (JoS)*, IOP Science, vol. 37, p. 104001-8, Oct 2016.

122. Sonal Singh, Mamta Khosla, Girish Wadhwa, B. Raj, "Design and analysis of double-gate junctionless vertical TFET for gas sensing applications", *Applied Physics A*, Springer, vol. 127, no. 16, 2 Jan 2021.

123. Inderjit Singh, B. Raj, Mamta Khosla, B. Rajesh Kumar Kaushik, "Potential MRAM technologies for low power SoCs", *SPIN World Scientific Publisher, SCIE*, vol. 10, no. 4, pp. 2050027, Dec 2020.

124. Shailendra Singh, B. Raj, "Parametric variation analysis on hetero-junction Vertical t-shape TFET for supressing ambipolar conduction", *Indian Journal of Pure and Applied Physics*, vol. 58, pp. 478–485, Jun 2020.

125. Shailendra Singh, Girish Wadhwa, Balwinder Raj, "Design and analysis of dual source vertical tunnel field effect transistor for high performance", *Transactions on Electrical and Electronics Materials*, Springer, vol. 21, pp. 74–82, Oct 2019.

126. Manjit Kaur, Neena Gupta, Sanjeev Kumar, B. Raj, Arun Kumar Singh, "RF performance analysis of intercalated graphene nanoribbon based global level interconnects", *Journal of Computational Electronics*, Springer, vol. 19, pp. 1002–1013, Jun 2020.

127. Girish Wadhwa, B. Raj, "Design and performance analysis of junctionless TFET biosensor for high sensitivity", *IEEE Nanotechnology*, vol. 18, pp. 567–574, 2019.

128. Jeetendra Singh, B. Raj, "Enhanced nonlinear memristor model encapsulating stochastic dopant drift", *JNO*, ASP, vol. 14, pp. 958–963, 2019.

129. B. Raj, A. K. Saxena, S. Dasgupta, "Analytical modeling of quasi planar nanoscale double gate FinFET with source/drain resistance and field dependent carrier mobility: A quantum mechanical study", *Journal of Computer (JCP)*, Academy Publisher, Finland, vol. 4, no. 9, pp. 1–8, 2009.

130. S. Bhushan, S. Khandelwal, B. Raj, "Analyzing different mode FinFET based memory cell at different power supply for leakage reduction", *Seventh International Conference on Bio-Inspired Computing: Theories and Application, (BIC-TA 2012) Advances in Intelligent Systems and Computing*, vol. 202, pp. 89–100, 2013.

131. Jeetendra Singh, B. Raj, "Temperature dependent analytical modeling and simulations of nanoscale memristor", *Journal: Engineering Science and Technology, an International Journal*, Elsevier, vol. 21, pp. 862–868, Oct 2018.

132. Shradhya Singh, Shashi Bala, Balwant Raj, B. Raj, "Improved sensitivity of dielectric modulated junctionless transistor for nanoscale biosensor design", *Sensor Letter*, ASP, vol. 18, pp. 328–333, Apr 2020.

133. Vivek Kumar, Santosh Kumar Vishvakarma, B. Raj, "Design and performance analysis of ASIC for IoT applications", *Sensor Letter*, ASP, vol. 18, pp. 31–38, Jan 2020.

134. Akanksha Jaiswal, R. K. Sarin, B. Raj, Shikha Sukhija, "A novel circular slotted microstrip-fed patch antenna with three triangle shape defected ground

structure for multiband applications", *Advanced Electromagnetic (AEM)*, vol. 7, no. 3, pp. 56–63, Aug 2018.
135. Girish Wadhwa, B. Raj, "Label free detection of biomolecules using charge-plasma-based gate underlap dielectric modulated junctionless TFET", *Journal of Electronic Materials (JEMS)*, Springer, vol. 47, no. 8, pp. 4683–4693, Aug 2018.
136. Gurmohan Singh, R. K. Sarin, B. Raj, "**Design and performance analysis of a new efficient coplanar quantum-dot cellular automata adder**", *Indian Journal of Pure & Applied Physics (IJPAP)*, vol. 55, pp. 97–103, Feb 2017.
137. Amandeep Singh, Mamta Khosla, Balwinder Raj, "Design and analysis of electrostatic doped schottky barrier CNTFET based low power SRAM," *International Journal of Electronics and Communications, (AEÜ)*, Elsevier, vol. 80, pp. 67–72, 2017.
138. Parminder Kaur, Vikas Pandey, B. Raj, "Comparative study of efficient design, control and monitoring of solar power using IoT", *Sensor Letter*, ASP vol. 18, pp. 419–426, May 2020.
139. Anil Kumar Bhardwaj, Sumeet Gupta, B. Raj, "Development & analysis of compact model for double gate schottky barrier CNTFET", *Journal of Nanoelectronics and Optoelectronics*, ASP, vol. 15, pp. 1199–1208, Aug 2020.
140. Girish Wadhwa, Priyanka Kamboj, Jeetendra Singh, B. Raj, "Design and investigation of junctionless DGTFET for biological molecule recognition", *Transactions on Electrical and Electronic Materials*, Springer, vol. 22, pp. 282–289, 2021.

Chapter 6

Perovskite-based tandem solar cells

*Shahraan Hussain, Divya Sharma,
Arun Kumar, and Balwinder Raj*

6.1 INTRODUCTION

In recent decades, photovoltaic technology has significantly advanced, leading to cost- and energy-efficient solutions. One of the promising approaches that has led to significant breakthroughs in this field is tandem-based solar cells. Tandem means an arrangement pattern of something, one behind another or stacked over one another. In order to harvest most of the bandgap falling onto the surface of solar cells, researchers have preferred this methodology, and results have been promising. Tandem cells strive to enhance the conversion efficiency of solar cell stacks one over another, varying the bandgap for complete or partially complete absorption bands of energy incidents on it [1–3]. Designing a solar cell involves complex engineering and scientific considerations. Some of the basic requirements are as follows:

Semiconductor material:
Consideration of material while designing a solar cell is a very important step. Currently, perovskite material is a popular and preferred material in the design of solar cells. Due to its exceptional properties, such as its high absorption coefficient, versatility, and being light and flexible, as well as low costs of fabrication, it has become a game-changing material in the field of solar cell design, particularly the tandem solar cell [4–6].

Bandgap energy:
Selection of the right bandgap material is very important when designing a solar cell. The bandgap should match the incident spectrum, otherwise the material may not harvest complete energy.

Absorption coefficient:
Absorption coefficient refers to the capacity of the material-absorbing photon. Therefore, it is highly preferred to have a high absorption coefficient material when designing a solar panel.

DOI: 10.1201/9781003487692-6

Encapsulation:
Protecting solar cells from external environmental factors such as moisture, UV radiation, and dust is very important for optimal output of the solar cell [7–10].

Durability and reliability:
To ensure long-term durability and reliability, researchers and engineers should consider various environmental conditions such as temperature fluctuation, UV exposure, and mechanical stress.

Cost-efficient:
In order to mass produce solar cells, material choice and design should be cost-efficient.

The following are a few important terms that should be understood prior to designing a solar cell.

6.1.1 Open circuit voltage

The potential difference between two sources after the termination of an external source, considering maximum value, is open circuit voltage V_{oc}. According to Hanna, the open circuit voltage v_{oc} is the maximum difference between quasi-Fermi levels of electrons, i.e. E_{FN}, and the quasi-Fermi level of hole, i.e. E_{FP} under ohmic contact.

$$qV_{oc} = E_{FN} - E_{FP} \tag{6.1}$$

6.1.2 Short-circuit current and current density

When an external circuit is short-circuited, the current that passes through the solar cell is called a short-circuit current I_{SC} as shown in Figure 6.1. Short-circuit current per unit area is known as current density, denoted by

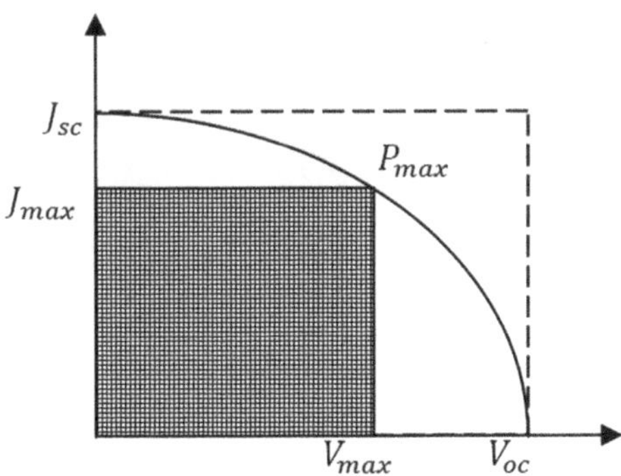

Figure 6.1 V-I characteristic diagram of a solar cell.

J_{SC}. Mathematically, J_{SC} is expressed as:

$$J_{SC} = \frac{I_{SC}}{S} \tag{6.2}$$

Fill factor:

Open circuit voltage and short-circuit current are the maximum voltage and current in solar cells, respectively. However, at both these points, the power of the solar cell is zero. The ratio between short-circuit current and open circuit voltage is referred to as the fill factor (FF), through which the performance of the solar cell can be determined. FF determines the maximum power of the solar cell. The value of FF generally lies between 0.5 and 0.8. Mathematically, it can be expressed as

$$FF = \frac{V_{max} \times J_{max}}{V_{OC} \times J_{SC}} = \frac{P_{max}}{V_{OC} \times J_{SC}} \tag{6.3}$$

where

$$P_{max} = V_{max} \times J_{max} \tag{6.4}$$

6.1.3 Power conversion efficiency

The most common parameter to calculate the performance of any device is efficiency, therefore to compare the performance of solar cells, efficiency needs to be identified, as presented in Figure 6.2. The efficiency of solar

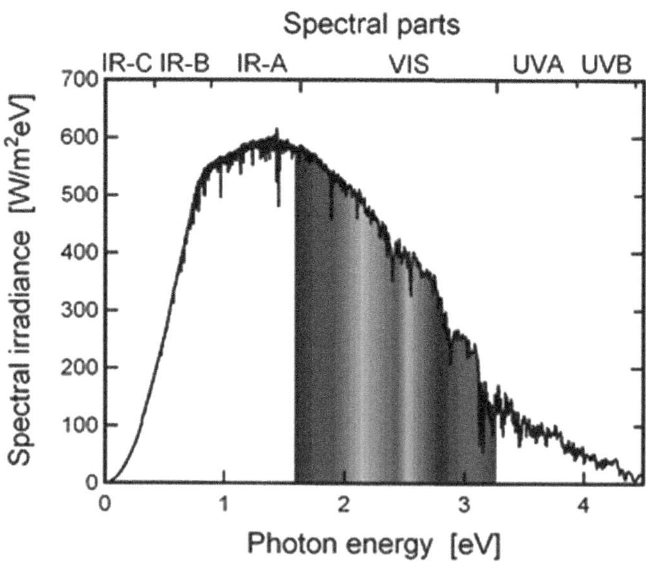

Figure 6.2 Photon energy vs spectral intensity graph of incident light (infrared-visible-ultraviolet).

cells depends upon the intensity and spectrum of light that is falling upon the solar cells [11–16].

Solar energy per unit area is denoted as P_{in}. P_{in} is the major component of calculating PCE. Mathematically, PCE is defined as

$$\eta = \frac{V_{OC} \times J_{SC} \times FF}{P_{in}} \tag{6.5}$$

Table 6.1 shows the evolution of solar cells and their efficiency.

A single-junction solar cell consists of three layers where the absorber layer is sandwiched between two contact layers as given in Figure 6.3. When a photon incident occurs on the absorber layer, it creates an electric carrier, i.e. an electron and hole [17–24]. As per the biasing created, the carrier moves toward its respective contact. Incident light consists of different energies, e.g. infrared-visible ultraviolet. The absorber layer is a semiconductor layer with a characteristic bandgap (Eg).

Energy (E ≥ Eg) greater than bandgap energy is absorbed and creates an electrical carrier [25–32]. Energy (E ≤ Eg) less than bandgap energy is not absorbed and is not utilized. For energy greater than bandgap energy (E > Eg), a fraction of energy is lost via thermal relaxation [33–39].

Table 6.1 Generation and material-wise efficiency of solar cell

Generation	Nomenclature	Material	Efficiency
First	Inorganic	Si	26.7%
		GaAs	28.8%
Second	Inorganic thin-film	CIGS	22.6%
		CdTe	22.1%
Third	Organic thin-film	Dye-sensitized	11.9%
		Organic	11.5%
Fourth	Organic-inorganic perovskite hybrid	Perovskite	22.7%

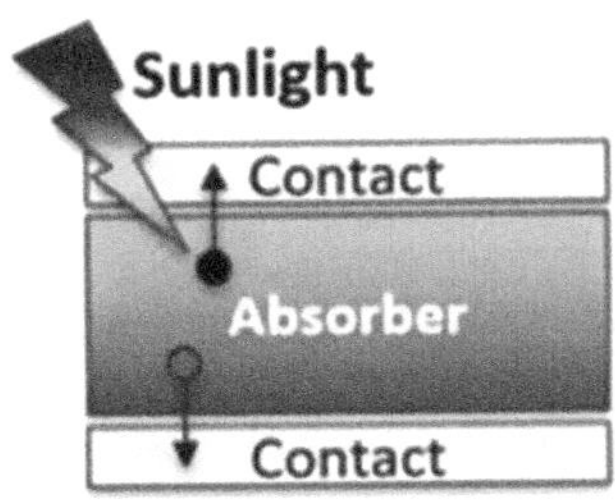

Figure 6.3 Sunlight incident on a solar cell, creating an electrical carrier.

6.2 FOURTH GENERATION: ORGANIC-INORGANIC PEROVSKITE HYBRID SOLAR CELL

Perovskite is currently considered a most promising material. In the past, perovskite attracted attention for its use in different technologies, such as optoelectronic devices, energy storage, and pollutant degradation. Due to its low cost, high efficiency, low preparation temperature, and excellent diffusion length, i.e. more than 1 μm, perovskite is a promising material for solar cell designing material [40–47].

The general formula of perovskite is ABX_3, which may be regarded as a derived form of ReO_3 structure, as shown in Figure 6.4. In ABX_3, 'A' is an organic-inorganic cation, such as $((CH_3)_2+(formamidinium))$ or $CH_3NH_3+(methylammonium)$; 'B' is a divalent metal cation, such as Pb_2 or Sn_2; and 'X' is a halide anion, such as Cl^-, Br^-, or I^-, which bonds to both. The size of the 'A' atom is larger than 'B'. The ideal ABX_3 cubic structure has a B-cation in six-fold coordinates, surrounded by an octahedron of anion 'X', and a 12-fold cuboctahedron coordinate structure of the 'A' cation, as shown in Figure 6.4(a) and (b).

6.3 TANDEM SOLAR CELL

Tandem solar cells are multi-junction solar cells with multiple p-n junctions made from different materials that absorb different wavelengths of light and convert them into electrical energy. Single junctions can absorb a specific range of wavelengths to produce electrical energy, whereas multi-junctions can utilize a maximum range of wavelengths to produce optimum efficient solar cells [48–55].

Traditional solar cells have a theoretical efficiency of 33.16%, whereas multi-junction tandem solar cells have a theoretical efficiency of 86.8%. As a result, researchers have become more interested in the higher efficiency concept, and the tandem solar cell is attracting more research attention.

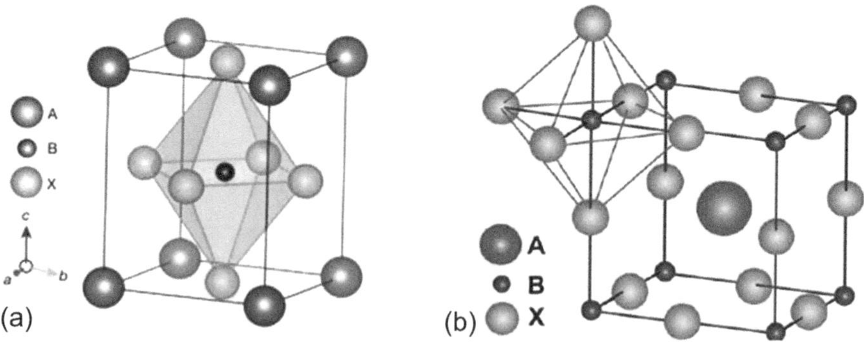

Figure 6.4 (a) Basic cube. (b) 3-D cubic ABX_3 perovskite structure.

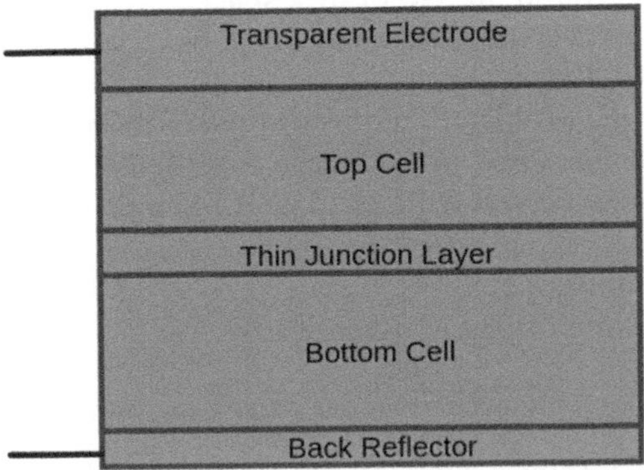

Figure 6.5 Common structure of 2-T multi-junction tandem solar cell.

The common structure of a tandem solar cell is presented in Figure 6.5. There are four types of tandem solar cell:

- 4-Terminal
- 2-Terminal
- Series parallel tandem (SPT)
- Reflective

6.4 SIMULATION

Simulation software name: SCAPS-1D
Software details:

- Version: 3.3.10
- Operating system: Windows XP
- VirtualBox 6.1

Hardware details:

- RAM; 8 GB
- Processor: Intel i-3 7th generation
- 128 GB SSD

Basics of SCAPS-1D:

- SCAPS developed on Lab Windows/CVI of the National Instrument platform.
- Windows OS.
- OS that support SCAPS are Windows 95, 98, NT, 2000, XP, Vista, Windows 7

- It occupies around 50 MB of disk space.
- Dedicated panels are there to perform basic actions in SCAPS:
  1. Run SCAPS.
  2. Define the problem, and insert properties of different materials.
  3. Set circumstances in which the solar cell will simulate.
  4. Indicate what you will calculate, i.e. which measurements you will simulate.
  5. Start simulation.
  a. Check simulation curves to study.

Scripting:

- A text file contains a command to follow, which is equivalent to a different mouse click role on SCAPS-1D simulation software is a script file [56–63].
- While running the script file, SCAPS will not respond to any other interventions.
- Once the script is executed, the normal interactive panel of SCAPS will be displayed.
- To design a tandem solar cell, a script should be used.
- To begin the process, click on script setup.
- Once the script setup is done, execute the script.
- One can load the file using the load button.
- The file has an extension of the script.
- After completion of the code, one should save the code along with the mentioned extension [64–73].

All command lines in the script will have three parts:
<command|argument|value>
The following are available script commands:

- load
- math
- calculate
- run
- run dll
- action
- save
- clear
- extract
- plot
- set
- Get
- loop
- show
- run system

For example:
  Load definition file name
  Each command has 'n' number of arguments followed by a value. Please refer to the documentation for more details [74–86].

6.5 RESULT AND DISCUSSION

6.5.1 Proposed tandem solar cell structure

The proposed tandem cell structure is shown in Figure 6.10, where the top layer consists of Cu2O, MAGeI3, and TiO2. The bottom layer consists of three layers where Fe-Si is sandwiched between C-Si [87–101]. Between the intermediate layers exists a layer that works as a tunnel junction. The intermediate layer consists of

  a–:(n^+) and μc–Si1–xOx:H(P^+)

In Figures 6.6, the single-cell structure is shown where one layer is stacked over the other. The width of each layer is defined in the properties mentioned in property panel windows, as shown in Figures 6.7 to 6.9.

Since SCAPS does not allow a simulation of the tandem structure using a GUI (graphical user interface) panel, scripting should be used to get all possible outputs. Figure 6.7 shows a scripting panel that is used to design and simulate the proposed tandem solar cell [102–116].

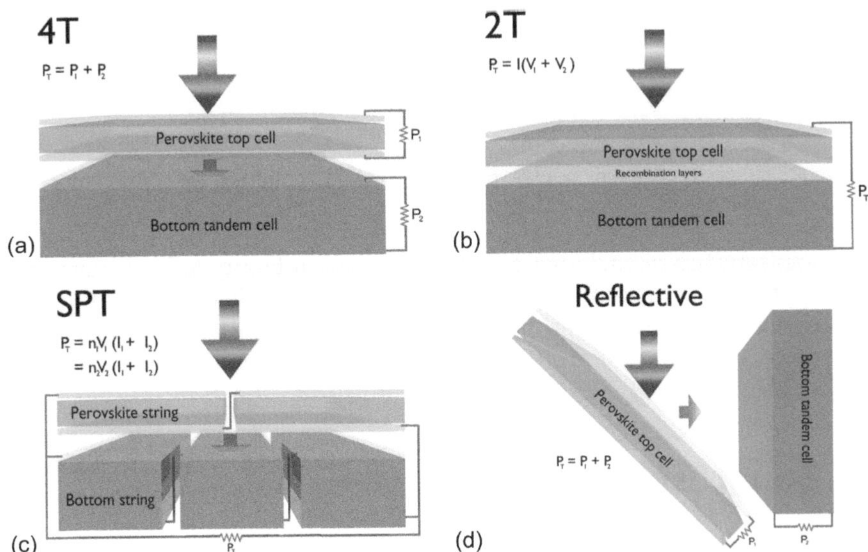

Figure 6.6 (a) 4-T tandem solar cell with independent connection for both cells, (b) series 2-Terminal tandem cell, (c) series parallel tandem, (d) IR reflector reflective tandem cell.

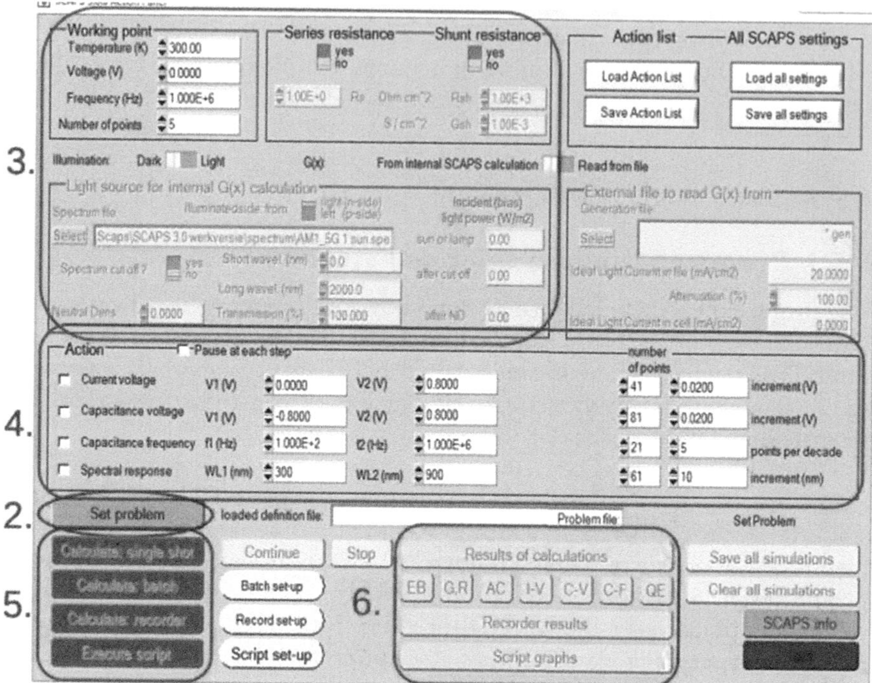

Figure 6.7 Start-up panel of SCAPS-1D simulation software. Numbers 2–6 explained below.

Figure 6.8 Snippet of main display panel of SCAPS, emphasizing script setup and execution.

The properties of each material are provided in Figure 6.7, with material name, material thickness, value, electron affinity, dielectric permittivity, etc. Using these material properties, the simulation software will calculate parameters such as fill factor, power conversion efficiency, Jsc, and Voc [117–125].

6.6 CONCLUSION

Attention has recently been given to tandem solar cells because of their optimum bandwidth for delivering more electrical energy. The junction layer directly affects the transparency of the solar cell of the bottom layer. Therefore, material selection should consider the transparency coefficient at

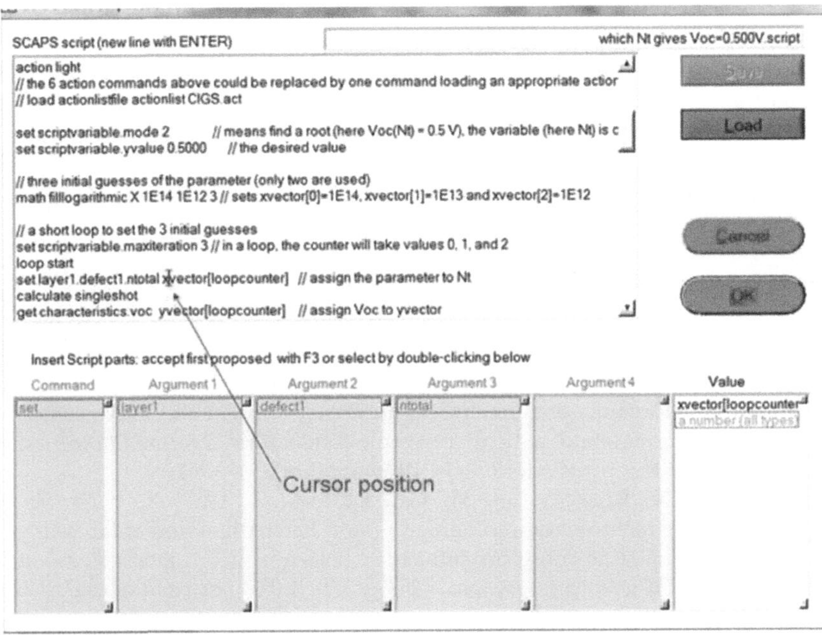

Figure 6.9 Script editor panel, showing a few lines of demo script.

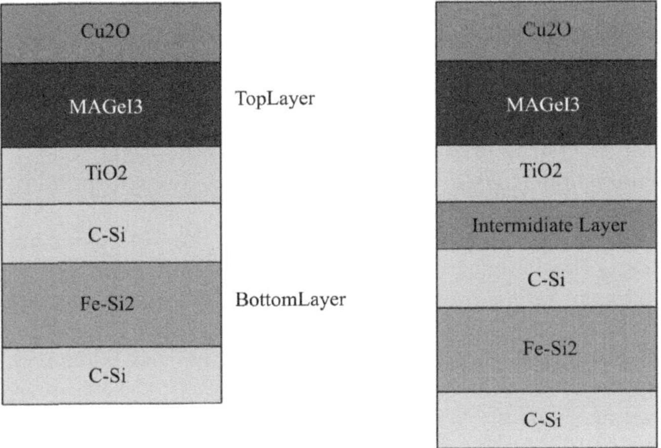

Figure 6.10 Proposed structure of tandem solar cell.

the top. Perovskite is also currently being considered, as it is abundant on earth and has a finished 27.25% efficiency. In addition, it does not damage the earth's environmental balance.

To design and simulate a solar cell, SCAPS-1D is a good option. It is very easy to design top and bottom cells using GUI, or one can design a number of individual solar cells and incorporate them in multi-junction solar cells. Researchers can also incorporate a range of data to check the simulation and select the best input value for the desired output. Scripting in SCAPS-1D simulation software is also very easy, and documentation explains all the commands properly. To design a tandem solar cell, a separate chapter is mentioned in the documentation that researchers can consider as a reference. The versions released after version 3 support multi-junction scripting.

REFERENCES

1. J. H. Heo, S. H. Im, J. H. Noh, T. N. Mandal, C.-S. Lim, J. A. Chang, Y. H. Lee, H.-J. Kim, A. Sarkar, M. K. Nazeeruddin, M. Grätzel, S. I. Seok, Efficient inorganic–organic hybrid hetero junction solar cells containing perovskite compound and polymerichole conductors", *Nature Photonics*, vol. 7, p. 486, 2013. https://doi.org/10.1038/nphoton.2013.80
2. H. Chen, X. Pan, W. Liu, M. Cai, D. Kou, Z. Huo, X. Fang, S. Dai, "Efficient panchromatic inorganic-organic hetero junction solar cells with consecutive charge transport tunnels in hole transport material", *Chemical Communications*, vol. 49, pp. 7277–7279, 2013. https://doi.org/10.1039/C3CC42297F
3. M. A. Green, K. Emery, "Solar cell efficiency tables", *Progress in Photovoltaics: Research and Applications*, vol. 1, pp. 25–29, 1993. https://doi.org/10.1002/pip.4670010306
4. B. Cai, Y. Xing, Z. Yang, W.-H. Zhang, J. Qiu, "High performance hybrid solar cellssensitized by organolead halide perovskites", *Energy & Environmental Science*, vol. 6, pp. 1480–1485, 2013. https://doi.org/10.1039/C3EE40343B
5. J. M. Ball, M. M. Lee, A. Hey, H. J. Snaith, "Low-temperature process edmeso-superstructured to thin-film perovskite solar cells", *Energy & Environmental Science*, vol. 6, pp. 1739–1743, 2013. https://doi.org/10.1039/C3EE40810H
6. J. A. Christians, R. C. M. Fung, P. V. Kamat, "An inorganic hole conductor for organo-leadhalide perovskite solar cells", *Improved Hole Conductivity with Copper Iodide, Journal of the American Chemical Society*, vol. 136, pp. 758–764, 2014.
7. A. S. Subbiah, A. Halder, S. Ghosh, N. Mahuli, G. Hodes, S. K. Sarkar, "Inorganic hole conducting layers for perovskite-based solar cells", *The Journal of Physical Chemistry Letters*, vol. 5, pp. 1748–1753, 2014. https://doi.org/10.1021/jz500645n
8. S. Park, J. H. Heo, J. H. Yun, et al., "Effect of multi-armed triphenylamine-based holetransporting materials for high performance perovskite solar cells", Chemical Science, vol. 7, pp. 5517–5522, 2016. https://doi.org/10.1039/C6SC00876C
9. J.-Y. Jeng, K.-C. Chen, T.-Y. Chiang, P.-Y. Lin, T.-D. Tsai, Y.-C. Chang, T.-F. "Nickel oxide electrode interlayer in CH3NH3PbI3 perovskite/PCBM planar-heterojunction hybrid solar cells", *Advanced Materials*, vol. 26, pp. 4107–4113, 2014. https://doi.org/10.1002/adma.201306217

10. Z. Zhu, Y. Bai, T. Zhang, Z. Liu, X. Long, Z. Wei, Z. Wang, L. Zhang, J. Wang, F. Yan, S. Yang, "High-performance hole-extraction layer of Sol–Gel-Processed NiO nanocrystalsfor inverted planar perovskite solar cells", *Angewandte Chemie*, vol. 126, pp. 12779–12783, 2014. https://doi.org/10.1002/anie.201405176

11. H. Li, K. Fu, A. Hagfeldt, et al. "A simple 3,4-ethylenedioxythiophene based hole-transporting material for perovskite solar cells", *Angewandte Chemie International Edition*, vol. 53, pp. 4085–4088, 2014. https://doi.org/10.1002/anie.201310877

12. X. Li, D. Bi, C. Yi, et al. "A vacuum flash assisted solution process for high-efficiency large-area perovskite solar cells", *Science*, vol. 80, pp. 1–10, 2016. https://doi.org/10.1126/science.aaf8060

13. H. Li, K. Fu, P. Boix, et al. "Hole-transporting small molecules based on thiophene cores for high efficiency perovskite solar cells", *ChemSusChem*, vol. 7, pp. 3420–3425, 2014. https://doi.org/10.1002/cssc.201402587

14. N. Kumari, S. R. Patel, J. V. Gohel, "Superior efficiency achievement for FAPbI3-perovskite thin film solar cell by optimization with response surface methodology technique and partial replacement of Pb by Sn", *Optik (Stuttg)*, vol. 176, pp. 262–277, 2019. https://doi.org/10.1016/j.ijleo.2018.09.066

15. R. Rajeswari, M. Mrinalini, S. Prasanthkumar, L. Giribabu, "Emerging of inorganic holetransporting materials for perovskite solar cells", *The Chemical Record*, vol. 17, pp. 681–699, 2017. https://doi.org/10.1002/tcr.201600117

16. Z. H. Bakr, Q. Wali, A. Fakharuddin, L. Schmidt-Mende, T. M. Brown, R. Jose, "Advances inhole transport materials engineering for stable and efficient perovskite solar cells", *Nano Energy*, vol. 34, pp. 271–305, 2017. https://dx.doi.org/10.1016/j.nanoen.2017.02.025

17. A. Krishna, A. C. Grimsdale, "Hole transporting materials for mesoscopic perovskite solarcells - towards a rational design?", *Journal of Materials Chemistry A*, vol. 5, pp. 16446–16466, 2017. https://doi.org/10.1039/C7TA01258F

18. D. Liu, L. Yongsheng, "Recent progress of dopant-free organic hole-transporting materials in perovskite solar cells", *Journal of Semiconductors*, vol. 38, p. 011005, 2017. https://doi.org/10.1088/1674-4926/38/1/011005

19. H. Chen, S. Yang, "Carbon-based perovskite solar cells without hole transport materials: The front runner to the market?", *Advanced Materials*, vol. 29, p. 1603994-n/a, 2017. https://doi.org/10.1002/adma.201603994

20. F. Zhang, C. Yi, P. Wei, et al. "A novel dopant-free triphenylamine based molecular "Butterfly" hole-transport material for highly efficient ands table perovskite sola rcells", *Advanced Energy Materials*, vol. 6, pp. 1–7, 2016. https://doi.org/10.1002/aenm.201600401

21. Candy Goyal, Jagpal Singh Ubhi, B. Raj, "Low leakage zero ground noise nanoscale full adder using source biasing technique", *Journal of Nanoelectronics and Optoelectronics*, American Scientific Publishers, vol. 14, pp. 360–370, Mar 2019.

22. Gurmohan Singh, R. K. Sarin, B. Raj, "A novel robust exclusive-OR function implementation in QCA nanotechnology with energy dissipation analysis", *Journal of Computational Electronics*, Springer, vol. 15, no. 2, pp. 455–465, Jun 2016.

23. Tanu Wadhera, Deepti Kakkar, Girish Wadhwa, B. Raj, "Recent advances and progress in development of the field effect transistor biosensor: A review",

Journal of Electronic Materials, Springer, vol. 48, no. 12, pp. 7635–7646, Dec 2019.

24. Girish Wadhwa, Priyanka Kamboj, Balwinder Raj, "Design optimisation of junctionless TFET biosensor for high sensitivity", *Advances in Natural Sciences: Nanoscience and Nanotechnology*, vol. 10, p. 045001, 2019.

25. Anjana Bhardwaj, Pradeep Kumar, B. Raj, Sunny Anand, "Design and performance optimization of doping-less vertical nanowire TFET using gate stack technique", *Journal of Electronic Materials (JEMS)*, Springer, vol. 41, no. 7, pp. 4005–4013, 2022.

26. Jeetendra Singh, B. Raj, "Tunnel current model of asymmetric MIM structure levying various image forces to analyze the characteristics of filamentary memristor", *Applied Physics A*, Springer, vol. 125, no. 3, p. 203.1, Feb 2019.

27. Priya Bansal, B. Raj, "Memristor modeling and analysis for linear dopant drift kinetics," *Journal of Nanoengineering and Nanomanufacturing*, American Scientific Publishers, vol. 6, pp. 1–7, 2016.

28. Amandeep Singh, Mamta Khosla, B. Raj, "Circuit compatible model for electrostatic doped schottky barrier CNTFET, "*Journal of Electronic Materials*, Springer, vol. 45, no. 12, pp. 4825–4835, 2016.

29. D. Vaithiyanathan, B. Raj, "Performance analysis of charge plasma induced graded channel si nanotube", *Journal of Engineering Research (JER)*, EMSME Special Issue, pp. 146–154, Aug 2021.

30. Abhishek Singh Tomar, Vijay Kumar Magraiya, B. Raj, "Scaling of access and data transistor for high performance DRAM cell design", *Quantum Matter*, vol. 2, pp. 412–416, Oct 2013.

31. Neeraj Jain, B. Raj, "Parasitic capacitance and resistance model development and optimization of raised source/drain SOI FinFET structure for analog circuit applications", *Journal of Nanoelectronics and Optoelectronins*, ASP, USA, vol. 13, pp. 531–539, Ap 2018.

32. Shradhya Singh, S. K. Vishvakarma, B. Raj, "Analytical modeling of split-gate junction-less transistor for a biosensor application", *Sensing and Bio-Sensing*, Elsevier, vol. 18, pp. 31–36, Apr 2018.

33. Maisagalla Gopal, Balwinder Raj, "Low power 8T SRAM cell design for high stability video applications", *ITSI Transaction on Electrical and Electronics Engineering*, vol. 1, no. 5, pp. 91–97, 2013.

34. Balwinder Raj, Jatin Mitra, Deepak Kumar Bihani, V. Rangharajan, A. K. Saxena, S. Dasgupta, "Analysis of noise margin, power and process variation for 32 nm FinFET based 6T SRAM cell", *Journal of Computer (JCP)*, Academy Publisher, Finland, vol. 5, no. 6, pp. 1–8, 2010.

35. Divya Sharma, Rajesh Mehra, B. Raj, "Comparative analysis of photovoltaic technologies for high efficiency solar cell design", *Superlattices and Microstructures*, Elsevier, vol. 153, p. 106861, May 2021.

36. Pawandeep Kaur, Avtar Singh Buttar, B. Raj, "A comprehensive analysis of nanoscale transistor based biosensor: A review", *Indian Journal of Pure and Applied Physics*, vol. 59, pp. 304–318, Apr 2021.

37. Divya Yadav, Balwant Raj, B. Raj, "Design and simulation of low power microcontroller for IoT applications", *Journal of Sensor Letters*, ASP, vol. 18, pp. 401–409, May 2020.

38. Shailendra Singh, B. Raj,"A 2-D analytical surface potential and drain current modeling of double-gate vertical t-shaped tunnel FET", *Journal of Computational Electronics*, Springer, vol. 19, pp. 1154–1163, Apr 2020.

39. Jeetendra Singh, B. Raj, "An accurate and generic window function for nonlinear memristor model", *Journal of Computational Electronics*, Springer, vol. 18, no. 2, pp. 640–647, Jun 2019.

40. Manjit Kaur, Neena Gupta, Sanjeev Kumar, B. Raj, Arun Kumar Singh, "Comparative RF and crosstalk analysis of carbon based nano interconnects", *IET Circuits, Devices & Systems*, vol. 15, no. 6, pp. 493–503, Feb 2021.

41. Nehru Kandasamy, Firdous Ahmad, D. Ajitha, B. Raj, Nagarjuna Telagam, "Quantum dot cellular automata based scan flip flop and boundary scan register", *IETE Journal of Research*, vol. 66, pp. 535–548, 2020.

42. S. K. Sharma, B. Raj, M. Khosla, "Enhanced photosensivity of highly spectrum selective cylindrical gate In1-xGaxAs nanowire MOSFET photodetector", *Modern Physics Letter-B*, vol. 33, no. 12, p. 1950144, 2019.

43. Jeetendra Singh, B. Raj, "Design and investigation of 7T2M NVSARM with enhanced stability and temperature impact on store/restore energy", *IEEE Transactions on Very Large Scale Integration Systems*, vol. 27, no. 6, pp. 1322–1328, Jun 2019.

44. Anil Kumar Bhardwaj, Sumeet Gupta, B. Raj, Amandeep Singh, "Impact of double gate geometry on the performance of carbon nanotube field effect transistor structures for low power digital design", *Computational and Theoretical Nanoscience*, ASP, vol. 16, pp. 1813–1820, 2019.

45. Neeraj Jain, B. Raj, "Thermal stability analysis and performance exploration of asymmetrical dual-k underlap spacer (ADKUS) SOI FinFET for security and privacy applications", *Indian Journal of Pure & Applied Physics (IJPAP)*, vol. 57, pp. 352–360, May 2019.

46. Amandeep Singh, Mamta Khosla, B. Raj, "Design and analysis of dynamically configurable electrostatic doped carbon nanotube tunnel FET", *Microelectronics Journal*, Elesvier, vol. 85, pp. 17–24, Mar 2019.

47. Neeraj Jain, Balwinder Raj, "Dual-k spacer region variation at the drain side of asymmetric SOI FinFET structure: Performance analysis towards the analog/rf design applications", *Journal of Nanoelectronics and Optoelectronics*, American Scientific Publishers, vol. 14, pp. 349–359, Mar 2019.

48. Jeetendra Singh, Sanjeev Sharma, B. Raj, Mamta Khosla, "Analysis of barrier layer thickness on performance of In1-xGaxAs based gate stack cylindrical gate nanowire MOSFET," *JNO*, ASP, vol. 13, pp. 1473–1477, Oct 2018.

49. Neeraj Jain, B. Raj, "Analysis and Performance Exploration of High-k SOI FinFETs Over the Conventional Low-k SOI FinFET toward Analog/RF Design", Journal of Semiconductors (JoS), IOP Science, vol. 39, no. 12, p. 124002-1-7, Dec 2018.

50. Candy Goyal, Jagpal Singh Ubhi, B. Raj, "A reliable leakage reduction technique for approximate full adder with reduced ground bounce noise', *Journal of Mathematical Problems in Engineering*, Hindawi, vol. 2018, Article ID 3501041, p. 16, 15 Oct 2018.

51. Jeetendra Singh, B. Raj, Mamta Khosla, "Design and performance analysis of nano-scale memristor-based nonvolatile SRAM", *Journal of Sensor Letter*", American Scientific Publishers, vol. 16, pp. 798–805, Oct 2018.

52. Girish Wadhwa, B. Raj, "Parametric variation analysis of charge-plasma-based dielectric modulated JLTFET for biosensor application", *IEEE Sensor Journal*, vol. 18, no. 15, pp. 6070–6077, 2018.

53. Jeetendra Singh, B. Raj, "Comparative analysis of memristor models for memories design", *JoS*, IoP, vol. 39, no. 7, p. 074006-1-12, Jul 2018.

54. Divya Yadav, Shailesh Singh Chouhan, Santosh Kumar Vishvakarma, B. Raj, "Application specific microcontroller design for IoT based WSN", *Sensor Letter*, ASP, vol. 16, pp. 374–385, May 2018.

55. Gurmohan Singh, R. K. Sarin, B. Raj, "Fault-tolerant design and analysis of quantum-dot cellular automata based circuits", *IEEE/IET Circuits, Devices & Systems*, vol. 12, pp. 638–664, 2018.

56. Jeetendra Singh, B. Raj, "Modeling of mean barrier height levying various image forces of metal insulator metal structure to enhance the performance of conductive filament based memristor model", *IEEE Nanotechnology*, vol. 17, no. 2, pp. 268–267, Mar 2018.

57. Aakash Jain, Sanjeev Sharma, B. Raj, "Analysis of triple metal surrounding gate (TM-SG) III-V nanowire MOSFET for photosensing application", *Opto-electronics Journal*, Elsevier, vol. 26, no. 2, pp. 141–148, May 2018.

58. Aakash Jain, Sanjeev Sharma, B. Raj, "Design and analysis of high sensitivity photosensor using cylindrical surrounding gate MOSFET for low power sensor applications", *Engineering Science and Technology, an International Journal*, Elsevier, vol. 19, no. 4, pp. 1864–1870, Dec 2016.

59. Amandeep Singh, Mamta Khosla, B. Raj, "Analysis of electrostatic doped schottky barrier carbon nanotube FET for low power applications," *Journal of Materials Science: Materials in Electronics*, Springer, vol. 28, pp. 1762–1768, 2017.

60. G. Saiphani Kumar, Amandeep Singh, B. Raj, "Design and analysis of gate all around CNTFET based SRAM cell design", *Journal of Computational Electronics*, Springer, vol. 17, no. 1, pp. 138–145, Mar 2018.

61. Gurinder pal Singh, B. S. Sohi, Balwinder Raj, "Material properties analysis of graphene base transistor (GBT) for VLSI analog circuits", *Indian Journal of Pure & Applied Physics (IJPAP)*, vol. 55, pp. 896–902, Dec 2017.

62. Amandeep Singh, Mamta Khosla, B. Raj, "Comparative analysis of carbon nanotube field effect transistor and nanowire transistor for low power circuit design," *Journal of Nanoelectronics and Optoelectronics*, American Scientific Publishers, USA, vol. 11, pp. 388–393, Jun 2016.

63. Sunil Kumar, B. Raj, "Estimation of stability and performance metric for inward access transistor based 6T SRAM cell design using n-type/p-type DMDG-GDOV TFET", *IEEE VLSI Circuits and Systems Letter*, vol. 3, no. 2, pp. 25–39, Jun 2017.

64. Shashikant Sharma, Anjan Kumar, Manisha Pattanaik, B. Raj, "Forward body biased multimode multi-threshold CMOS technique for ground bounce noise reduction in static CMOS adders", *International Journal of Information and Electronics Engineering*, pp. 567–572, vol. 3, no. 3, 2013.

65. Hamendra Singh, Pankaj Kumar, Balwinder Raj, "Performance Analysis of Majority Gate SET Based 1-bit Full Adder", *International Journal of Computer and Communication Engineering* (IJCCE), IACSIT Press Singapore, ISSN: 2010-3743, Vol. 2, no. 4, 2013.

66. Anil Kumar Bhardwaj, Sumeet Gupta, B. Raj, "Investigation of parameters for schottky barrier (SB) height for schottky barrier based carbon nanotube field effect transistor device", *Journal of Nanoelectronics and Optoelectronics*, ASP, vol. 15, pp. 783–791, Jul 2020.

67. Priya Bansal, B. Raj, "Memristor: A versatile nonlinear model for dopant drift and boundary issues," *JCTN*, American Scientific Publishers, vol. 14, no. 5, pp. 2319–2325, May 2017.

68. Neeraj Jain, B. Raj, "An analog and digital design perspective comprehensive approach on Fin-FET (Fin-Field Effect transistor) technology: A review", *Reviews in Advanced Sciences and Engineering (RASE)*, ASP, vol. 5, pp. 1–14, 2016.

69. Sanjeev Sharma, B. Raj, Mamta Khosla, "Subthreshold performance of In1-xGaxAs based dual metal with gate stack cylindrical/surrounding gate nanowire MOSFET for low power analog applications", *Journal of Nanoelectronics and Optoelectronics*, American Scientific Publishers, USA, vol. 12, pp. 171–176, 2017.

70. Balwinder Raj, A. K. Saxena, S. Dasgupta, "Analytical modeling for the estimation of leakage current and subthreshold swing factor of nanoscale double gate FinFET device", *Microelectronics International, UK*, vol. 26, pp. 53–63, 2009.

71. Shailendra Singh, Girish Wadhwa, Balwinder Raj, "An analytical modeling for dual source vertical tunnel field effect transistor", *International Journal of Recent Technology and Engineering (IJRTE)*, vol. 8, no. 2, pp. 603–608, Jul 2019.

72. Shailendra Singh, B. Raj, "Design and analysis of hetrojunction vertical T-shaped tunnel field effect transistor", *Journal of Electronics Material*, Springer, vol. 48, no. 10, pp. 6253–6260, Oct 2019.

73. Candy Goyal, Jagpal Singh Ubhi, B. Raj, "A low leakage CNTFET based inexact full adder for low power image processing applications", *International Journal of Circuit Theory and Applications*, Wiley, vol. 47, no. 9, pp. 1446–1458, Sept 2019.

74. B. Raj, A. K. Saxena, S. Dasgupta, "A compact drain current and threshold voltage quantum mechanical analytical modeling for FinFETs", *Journal of Nanoelectronics and Optoelectronics (JNO)*, USA, vol. 3, no. 2, pp. 163–170, 2008.

75. Girish Wadhwa, B. Raj, "An analytical modeling of charge plasma based tunnel field effect transistor with impacts of gate underlap region", *Superlattices and Microstructures*, Elsevier, vol. 142, p. 106512, Jun 2020.

76. Shailendra Singh, B. Raj, "Modeling and simulation analysis of SiGe hetrojunction double GateVertical t-shaped tunnel FET", *Superlattices and Microstructures*, Elsevier vol. 142, p. 106496, Jun 2020.

77. Amandeep Singh, Dinesh Kumar Saini, Dinesh Agarwal, Sajal Aggarwal, Mamta Khosla, B. Raj, "Modeling and simulation of carbon nanotube field effect transistor and its circuit application," *Journal of Semiconductors (JoS)*, IOP Science, vol. 37, p. 074001-6, Jul 2016.

78. Neeraj Jain, Balwinder Raj, "Device and circuit co-design perspective comprehensive approach on FinFET technology: A review", *Journal of Electron Devices*, vol. 23, no. 1, pp. 1890–1901, 2016.

79. Sunil Kumar, B. Raj, "Analysis of I_{ON} and ambipolar current for dual-material gate-drain overlapped DG-TFET," *Journal of Nanoelectronics and Optoelectronics*, American Scientific Publishers, USA, vol. 11, pp. 323–333, Jun 2016.
80. Naveed Anjum, Tarun Bali, B. Raj, "Design and simulation of handwritten multiscript character recognition", *International Journal of Advanced Research in Computer and Communication Engineering*, vol. 2, no. 7, pp. 2544–2549, Jul 2013.
81. Sanjeev Sharma, B. Raj, Mamta Khosla, "A gaussian approach for analytical subthreshold current model of cylindrical nanowire FET with quantum mechanical effects", *Microelectronics Journal*, Elsevier, vol. 53, pp. 65–72, Apr 2016.
82. Karmjit Singh, B. Raj, "Performance and analysis of temperature dependent multi-walled carbon nanotubes as global interconnects at different technology nodes," *Journal of Computational Electronics*, Springer, vol. 14, no. 2, pp. 469–476, Jun 2015.
83. Sunil Kumar, B. Raj, "Compact channel potential analytical modeling of DG-TFET based on evanescent–mode approach," *Journal of Computational Electronics*, Springer, vol. 14, no. 2, pp. 820–827, Jul 2015.
84. Karmjit Singh, B. Raj, "Temperature dependent modeling and performance evaluation of multi-walled CNT and single-walled CNT as global interconnects," *Journal of Electronic Materials*, Springer, vol. 44, no. 12, pp. 4825–4835, Dec 2015.
85. V. K. Sharma, M. Pattanaik, B.Raj, "INDEP approach for leakage reduction in nanoscale CMOS circuits", *International Journal of Electronics*, Taylor & Francis, vol. 102, no. 2, pp. 200–215, 2014.
86. Karmjit Singh, B. Raj, "Influence of temperature on MWCNT bundle, SWCNT bundle and copper interconnects for nanoscaled technology nodes," *Journal of Materials Science: Materials in Electronics*, Springer, vol. 26, no. 8, pp. 6134–6142, 2015.
87. Naveed Anjum, Tarun Bali, B. Raj, "Design and simulation of handwritten gurumukhi and devanagri numerical recognition", *International Journal of Computer Applications*, Foundation of Computer Science, New York, USA, vol. 73, no. 12, pp. 16–21, 2013.
88. S. Khandelwal, V. Gupta, B. Raj, R. D. "Gupta, process variability aware low leakage reliable nano scale DG-FinFET SRAM cell design technique", *Journal of Nanoelectronics and Optoelectronics*, vol. 10, no. 6, pp. 810–817, Dec 2015.
89. Girish Wadhwa, Priyanka Kamboj, Jeetendra Singh, B. Raj, "Design and investigation of junctionless DGTFET for biological molecule recognition", *Transactions on Electrical and Electronic Materials*, Springer, vol. 22, pp. 282–289, 2021.
90. Tulika Chawla, Mamta Khosla, B. Raj, "Optimization of double-gate dual material GeOI-vertical TFET for VLSI circuit design", *IEEE VLSI Circuits and Systems Letter*, vol. 6, no. 2, pp. 13–25, Aug 2020.
91. Sachin Kumar Verma, Shailendra Singh, Girish Wadhwa, B. Raj, "Detection of biomolecules using charge-plasma based gate underlap dielectric modulated dopingless TFET", *Transactions on Electrical and Electronic Materials (TEEM)*, Springer, vol. 21, pp. 528–535, Jun 2020.
92. Neeraj Jain, B. Raj, "Impact of underlap spacer region variation on electrostatic and analog/RF performance of symmetrical high-k SOI FinFET at 20

nm channel length", *Journal of Semiconductors (JoS)*, IOP Science, vol. 38, no. 12, p. 122002, Dec 2017.

93. Shailendra Singh, B. Raj, "Analytical modeling and simulation analysis of T-shaped III-V heterojunction Vertical T-FET", *Superlattices and Microstructures*, Elsevier, vol. 147, p. 106717, Nov 2020.

94. Gurmohan Singh, R. K. Sarin, B. Raj, "Design and analysis of area efficient QCA based reversible logic gates", *Journal of Microprocessors and Microsystems*, Elsevier, vol. 52, pp. 59–68, May 2017.

95. Amandeep Singh, Mamta Khosla, B. Raj, "Compact model for ballistic single wall CNTFET under quantum capacitance limit," *Journal of Semiconductors (JoS)*, IOP Science, vol. 37, p. 104001-8, Oct 2016.

96. Sonal Singh, Mamta Khosla, Girish Wadhwa, B. Raj, "Design and analysis of double-gate junctionless vertical TFET for gas sensing applications", *Applied Physics A*, Springer, vol. 127, no. 16, 2 Jan 2021.

97. Inderjit Singh, B. Raj, Mamta Khosla, B. Rajesh Kumar Kaushik, "Potential MRAM technologies for low power SoCs", *SPIN World Scientific Publisher, SCIE*; vol. 10, no. 4, p. 2050027, Dec 2020.

98. Shailendra Singh, B. Raj, "Parametric variation analysis on hetero-junction Vertical t-shape TFET for supressing ambipolar conduction", *Indian Journal of Pure and Applied Physics*, vol. 58, pp. 478–485, Jun 2020.

99. Shailendra Singh, Girish Wadhwa, Balwinder Raj, "Design and analysis of dual source vertical tunnel field effect transistor for high performance", *Transactions on Electrical and Electronics Materials*, Springer, vol. 21, pp. 74–82, Oct 2019.

100. Manjit Kaur, Neena Gupta, Sanjeev Kumar, B. Raj, Arun Kumar Singh, "RF performance analysis of intercalated graphene nanoribbon based global level interconnects", *Journal of Computational Electronics*, Springer, vol. 19, pp. 1002–1013, Jun 2020.

101. Girish Wadhwa, B. Raj, "Design and performance analysis of junctionless TFET biosensor for high sensitivity", *IEEE Nanotechnology*, vol. 18, pp. 567–574, 2019.

102. Jeetendra Singh, B. Raj, "Enhanced nonlinear memristor model encapsulating stochastic dopant drift", *JNO*, ASP, vol. 14, pp. 958–963, 2019.

103. V. K. Sharma, M. Pattanaik, B. Raj, "ONOFIC approach: Low power high speed nanoscale VLSI circuits design", *International Journal of Electronics*, Taylor & Francis, vol. 101, no. 1, pp. 61–73, 2014.

104. S. Khandelwal, Balwinder Raj, R. D. Gupta, "FinFET based 6T SRAM cell design: Analysis of performance metric, process variation and temperature effect", *Journal of Computational and Theoretical Nanoscience*, ASP, USA, vol. 12, pp. 2500–2506, 2015.

105. Sumit Singh, Shekhar Yadav, Jagdeep Rahul, Anurag Srivastava; B. Raj, "Impact of HfO$_2$ in graded channel dual insulator double gate MOSFET", *Journal of Computational and Theoretical Nanoscience*, American Scientific Publishers, vol. 12, no. 6, pp. 950–953, Apr 2015.

106. Vijay Kumar Sharma, Manisha Pattanaik, B. Raj, "PVT variations aware low leakage INDEP approach for nanoscale CMOS circuits", *Microelectronics Reliability*, Elsevier, vol. 54, pp. 90–99, 2014.

107. B. Raj, A. K. Saxena, S. Dasgupta, "Quantum mechanical analytical modeling of nanoscale DG FinFET: Evaluation of potential, threshold

voltage and source/drain resistance", *Elsevier's Journal of Material Science in Semiconductor Processing*, Elsevier, vol. 16, no. 4, pp. 1131–1137, 2013.

108. Maisagalla Gopal, Siva Sankar D Prasad, Balwinder Raj, "8T SRAM cell design for dynamic and leakage power reduction", *International Journal of Computer Applications*, Foundation of Computer Science, New York, USA, vol. 71, no. 9, pp. 43–48, Jun 2013.

109. Manisha Pattanaik, B. Raj, Shashikant Sharma, Anjan Kumar, "Diode based trimode multi-threshold CMOS technique for ground bounce noise reduction in static CMOS adders", *Advanced Materials Research*, Trans Tech Publications, Switzerland, vol. 548, pp. 885–889, 2012.

110. Balwinder Raj, A. K. Saxena, S. Dasgupta, "Nanoscale FinFET based SRAM cell design: Analysis of performance metric, process variation, underlapped FinFET and temperature effect", *IEEE Circuits and System Magazine*, vol. 11, no. 2, pp. 38–50, 2011.

111. V. K. Sharma, M. Pattanaik, Balwinder Raj, "Leakage current ONOFIC approach for deep submicron VLSI circuit design", *International Journal of Electrical, Computer, Electronics and Communication Engineering, World Academy of Sciences, Engineering and Technology*, vol. 7, no. 4, pp. 239–244, 2013.

112. Tulika Chawla, Mamta Khosla, B. Raj, "Design and simulation of triple metal double-gate germanium on insulator vertical tunnel field effect transistor", *Microelectronics Journal*, Elsevier, vol. 114, p. 105125, Aug 2021.

113. Parminder Kaur, Sandeep Singh Gill, B. Raj, "Comparative analysis of OFETs materials and devices for sensor applications", *Journal of Silicon*, Springer, vol. 14, pp. 4463–4471, 2022.

114. Sanjeev Kumar Sharma, Parveen Kumar, Balwant Raj, B. Raj, "In$_{1-x}$Ga$_x$As double metal gate-stacking cylindrical nanowire MOSFET for highly sensitive photo detector", *Journal of Silicon*, Springer, vol. 14, pp. 3535–3541, 2022.

115. B. Raj, A. K. Saxena, S. Dasgupta, "Analytical modeling of quasi planar nanoscale double gate FinFET with source/drain resistance and field dependent carrier mobility: A quantum mechanical study", *Journal of Computer (JCP)*, Academy Publisher, Finland, vol. 4, no. 9, pp. 1–8, 2009.

116. S. Bhushan, S. Khandelwal, B. Raj, "Analyzing different mode FinFET based memory cell at different power supply for leakage reduction", *Seventh International Conference on Bio-Inspired Computing: Theories and Application, (BIC-TA 2012) Advances in Intelligent Systems and Computing*, vol. 202, pp. 89–100, 2013.

117. Jeetendra Singh, B. Raj, "Temperature dependent analytical modeling and simulations of nanoscale memristor", *Journal: Engineering Science and Technology, an International Journal*, Elsevier, vol. 21, pp. 862–868, Oct 2018.

118. Shradhya Singh, Shashi Bala, Balwant Raj, B. Raj, "Improved sensitivity of dielectric modulated junctionless transistor for nanoscale biosensor design", *Sensor Letter*, ASP, vol. 18, pp. 328–333, Apr 2020.

119. Vivek Kumar, Santosh Kumar Vishvakarma, B. Raj, "Design and performance analysis of ASIC for IoT applications", *Sensor Letter*, ASP, vol. 18, pp. 31–38, Jan 2020.

120. Akanksha Jaiswal, R. K. Sarin, B. Raj, Shikha Sukhija, "A novel circular slotted microstrip-fed patch antenna with three triangle shape defected ground structure for multiband applications", *Advanced Electromagnetic (AEM)*, vol. 7, no. 3, pp. 56–63, Aug 2018.
121. Girish Wadhwa, B. Raj, "Label free detection of biomolecules using charge-plasma-based gate underlap dielectric modulated junctionless TFET", *Journal of Electronic Materials (JEMS)*, Springer, vol. 47, no. 8, pp. 4683–4693, Aug 2018.
122. **Gurmohan Singh, R. K. Sarin, B. Raj, "Design and performance analysis of a new efficient coplanar quantum-dot cellular automata adder"**, *Indian Journal of Pure & Applied Physics (IJPAP)*, vol. 55, pp. 97–103, Feb 2017.
123. Amandeep Singh, Mamta Khosla, Balwinder Raj, "Design and analysis of electrostatic doped schottky barrier CNTFET based low power SRAM," *International Journal of Electronics and Communications, (AEÜ)*, Elsevier, vol. 80, pp. 67–72, 2017.
124. Parminder Kaur, Vikas Pandey, B. Raj, "Comparative study of efficient design, control and monitoring of solar power using IoT", *Sensor Letter*, ASP vol. 18, pp. 419–426, May 2020.
125. Anil Kumar Bhardwaj, Sumeet Gupta, B. Raj, "Development & analysis of compact model for double gate schottky barrier CNTFET", *Journal of Nanoelectronics and Optoelectronics*, ASP, vol. 15, pp. 1199–1208, Aug 2020.

Chapter 7

Aerodynamics of wind turbines operating in Algerian desert environments

Arezki Smaili, Abdelhamid Bouhelal,
and Mohammed Amokrane Mahdi

7.1 INTRODUCTION

Over the past few decades, the world's demand for energy has grown significantly due to various factors such as population growth, urbanization, industrialization, and changing lifestyles. This increase in demand has led to an over-reliance on finite fossil fuels, resulting in environmental and economic problems such as pollution, climate change, and geopolitical conflicts. In response to these challenges, there is an urgent need to shift toward cleaner and more sustainable sources of energy. Renewable energy sources, such as wind, solar, and hydro power, are seen as viable solutions to address these issues.

Among the various renewable energy sources, wind energy has emerged as a promising solution to meet the world's growing energy demands. Wind power is clean, abundant, and widely distributed, making it a highly attractive option for energy production. In recent years, the global wind energy industry has experienced tremendous growth, with a cumulative installed capacity of over 837 GW by the end of 2021, according to the Global Wind Energy Council's April 2022 report.

The growth of wind energy can be attributed to a number of factors, including technological advancements, declining costs, and increasing demand for clean and sustainable energy. Improvements in the design and efficiency of wind turbines have increased the potential for wind energy to provide a significant share of global electricity demand. In addition, increasing awareness of the need to reduce greenhouse gas emissions has led to greater investment in wind energy, making it one of the fastest-growing renewable energy sources.

The wind energy sector is characterized by a significant level of innovation and technological development, with new wind turbine designs and control strategies being developed to optimize energy generation. There is also a growing interest in developing wind energy in regions with high wind resources, including coastal areas and deserts. Deserts, in particular, are considered to be ideal locations for wind energy development due to the high wind speeds and consistent wind patterns.

DOI: 10.1201/9781003487692-7

The potential of wind energy in the Sahara desert of Algeria is particularly promising. Algeria is one of the largest countries in Africa and has an abundance of land, with a large portion of the country being covered by the Sahara desert. The desert is known for its strong and consistent winds, making it an ideal location for wind energy production. The Laboratory of Green and Mechanical Development (LGMD) at Ecole Nationale Polytechnique (ENP), Algiers, has been conducting research in this field, with a focus on the aerodynamics of wind turbines in desert climates.

According to a study by the World Bank, published in its African Energy Outlook 2019 report, Algeria is one of the countries in the African continent with the highest onshore wind potential, which amounts to 7700 GW, i.e., more than nine times the world's currently installed wind capacity (until the end of 2021).

The Sahara desert of Algeria represents a unique environment for the deployment of wind turbines due to its harsh climate, characterized by higher ambient air temperatures, sandstorms, and lower humidity. Understanding the aerodynamics of wind turbines that would be installed in such desert environments is a critical step for the development of efficient and sustainable wind energy systems. One of the main challenges of wind energy production in desert climates is the impact of sand particles on the performance of wind turbines. The accumulation of sand on the blades can alter their aerodynamic shape and increase the drag force, thus reducing their efficiency. Increasing air temperature may also significantly impact wind turbine operations in desert climates. High temperatures can cause a reduction in air density, which in turn affects the lift and drag forces acting on the blades. The reduced air density can also increase the resistance to the rotation of the blades, consequently reducing the efficiency of the turbine.

The combination of sand particles and high temperatures in desert climates presents a unique challenge for wind energy production. However, studies conducted by LGMD researchers have provided valuable insights into the specific challenges and limitations of wind energy production in this environment. Advanced methods based on computational fluid dynamics (CFD) simulations have been proposed to predict the performance of wind turbines in desert climates.

The objective of this chapter is to provide a comprehensive review of the aerodynamics of wind turbines in the Sahara desert of Algeria, with a focus on the impact of sand particles and ambient air temperature upon the resulting wind turbine performances. The specific objectives of this chapter can be summarized as follows:

1. To provide insights into the recent studies conducted by LGMD on the aerodynamics of wind turbines, including prediction of performances, near-wake studies, farm optimization, and more.

2. To highlight the potential of the Sahara desert in Algeria as a source of wind energy and the unique challenges and limitations associated with wind energy production in desert climates.
3. To summarize recent studies that have investigated the effects of sand on wind turbine aerodynamics, including the characterization of sand particle size and distribution, and the development of accurate numerical methods for predicting the effect of sand on the aerodynamic performance of wind turbines.
4. To examine the impact of air temperature on wind turbine aerodynamics and explore methods for predicting this effect, such as the use of cooling systems and the optimization of blade shape and surface area.
5. To highlight the need for further studies to fully understand and optimize wind turbine aerodynamics in desert climates, and to identify advanced methods, such as CFD simulations and field experiments, that can be used to study the performance of wind turbines in desert climates.

Finally, the chapter aims to provide valuable information and insights for policymakers, engineers, and energy experts as they work to develop sustainable and efficient wind energy systems in desert regions. The findings from this review will also be useful for researchers working in the field of wind energy, as well as for students and academics interested in renewable energy and sustainable development.

7.2 LGMD'S RECENT STUDIES ON THE AERODYNAMICS OF WIND TURBINES

7.2.1 Wind turbine performance predictions

The evolution of wind energy technology has been driven by advancements in the field of aerodynamics, which plays a crucial role in the efficient functioning of wind turbines. Wind energy aerodynamics may be separated into two categories: rotor aerodynamics and wind farm aerodynamics [1]. The former encompasses the complex interplay between the rotor blades and the surrounding air, while the latter focuses on the aerodynamic phenomena that occur in the wake of multiple turbines within a wind farm. In-depth knowledge of the aerodynamic performance of rotor blades can lead to the development of more efficient wind turbines that are capable of generating higher power outputs. On the other hand, the study of wake aerodynamics can be used to optimize wind farm design and layout to increase the overall efficiency of the wind energy system. As such, a comprehensive understanding of both rotor and wind farm aerodynamics is essential for the successful development of wind energy technology.

Wind turbines have become increasingly important in the pursuit of clean and renewable energy. As such, researchers have devoted considerable time and effort to understanding the aerodynamic principles that govern their operation. One of the most important aspects of wind turbine aerodynamics is rotor aerodynamics, which can be modeled using several different methods. In this section, we will discuss these methods and provide a brief description of each.

A wind turbine rotor must be precisely modeled in order to investigate its aerodynamics, both in operating conditions and in relation to the wind. Numerous techniques are available, such as the blade element momentum (BEM) methods, the full rotor geometry CFD methods (full Navier-Stokes) [1], the actuator approaches (actuator disc model (ADM), actuator line model (ALM), and actuator surface method (ASM) [2]), and the vortex methods (including the lifting line, lifting surface, and the panel method).

Using an iterative algorithm to achieve a force balance acting on the blade components and the forces acting on the flow field, the BEM technique models the rotor as a collection of blade elements. This allows for the determination of numerous parameters, including mechanical power and thrust. The underlying presumptions of this approach are that forces can be averaged over distinct annuli and that there is no radial dependency [3, 4].

Since the classical BEM theory requires aerodynamic coefficients as inputs to forecast wind turbine rotor performance, an alternate strategy based on artificial intelligence was presented by A. Bouhelal et al. [5]. The researchers employed an artificial neural network (ANN) in their investi gation to forecast the aerodynamic coefficients of airfoils. The ANN was designed to account for multiple features, incorporating hidden layer count, neurons per layer, function of activation, input models, and learning algorithms [6–12]. The BEM technique was then combined with this improved ANN model to properly forecast wind turbine performance, even in situations where airfoil data was unavailable. All things considered, this method offers a quick and effective way to forecast wind turbine performance and optimize wind turbine design, which helps to create more sustainable and effective energy sources [13–21].

Rotor blades are modeled as vortex models, which trail and shed vorticity in the wake by means of lifting lines or surfaces. The turbine is simulated by the ADM as a disk that exerts forces on the fluid inside the flow. While the ASM determines the forces on a 2-D airfoil as a function of the chord, the ALM represents wind turbine blades as a set of blade components along each blade axis [22–29].

Finally, the full Navier-Stokes method is a full rotor geometry CFD method that models the flow through the rotor using the Navier-Stokes equations. Although this method is the most computationally expensive, it offers the most detailed image of the flow field surrounding the rotor. Because of this, it is usually only employed in research or in the creation of extremely specialized wind turbines [1, 6].

A. Bouhelal et al. [6] conducted a comprehensive Navier-Stokes study [6] to examine the effects of various Reynolds-averaged Navier-Stokes (RANS) turbulence models on the prediction of horizontal-axis wind turbine (HAWT) performance under various wind conditions. Measurements from the MEXICO project were used in the study to validate the CFD codes. The study assessed four RANS turbulence models, and it found that at low wind speeds, all models correctly predicted the aerodynamic performance and wake velocity of the HAWT. On the other hand, the k-ε model proved to be the most accurate predictor of aerodynamic performance at high wind speeds within a reasonable computational time. The study also demonstrated that near-wall effect modeling accurately predicted blade loads and velocity in the near wake. Overall, the study discovered that for high wind speeds, the high Reynolds models outperformed the low Reynolds models [30–37].

7.2.2 Near-wake predictions

Accurately forecasting the wake of an isolated wind turbine is essential for maximizing power output in a wind farm, as wind turbines situated in the wake of upstream rotors generate 40–60% less power than when isolated. The wake of a wind turbine is separated into two parts: close and far wake. The far wake is located far from the wind turbine and immediately affects downstream turbines, whereas the close wake is a downstream induction zone one to five times the rotor diameter. In order to develop fast engineering wake models, improve nacelle anemometry technology for estimating wind farm output power, evaluate annual energy production, optimize wind farms, optimize the interaction between aerodynamics and hydrodynamics for offshore wind turbines, and assess maintenance testing and system safety, among other uses for wind turbines, accurate near-wake predictions are essential [38–45].

M. Tata et al. [7] used both experimental and numerical approaches to investigate fluid flow in the near wake of a scaled horizontal-axis wind turbine (HAWT). The particle image velocimetry (PIV) technique was used in the experimental part to measure the velocity field in the near wake of a wind turbine with a 0.350 m rotor diameter at two different tip speed ratios (6.76 and 9.91) in a wind tunnel. Two turbulence models (k-ε and k-SST) were used in the numerical simulations, and the wind turbine rotor was modeled using the actuator disc approach combined with the blade element momentum theory. The primary goal of this study was to compare numerical simulation results to experimental measurements and to investigate the performance of various turbulence models. The study found that the k- and k- SST high Reynolds turbulence models accurately predicted flow in the outer part of the blade. Other turbulence models, such as low Reynolds and transient LES and DES, were recommended for use in the inner part of the blade near the nacelle.

Using particle image velocimetry (PIV) and an ADM, Hamlaoui et al. [8] investigated the wake flow characteristics downstream of the rotor blade of a small horizontal-axis wind turbine (HAWT) at a Reynolds number of Re = 47000. The objective of this study was to examine how the wake changed over time and how the nacelle affected the properties of the flow field. The estimated velocity field near the rotor blade root, particularly the axial and tangential velocity components, was shown to be greatly improved by the 3D corrections for stall delay phenomena consideration at a distance of 0.225 D downstream from the rotor. However, the nacelle dominated the flow field at a distance of 2.4 D from the rotor, which resulted in the loss of the structured distribution of the velocity field because of wake expansion. It was discovered that, downstream of the rotor, $1.8 < z/D < 2.4$ is the near-wake upper limit. The rotor blade tip was shown to be the primary source of turbulence, whereas the nacelle was found to be the source of turbulence dissipation. The study also examined turbulent kinetic energy and turbulence-specific dissipation fields. In order to examine downstream small HAWT performance and near-wake forecasts under low Reynolds number flow conditions, the study offered precise correlations. Small HAWT experimental measurements were used to validate the study, and the outcomes demonstrated a high degree of accuracy and agreement with the experimental findings.

Recently, A. Bouhelal et al. [9] examined the fluid behavior within the close wake of an actual turbine model (MEXICO rotor) with the nacelle present using full Navier-Stokes simulations. The simulations were performed using the transitional k-kl-ω turbulence model in conjunction with RANS equations. Comprehensive PIV measurements were compared with simulation cases with and without the nacelle, highlighting the impact of the nacelle and blade rotation on the inductive region and near wake.

A detailed investigation was conducted into the axial and radial flow behaviors at the induction region. The goal of the study was to demonstrate how, and possibly why, the nacelle affects near-wake flow and numerical prediction accuracy in different scenarios. Simulation results show that the blade rotation dominates the near-wake region, and at high wind speeds, nacelle geometry can increase the accuracy of both axial and radial flow prediction by up to 15%. The impacts of the nacelle may be ignored at low wind speeds. Additionally, it has been shown that the presence of the nacelle enhances root and tip vorticities by increasing flow separation at the trailing edges of the blade airfoils. The study comes to the conclusion that short nacelle diameters should be used to reduce flow separation on the blades and increase average velocity downstream of the rotor in order to optimize wind farm output power [46–53].

7.2.3 Wind farm optimization

Wind turbine placement optimization is critical for maximizing the efficiency and profitability of wind energy projects. A poorly optimized placement can lead to significant losses in power generation due to wake effects, which occur when a turbine reduces wind speed in its wake, thereby reducing the power output of downstream turbines [46–53]. Additionally, a suboptimal placement can result in higher maintenance costs, as turbines may be subjected to higher loads or excessive turbulence, leading to premature wear and tear. With the increasing demand for renewable energy sources and the global drive toward reducing carbon emissions, wind energy has become a vital component of the energy mix. As such, it is crucial to optimize the placement of wind turbines to achieve maximum power generation and ensure the long-term sustainability of wind energy projects. By using mathematical approaches such as the genetic algorithms, Monte Carlo, random number generation, and wake models, as well as developing computer programs to simulate and analyze data, researchers and engineers can determine the optimal placement of turbines in a wind farm and ultimately contribute to the growth of the renewable energy industry.

S. Zerganeet al. [10] proposed a new optimization method for wind turbine placement in a wind farm with the aim of maximizing power production and reducing wake effects. This method involves generating pseudo-random numbers and using the Jensen linear wake model in numerical simulations to determine the optimum positioning of wind turbines. A computer program was developed to carry out these simulations, and the results showed a significant improvement in power production compared to previous studies that used genetic algorithms and viral basis methods. The optimal number of wind turbines for a given wind farm size was also determined. The proposed method was found to be more suitable for predicting maximum power production and enhancing total power output. The authors suggest that future studies should compare this method with other optimization techniques and consider more complex wake models.

Studying and optimizing the placement of wind turbines in a wind farm pose several challenges. First, wind flow is complex and influenced by a range of factors such as topography, ground conditions, and atmospheric conditions, which can make it difficult to accurately predict the wind direction and velocity at any given point in a wind farm. Additionally, the placement of one turbine can affect the performance of other turbines due to the wake effect, which can reduce the energy output of downstream turbines. Moreover, wind turbines are costly to install and maintain, and their placement should be optimized to ensure maximum energy production while minimizing the cost of installation and maintenance. Finally, wind farm layout optimization involves trade-offs between conflicting objectives, such as maximizing energy production, minimizing wake effect, and reducing the cost of installation and maintenance, making it a complex multi-objective optimization problem. Therefore, developing effective optimization methods that can account for these challenges is critical for achieving maximum energy production and reducing the cost of wind energy.

7.3 WIND TURBINE AERODYNAMICS IN DESERT CLIMATES

7.3.1 Challenges and limitations of wind resource assessment in desert climates

Wind resource assessment in desert areas presents a unique set of challenges and limitations that must be overcome to fully realize the potential of wind energy in these regions. One of the main challenges of wind resource assessment in desert climates is the lack of reliable data. This is because conventional methods of wind measurement, such as using anemometers, are not always accurate in desert environments due to the presence of sand particles that can interfere with wind speed measurements. Moreover, the harsh desert conditions can cause significant wear and tear on the equipment, leading to inaccuracies in the data collected [54–61].

Temperature is another factor that can significantly impact wind resource assessment in desert climates. High temperatures can cause thermal gradients that can interfere with wind measurements, leading to inaccuracies in the data collected. Moreover, high temperatures can also cause significant wear and tear on equipment, further reducing the reliability of wind data collected in these environments.

Sandstorms are also a significant limitation to wind resource assessment in desert climates. Sandstorms can cause significant damage to wind turbines and can interfere with the performance of equipment used to collect wind data. Moreover, the presence of sand particles in the air can also reduce the accuracy of wind speed measurements, further compounding the challenges of wind resource assessment in desert environments [62–68].

7.3.2 Effect of sand particles on the efficiency of wind turbines

In the case of wind turbines installed in desert environments, experimental studies are often difficult and expensive to conduct due to the harsh and unpredictable conditions of the desert. Therefore, numerical simulations can provide a cost-effective and efficient alternative for investigating the impact of sand particles on the aerodynamic performance of wind turbines. Theatrically, two approaches can be used to model the behavior of multiphase flows containing air and sand particles: the Eulerian-Eulerian and the Eulerian-Lagrangian approaches. A. Bouhelal et al. [11] investigated the impact of sand particles on the aerodynamic performance of wind turbines in desert environments using a multiphase CFD model based on the Eulerian-Lagrangian approach [69–76]. A parametric study was performed, varying the particle concentration, particle volume fraction, and particle diameter. The results showed that the presence of sand particles could significantly reduce the aerodynamic performance of the rotor. The rotor performance degraded with the increase of the particle concentration and particle size, especially for the higher concentrations and volume fractions. The study concluded that the sand wind can degrade the output power of wind turbines, and the rotor is more susceptible to degradation for larger particles even at smaller concentrations. In the next section, a detailed description of the mathematical model and numerical method used will be presented [77–85].

7.3.2.1 Mathematical model

The equation of motion for a solid particle in a gas is given by integrating the force balance on the particle, written in a Lagrangian reference frame. As an assumption, only dynamic forces, such as gravity and drag forces, are considered, while other force types like heat forces, thermophoretic forces, and thermodynamic forces, are disregarded. The trajectory of a sand particle is given by

$$\frac{d\vec{u}_p}{dt} = \underbrace{F_D\left(\vec{u} - \vec{u}_p\right)}_{Drag\ force} + \underbrace{\frac{\vec{g}\left(\rho_p - \rho\right)}{\rho_p}}_{Gravity\ force} + \underbrace{\vec{F}}_{other\ forces} \tag{7.1}$$

where $\vec{u}$ and $\vec{u}_p$ are respectively the air and the sand particle velocities. $F_D\left(\vec{u} - \vec{u}_p\right)$ is the particle force of the drag per unit mass. F_D is given by

$$F_D = \frac{18\mu}{\rho_p d_p^2} \frac{C_D Re}{24} \tag{7.2}$$

where μ is the air viscosity, d_p is the diameter of the sand particles, C_D is the drag coefficient, and Re is the relative Reynolds number, which is defined as

$$Re \equiv \frac{\rho d_p \left| u_p - u \right|}{\mu} \tag{7.3}$$

To find the drag coefficient of sand particles, a correlation for smooth spherical particles has been used [1, 11]:

$$C_D = a_1 + \frac{a_2}{Re} + \frac{a_3}{Re^2} \tag{7.4}$$

where a_1, a_2, and a_3 are constants varied as a function of the Reynolds number.

Lift forces are also considered in some multiphase flow models. The most commonly used lift force for fluid-spherical particle interactions is Saffman's law given by

$$\vec{F} = \frac{2K\upsilon^{1/2}\rho d_{ij}}{\rho_p d_p (d_{lk}d_{kl})^{1/4}} \left(\vec{u} - \vec{u}_p\right) \tag{7.5}$$

where K is an empirical constant (=2.594), and d_{ij} is the fluid deformation tensor.

7.3.2.2 Numerical method

To account for the influence of solid particles on the fluid phase, a hybrid CFD-Lagrangian model is used.

The calculation of momentum transfer from the continuous to the discrete phase entails examining a particle's momentum shift as it travels through each control volume [86–93]. This change in momentum is determined using the formula:

$$F = \Sigma \left(\frac{18\mu C_D Re}{24\rho_p d_p^2}\left(u - u_p\right) + \frac{\vec{g}\left(\rho_p - \rho\right)}{\rho_p} + F_{MRF} \right) \dot{m}_p \text{''} t \tag{7.6}$$

where $\dot{m}_p$ is the mass flow rate of the particles, F_{MRF} is the forces of the multiple reference frame of the rotor, and "t is the time step.

This exchange of momentum manifests as an energy sink in the ongoing momentum balance in any further calculations of the continuous phase flow field.

The coupling procedure can be summarized as follows:

1. Before introducing the discrete phase, solve the continuous phase flow field.
2. Calculate the particle trajectories for each discrete phase injection to introduce the discrete phase.
3. Recalculate the continuous phase flow using the interphase momentum and mass exchange calculated during the previous particle calculation.
4. In the modified continuous phase flow field, recalculate the discrete phase trajectories.
5. Continue the previous two procedures until you arrive at a converged solution, where each extra computation yields the same results for the discrete phase particle trajectories and the continuous phase flow field.

7.3.2.3 Hypothesis of the study

In this study, several assumptions are made to simplify the calculation process, including the homogeneity and constant diameter of the particles, constant velocity of the fluid and sand, equal velocity of particles and air, constant concentration of particles, disregarding heat transfer and thermal effects, consideration of collision with the blade wall as an elastic collision, and disregarding collision between the particles themselves [94–98].

7.3.2.4 Initial/boundary conditions

The particle volume fraction, α_s, may be computed as a function of particle flow rate $(\dot{m}_p)$, velocity(u_p), and density as follows $\left(\rho_p = 2240\,kg\,/\,m^3\right)$:

$$\alpha_{sinlet} = \frac{\dot{m}_p}{\rho_p u_p A_{inlet}} \tag{7.7}$$

Five different sand particle dimensions have been included in the study to represent both fine and medium sand because the diameter of the particles in the analyzed location is unclear [99–106]. These dimensions correspond to the International Scale for Identification and Classification of Soils, ISO, and are $d_p = 100, 200, 300, 400,$ and $500\,\mu m$.

In the Adrar area, the average yearly velocity at a height of 10 meters is around 6.30 m/s. The Kabertene farm's wind turbines are erected with a 55-meter-tall tower. A power law, which has the following definition, may be used to represent the velocity profile as a function of height:

$$\frac{u(h)}{u_0} = \left(\frac{h}{h_0}\right)^{\alpha} \tag{7.8}$$

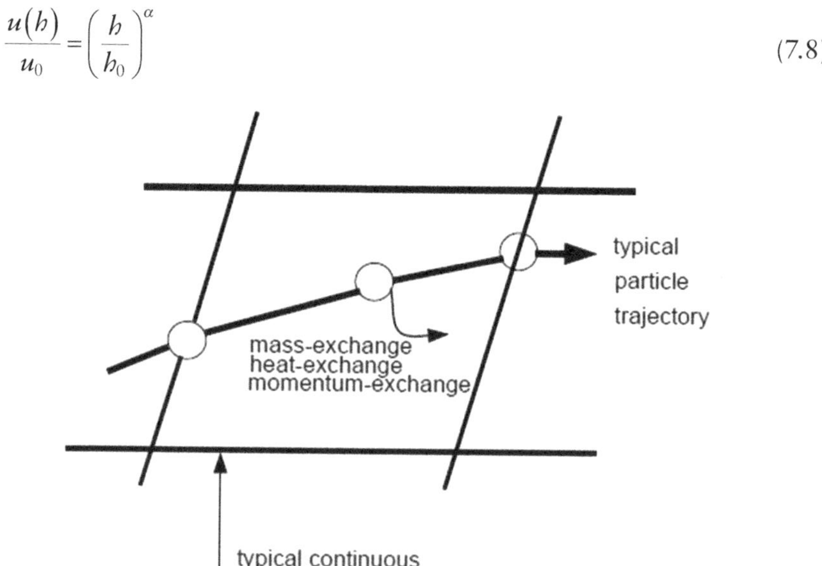

Figure 7.1 Illustration of the hybrid technique transfer of momentum between discrete and continuous phases.

where α is the surface roughness coefficient, which varies depending on the site features, and $u(h)$ is the wind speed at the necessary height h, and u_0 is the wind speed recorded at the reference height h_0.

The solver offers a number of ways to introduce particles into the computational domain. The following categories apply to these injection techniques:

- Single injection: With this method, a particle stream is injected from a single location.
- Group injection: As seen in Figure 7.2(a), this technique involves injecting particle streams in a line.
- Cone injection: As shown in Figure 7.2(b), this method entails injecting particle streams in a three-dimensional conical pattern.
- Surface injection: Particle streams are injected from a surface (one from each face) in this study using the surface injection method. In this instance, the computational domain's inlet serves as the injection surface, as seen in Figure 7.2.

Particles change momentum as they rebound after colliding with the wall boundary. This change in momentum is described by the coefficient of restitution, which is shown in Figure 7.3. It is assumed in this study that the particle rebound from the wall boundary is the result of an elastic collision. This indicates that following the rebound, the particles maintain all of their normal or tangential momentum [107–112].

The quantity of momentum that the particle retains after colliding with the boundary in a direction perpendicular to the wall is measured by a parameter called the normal coefficient of restitution. Mathematically speaking, the normal coefficient of restitution, or e_n, is defined as:

$$e_n = \frac{U_{p2,n}}{U_{p1,n}} \tag{7.9}$$

where $U_{p1,n}$ is the sand speed perpendicular to the wall before the collision, and $U_{p2,n}$ is the sand speed perpendicular to the wall after the collision.

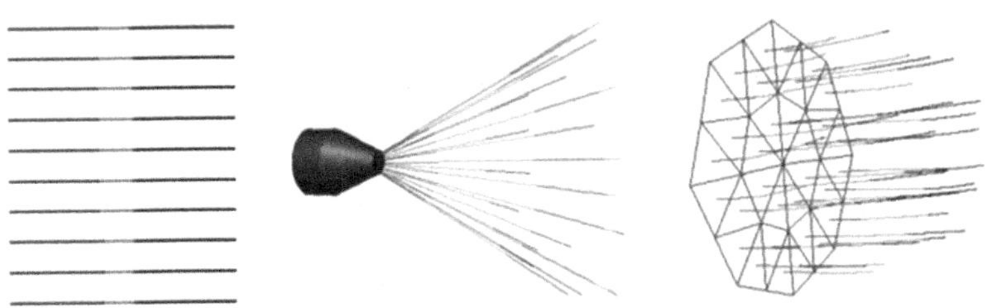

Figure 7.2 Different types of particle injections: (a) group injection, (b) 3D cone injection, and (c) surface injection.

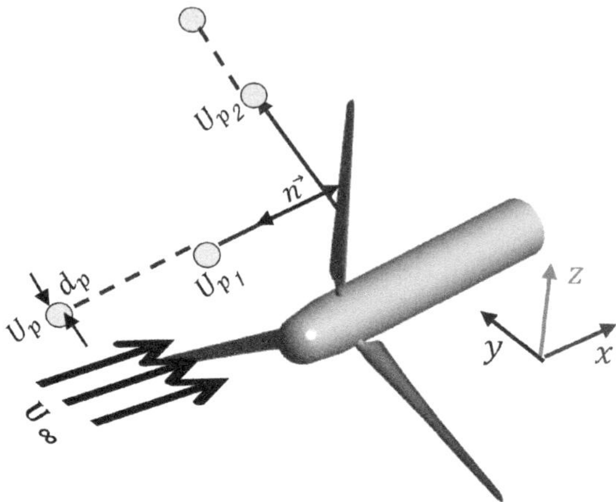

Figure 7.3 Illustration of sand particle reflection on the blade surface.

Similarly, the amount of momentum that the particle conserves in the direction parallel to the wall following the collision is measured by the tangential coefficient of restitution or e_t. Both the normal and tangential coefficients of restitution were set to 1 in order to guarantee that the particles interact with the blade surface in an elastic manner.

7.3.2.5 Results and discussion

The mean concentration of mass and mean volume fraction were calculated, respectively, by

$$\bar{\alpha}_s = \frac{\int_{i=1}^{N_{cell}} (\alpha_s.v)_i}{V_{tot}} \tag{7.10}$$

$$\bar{c} = \frac{\int_{i=1}^{N_{cell}} (c.v)_i}{V_{tot}} \tag{7.11}$$

where, $\bar{\alpha}_s$ and $\bar{c}$ are the average volume fraction and the average concentration of the particles.

The average volume percent and mean concentrations calculated for the five simulated situations are as follows: $\bar{\alpha}_s = [0.0001 - 0.0004 - 0.0009 - 0.0013 - 0.0015]$ and $\bar{c} = [0.3152 - 0.9455 - 1.8910 - 2.8365 - 3.1517] (kg/m^3)$.

Figure 7.4 shows the computed rotor torque curves as a function of five mean volume fractions and their corresponding sand mass concentrations for five different particle diameters.

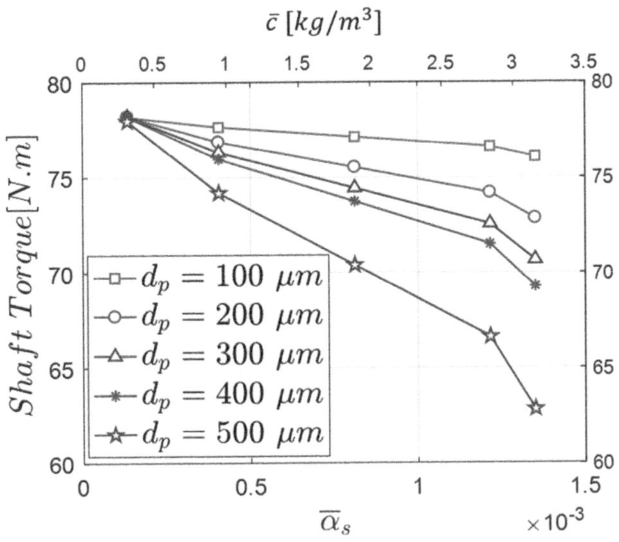

Figure 7.4 Rotor torque variation as a function of particle concentration for various particle diameters.

Based on Figure 7.4, it is evident that sand particles have a significant impact on the rotor torque. The study examined five different sizes of sand particles, and it was found that the aerodynamic torque decreases as the particle volume fraction and concentration increase. To evaluate the effect of sand particle concentration, a specific case was taken (d_p = 500 μm), while the influence of particle diameter was examined in the subsequent section. The results of the study revealed that the decrease in torque with concentration followed a semi-linear relationship [113–117]. The slope of the curve increased with increasing particle concentration. For low-volume fractions and concentrations, the rotor experienced a slight decrease in torque of 0.28 Nm, which was a small fraction of its total torque. These findings suggest that the presence of sand particles, even at low concentrations, can negatively impact the aerodynamic performance of rotors. The semi-linear relationship between torque and particle concentration suggests that as the concentration of particles increases, the detrimental effect on rotor performance becomes more severe. Therefore, it is essential to consider the potential impact of sand particles when designing and operating wind turbines in sandy environments.

Increasing sand concentration leads to significant declines in rotor performance, with losses of up to 15.4292 N.m observed at high concentrations. The interaction between the sand particles and the air flowing through the rotor is the primary cause of these losses. These findings underscore the importance of considering sand particle effects when designing and operating wind turbines in sandy environments, particularly at high concentrations. Further research is needed to develop effective mitigation measures.

Figure 7.5 shows that power loss due to sand particles is a function of particle size, with losses of up to 20% observed at higher particle diameters and concentrations. Power loss decreases with decreasing particle diameter. At a concentration of $\bar{\alpha}_s = 1.5 \times 10^{-3}$, power loss was 2.8%, 6.9%, 9.7%, 11.5%, and 19.7% for particle sizes of 0.1, 0.2, 0.3, 0.4, and 0.5 mm, respectively [118–120].

Figure 7.6 depicts the pressure factor variation at the first cross-section of the blades for concentrations of $\bar{\alpha}_s = 1.5 \times 10^{-3}$ and $\bar{c} = 0.3152\,kg/m^3$. It demonstrates that increasing particle size reduces pressure distribution, especially at the extrados section of the front edges of the blade airfoils. The influence of particles of sand on the motion of the field is immediately reflected in this drop in pressure. These studies indicate that the presence of particles of sand can significantly reduce wind turbine performance.

In Figure 7.7, the relationship between power loss and particle diameter under different volume fractions is presented. The curves of power loss increase with increasing particle diameter for all cases, and this evolution follows a semi-exponential law. Particle size has a significant impact on rotor performance, particularly at higher volume fractions. For fine particle sizes (0.1 mm), power loss does not exceed 4%, even at the highest concentrations. However, for larger particle sizes, power loss is sensitive even at the smallest volume fractions, except for a volume fraction of 0.3%, where the loss of energy is negligible for all cases studied.

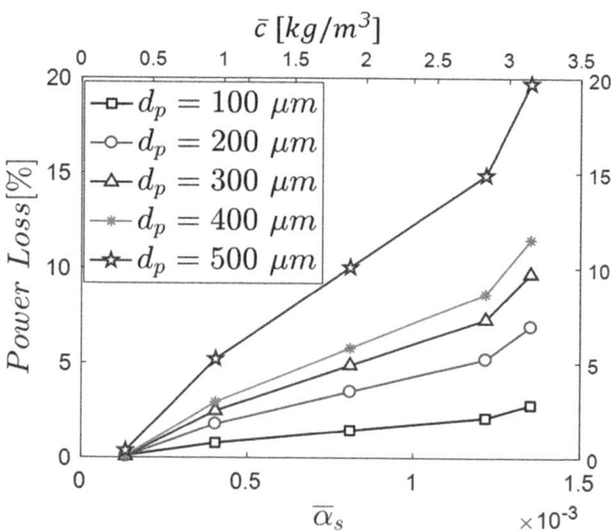

Figure 7.5 Power loss vs particle concentration for various particle diameters.

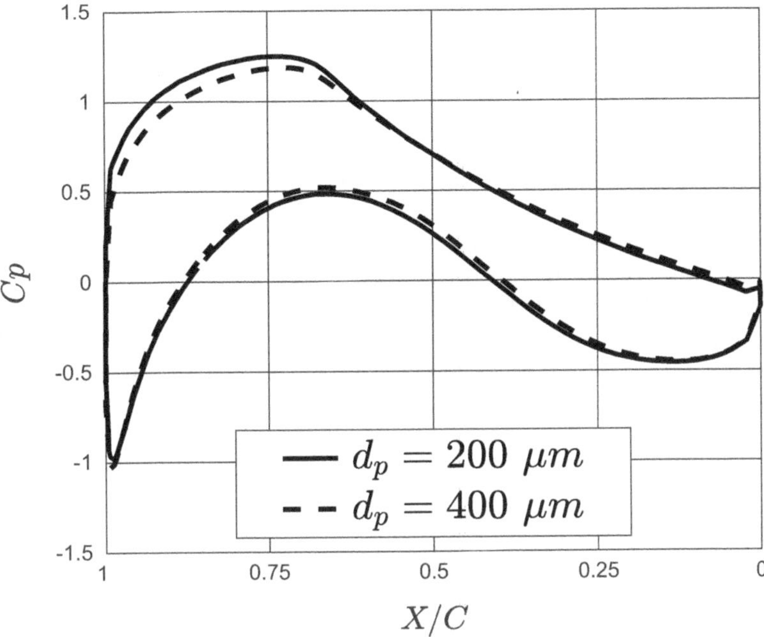

Figure 7.6 Pressure coefficient distribution at r/R=0.25 spanwise blade section for varied particle diameters, for $\bar{\alpha}_s = 1.5 \times 10^{-3}$.

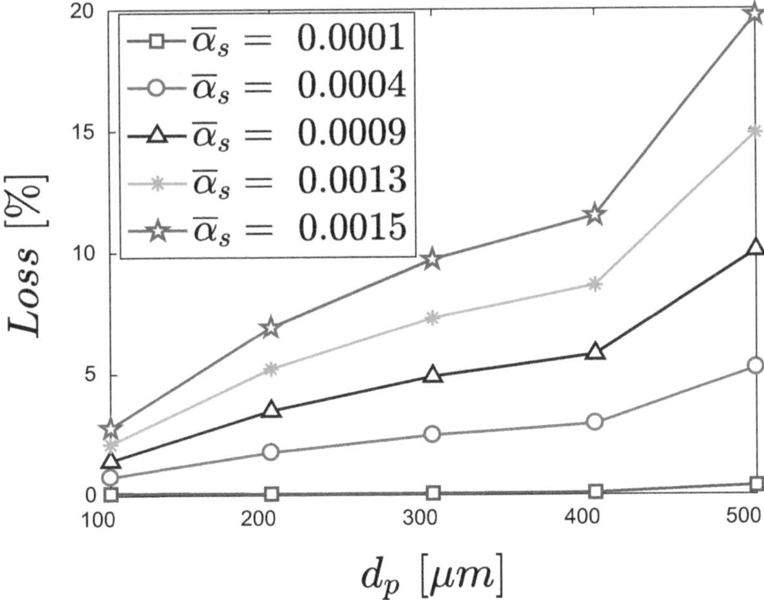

Figure 7.7 Power and torque loss vs particle diameter for various mean particle volume fractions.

7.4 EFFECT OF THE BLADE SURFACE ROUGHNESS

A. Bouhelal et al. [12] used a full Navier-Stokes CFD model to investigate the effects of blade roughness (due to sand accumulation) on the aerodynamic wind turbine performance. The research was carried out under a variety of wind conditions, with the flow over the rotor varying from fully attached flow to massively separated flow with varying roughness degrees. To close RANS equations, the k-ε RNG turbulence model was used, and a modification of the logarithmic law of the wall function was used to consider the numerical effect of roughness.

The study discovered that roughness significantly reduces rotor performance, with total power loss of the rotor reaching up to 35% for all test cases. For low and high wind speeds, the effect of small-size roughness was negligible. The presence of roughness heights induced early transition and increased the extent of transitional flow, as well as the intensity of turbulence near the wall. In each wind speed, there was a critical value of roughness beyond which the performance was unaffected.

Instead of employing complex solid-fluid interaction models, the study suggests using blade surface roughness as a simpler method for modeling the effect of sand particles on wind turbine performance. In the following section, we will go over how to put this method into action. Notably, this approach is less expensive than the previous Eulerian-Lagrangian simulations.

7.4.1 Roughness modeling

To incorporate the impact of roughness in the CFD code, a modification of the universal standard wall law function can be used as a boundary condition at the blade surface:

$$\frac{U_p u^*}{\tau_w / \rho} = \frac{1}{\kappa} \ln\left(E \frac{u^* y_p}{v} \right) - \Delta B \tag{7.12}$$

where U_p and y_p are the velocity and height at the wall-adjacent cell's center point P. E is the empirical constant for the smooth wall (E = 9.793). κ is the Von Karman constant (= 0.41), τ_w is the wall shear stress, and u^* is the wall friction velocity expressed as

$$u^* = C_\mu^{1/4} k_p^{1/2} \tag{7.13}$$

In the equation above, the variable k_p represents the turbulent kinetic energy at the wall-adjacent cell center, and the constant C_μ has a default value of 0.09.

As a function of dimensionless sand-grain roughness height k_s^+, the roughness function (ΔB) is defined as follows:

$$k_S^+ = \frac{u^* k_S}{\nu} \tag{7.14}$$

where k_s is the height of the equivalent sand-grain roughness. The roughness is classified into three regimes based on the value of k_S^+: smooth ($k_S^+ < 2.25$), transitional ($2.25 \leq k_S^+ < 90$), and fully rough ($k_S^+ > 90$). The fully rough regime is represented by the following formula:

$$\Delta B = \frac{1}{\kappa} \ln\left(1 - \frac{1}{2} k_S^+\right) \tag{7.15}$$

The relationship between the actual roughness lengths y_0 and the equivalent sand-grain roughness heights k_S can be expressed as follows:

$$k_S = 2 E y_0 \tag{7.16}$$

7.4.2 Numerical simulation

The purpose of the study was to investigate the influence of surface roughness on the flow around the MEXICO blade. The range of roughness heights to be investigated was determined using the findings of a prior experimental investigation that assessed the magnitude of roughness induced by dust collection in a dusty environment. The size of roughness might change from 0.01 mm to 0.7 mm over a period of one week to nine months, according to the findings of this study. In the current investigation, roughness heights were changed between 0.01 and 0.1 mm since it was observed that the influence of roughness beyond this interval is insignificant. The research also included a roughness height of 0.0005 mm to evaluate the influence of roughness on smaller values. The numerical examination under a smooth blade surface contains three scenarios with varied wind speeds, including the turbulent wake state at U=10 (m/s), the design circumstances at U=15 (m/s), and the separated flow conditions at U=24 (m/s). These examples demonstrate how rotor loading impacts numerical flow computations.

Table 7.1 shows the operational conditions for each of the computational cases, which were run with a constant blade pitch angle of –2.3° and a constant rotational speed of 425.1 rpm.

Table 7.1 Three investigated cases and their operational conditions

Wind speed (m/s)	Air density (Kg/m³)	Pressure (Pa)
10.05	1.197	101398
15.06	1.191	101345
24.05	1.195	101407

7.4.3 Results and discussion

Figure 7.8 depicts a comparison of normal and tangential forces at various roughness heights and wind speeds. At U=10 m/s, the first test case represents a mostly attached flow, with the normal force decreasing as the roughness height increases up to a critical value of 0.02 mm. Smaller roughness height values (0.0005 mm) have no discernible effect on forces. Furthermore, the roughness has a greater impact on the forces in the outer part of the blade. When it comes to tangential force, an increase in roughness height causes

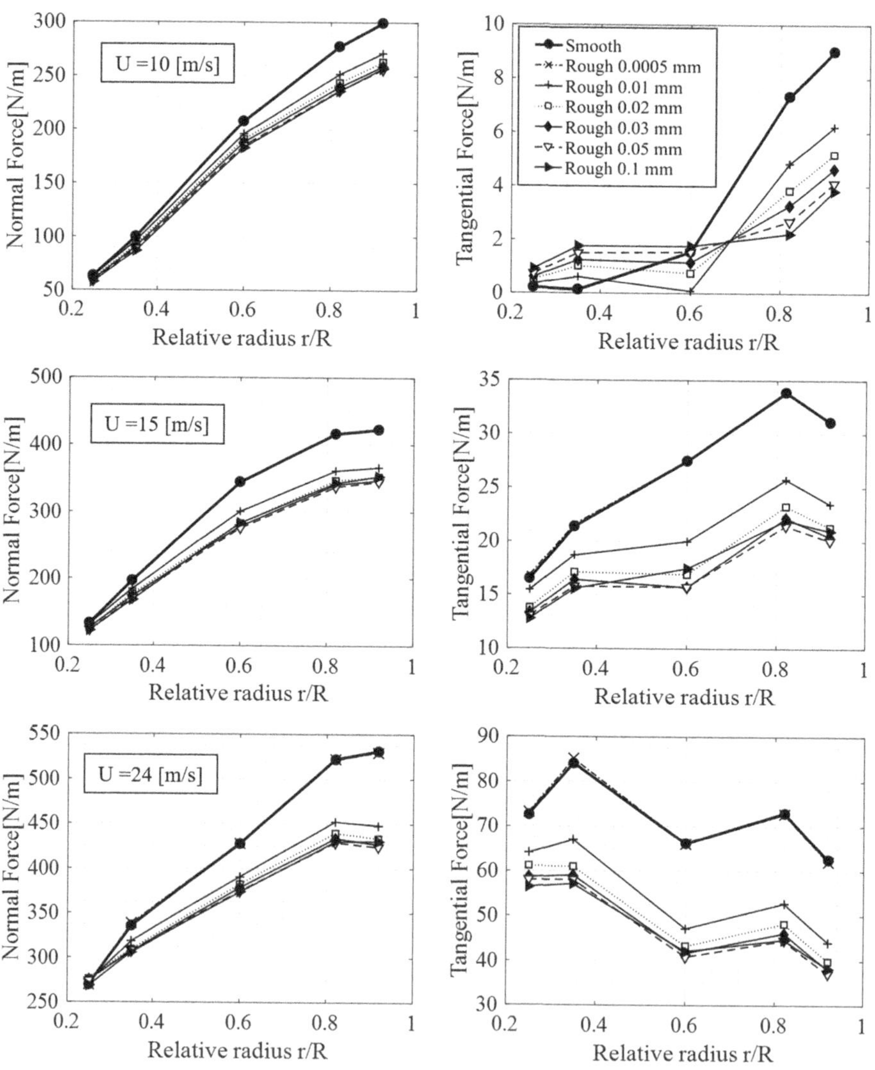

Figure 7.8 Normal and tangential forces are compared at various roughness heights.

a decrease on the outside of the blade (r/R > 0.6) and an increase on the inside.

The flow is classified as an onset of stall in the second test case at U=15 m/s, and an increase in roughness height results in a significant decrease in both normal and tangential forces, particularly in the outer part of the blade. This suggests that roughness has a significant impact on stall conditions, resulting in a decrease in overall blade performance.

Finally, the flow is classified as a separated flow condition in the third test case at U=24 m/s. As the roughness height increases, the tangential force decreases significantly in both the outer and inner parts of the blade. As a result, it is possible to conclude that roughness has a significant impact on flow separation and stall conditions. These findings suggest that the roughness of the blade surface must be carefully considered when designing wind turbines in order to optimize their aerodynamic performance.

Figure 7.9 depicts the pressure coefficient distribution at r/R=0.6 span for various roughness heights. The figure shows that the pressure coefficient decreases as the roughness heights increase for all three test cases. The effect of roughness is minimal in the first case, where the flow is attached, but as wind speed increases, the impact of roughness becomes more significant. Furthermore, the figure shows that roughness has a greater effect on the extrados of the blade section profile than on the intrados for all three test cases, suggesting that roughness influences flow separation.

Figure 7.10 depicts the distribution of turbulence intensity at r/R=0.6 span under different roughness heights to support this conclusion. The figure depicts how increasing roughness heights causes an increase in turbulence intensity, particularly on the leading edge of the three test cases. This observation supports the idea that roughness influences flow separation and can affect wind turbine aerodynamic performance. In general, these findings emphasize the importance of considering roughness effects when designing wind turbines to maximize performance.

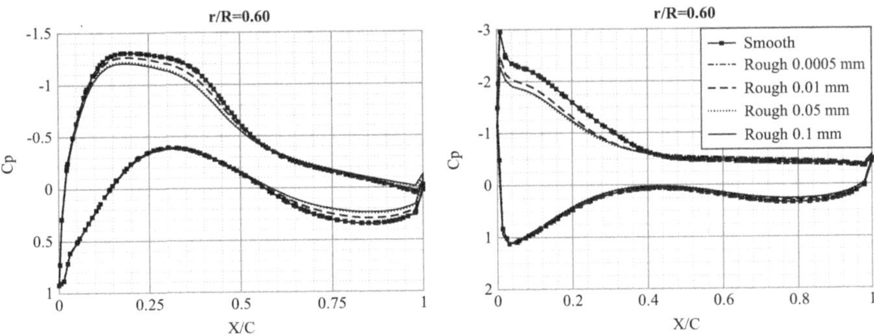

Figure 7.9 Pressure coefficient distribution at r/R=0.6 span for different roughness heights (left: U =10 m/s, right: U =24 m/s).

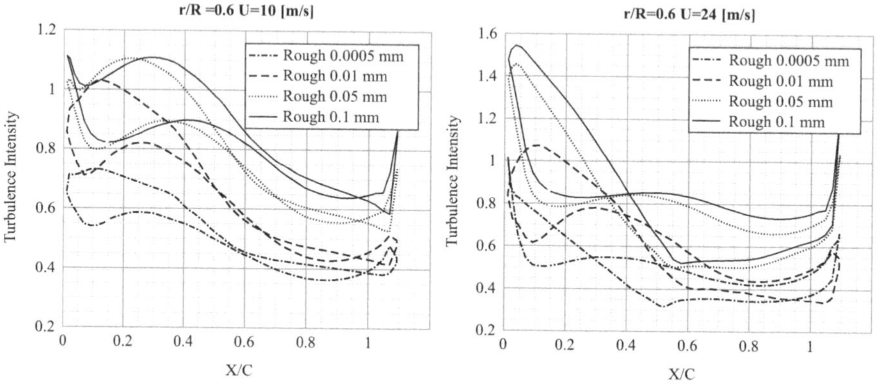

Figure 7.10 Turbulence intensity distribution at r/R=0.6 span and varied roughness heights (left: U =10 m/s, right: U =24 m/s).

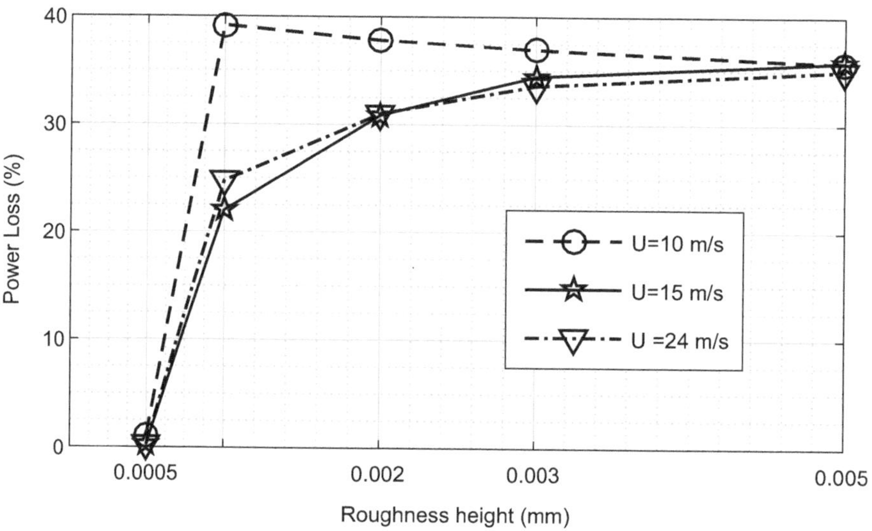

Figure 7.11 The effects of roughness heights on turbine power loss at various wind speeds.

Figure 7.11 depicts the effect of roughness heights on wind turbine power loss at various wind speeds. The results show that small roughness heights (0.0005 mm) have no effect in the first test case, where the flow is mostly attached. However, the power loss remains nearly constant at 38% for large roughness heights ranging from 0.001 mm to 0.005 mm. For small roughness heights, the impact of roughness is insignificant in cases two and three, where the flow is separated. The power loss increases rapidly as the roughness heights increase until it reaches a critical value beyond which the roughness has no effect. The rotor's total power loss can be as high as 35%.

7.5 EFFECT OF AMBIENT AIR TEMPERATURE ON WIND TURBINES OPERATION

The study conducted by A. Smaili et al. [13] delves into the thermal behavior of a wind turbine nacelle operating in extreme Saharan climate conditions, where high temperatures can wreak havoc on the electronic components housed within the nacelle. To tackle this issue, the authors propose a numerical method that accurately models the airflow in and around the nacelle through the use of Reynolds-averaged Navier-Stokes equations, while also accounting for heat transfer effects through the energy equation. The resulting simulations offer insight into the temperature fields within and surrounding the nacelle, which can aid in the design and implementation of effective cooling systems to maintain acceptable temperature levels.

Although this study presents a preliminary investigation into the thermal behavior of a wind turbine nacelle, the proposed numerical method shows promise for future, more detailed research. The linear trend variations observed between cooling capacity and temperature levels within the nacelle as a function of cold-plate temperature provide useful information for future design considerations. Additionally, the findings emphasize the crucial role of balancing the environmental heat load with the required cooling capacity during nacelle thermal control.

Moving forward, future research can build upon this study by incorporating practical and nacelle thermal design considerations to develop more comprehensive numerical models. Such models can potentially offer even greater accuracy in predicting temperature fields and inform the development of more efficient cooling systems for wind turbine nacelles operating in extreme climates. Overall, this study highlights the importance of understanding and managing the thermal behavior of a wind turbine nacelle to ensure optimal performance and longevity of these important renewable energy systems.

Recently, M.A. Mahdi et al. [14] performed numerical simulations to explore the effect of turbulent convection heat transfer from the environment on the thermal behavior of wind turbine nacelles working in the Algerian Sahara. Heat and velocity profiles across the nacelle's two-dimensional and three-dimensional arrangements, as well as changes in the nacelle's median temperature and necessary cooling capacity, were presented and addressed by the authors. They also proposed an iterative procedure for mesh and time-step dependence analysis, as well as double quantitative validation of numerical results. The study discovered that appropriate values of required cooling loads are dependent on accurate predictions of natural convection within the nacelle and suggested that future research should consider more practical aspects of nacelle thermal design, such as heat exchange by radiation, unsteady thermal loads, and external heat exchange with surroundings and flow fields around the nacelle.

7.5.1 Problem specification

The computational realm and boundaries employed in the investigation are depicted in Figure 7.12. To get the temperature field, isothermal conditions were implemented in the nacelle's interior wall (i.e., the cold surface at TC) and the power supply wall (i.e., the hot surface at TH). The velocity field is subject to non-slip conditions. To analyze the temperature and air circulation fields in the 2D case, the research considers two longitudinal paths (A and B) and two transverse lines (C and D), as well as the vertical upward-oriented path in the left-mid-plane across hot and cold plates. To explore the 3D case, two vertical Y Z-plane segments (I) and (II) crossing through Lines C and D are analyzed.

7.5.2 Numerical simulation

The authors examined a standard 850 kW commercial wind turbine and reduced the nacelle shape to a 13 m × 3 m horizontal circular cylinder in order to study the thermal behavior of a wind turbine nacelle. A coaxial cylindrical generator measuring 3.22 m × 1.4 m was positioned within the nacelle, 0.48 m from the rear of the nacelle. Throughout the trial, the hot plate temperature T_H was kept at a steady 100 °C safety limit. As a result, the temperature range $\Delta T = T_H - T_C$ was solely regulated by changing the cooling temperature T_C. Notably, the authors found that ΔT varying in the interval of 120 °C to 70 °C, the Rayleigh number based on the longest distance between the hot and cold plates, i.e., the length of line A, was $R_{aA} \sim 10^{12}$. This implies that the airflow within the nacelle would continue to be very turbulent. As a result, a time-dependent strategy was required to find the answer. To solve governing equations using a second-order upwind differencing scheme, the ANSYS FLUENT finite-volume approach with the PISO algorithm is employed. Except for the pressure equation in the 3D

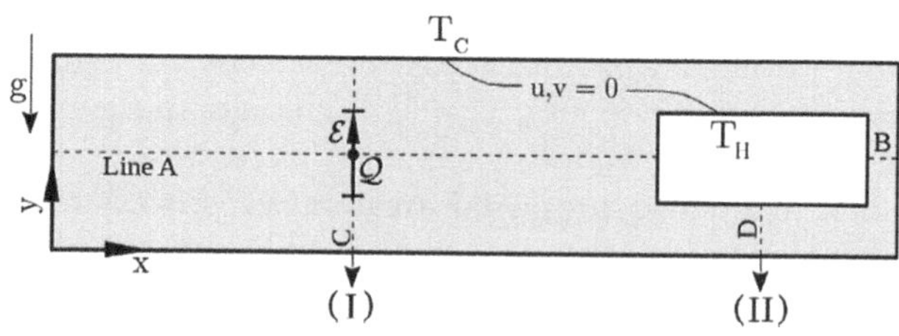

Figure 7.12 Computational domain and boundary conditions.

case, relaxation factors of 0.9 and 1 are used for the energy equation and other equations, respectively.

7.5.3 Results and discussion

The temperature distribution and streamlined outlines for the 2D case are shown in Figure 7.13 at $T_C = 0$ C and $T_H = 100$ C. The results are reasonable and show the numerical method's validity. The air on the left side of the nacelle exhibits thermal stratification, which is caused by a concentration of hot air in the top section due to a drop in density, confirming the correctness of the Boussinesq approximation.

The temperature patterns in the 3D nacelle are shown in Figure 7.14, revealing a more distinct thermal stratification phenomenon on the left side of the nacelle.

Figures 7.15 and 7.16 depict the necessary capacity for cooling and the mean temperature within the nacelle as an expression of cooling temperature, TC, for two physical scenarios: turbulent natural conduction 3D case (3D TNC), turbulent natural convection 2D case (2D TNC), and the pure conduction case (COND) in which buoyancy effects are ignored. Both factors are proportional to the cooling temperature. Figure 7.15 illustrates that the turbulence natural convection situation necessitates much more cooling

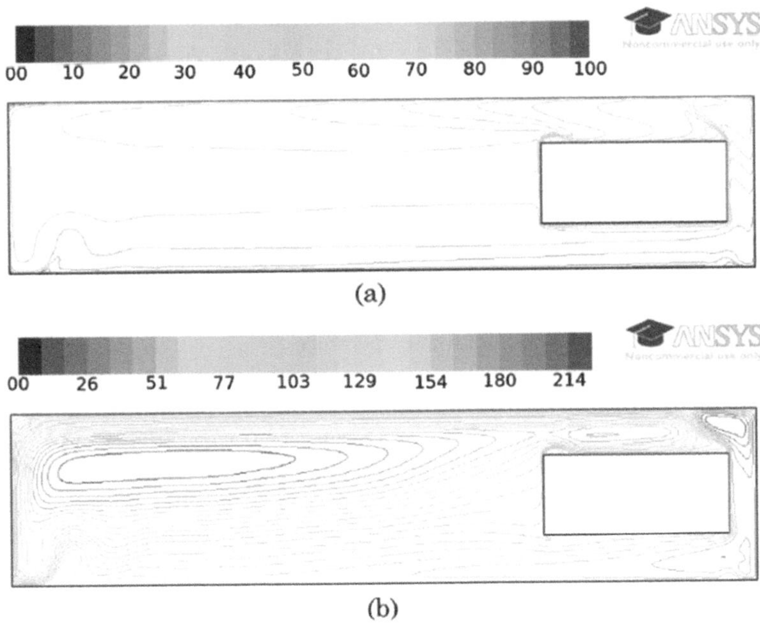

Figure 7.13 (a) The distribution of temperature in degrees Celsius (°C), and (b) the contour lines of streamlines in units of 10⁻³ kg/s within the 2D nacelle.

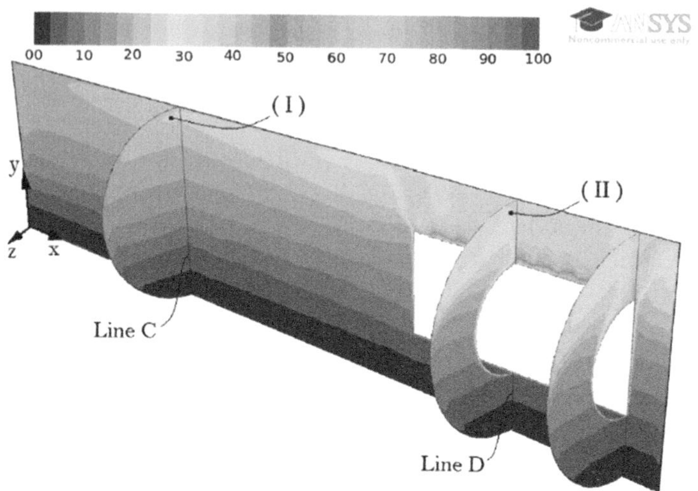

Figure 7.14 Temperature distribution within the 3D nacelle (oC).

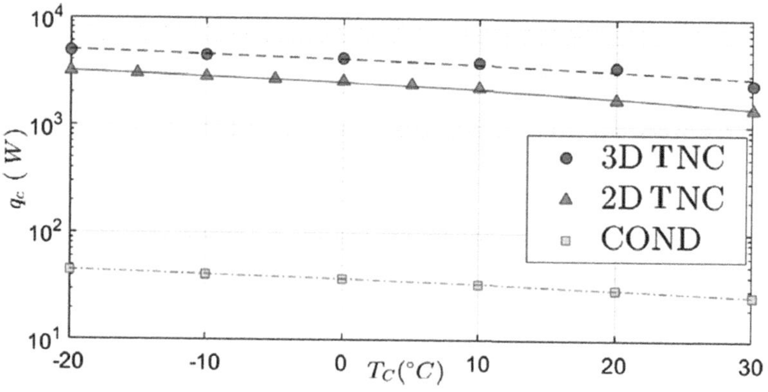

Figure 7.15 Cooling as a function of the temperature.

capacity than a pure conduction case, resulting in a higher average temperature. This finding highlights the importance of buoyant forces in the nacelle cooling procedure. Furthermore, it has been observed that the smaller the cooling temperature, the greater the energy required for cooling. This also serves as a qualitative validation of the numerical technique. The results also reveal that the influence of turbulent natural convection on nacelle thermal behavior rises with temperature span, T, making heat exchange via natural convection a crucial issue in nacelle cooling system thermal design. When the 2D and 3D findings are compared, the average relative difference is roughly 35%.

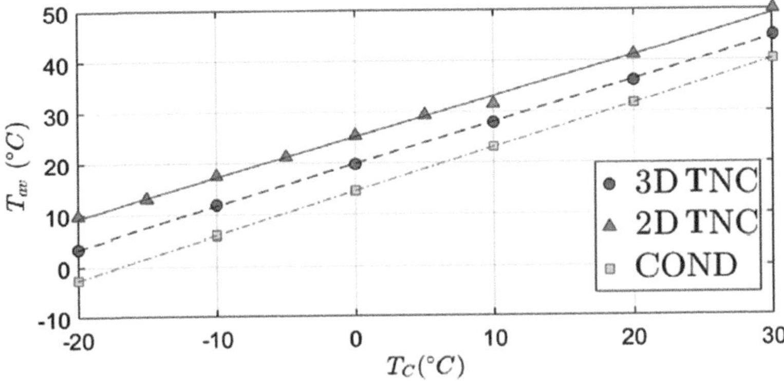

Figure 7.16 Mean temperature as a function of the temperature.

This might be attributed to the fact that turbulence in the 3D arrangement is projected to be significantly more substantial due to the huge dimensions of the computing domain, which would boost the exchange of heat and the thermal loads that the turbine cooling structure would combat. The mean temperature seems to be lower in the 3D case due to the massive amount of air within the nacelle in this situation.

7.6 CONCLUSION

Wind energy is a promising source of renewable energy that can contribute significantly to the world's growing energy demands while reducing the impacts of climate change. The Sahara desert in Algeria presents a unique opportunity for wind energy production, with high wind speeds and vast uninhabited land. However, wind energy production in desert climates presents challenges, such as the effect of sand particles and excessive ambient air temperature on wind turbine performance, which must be addressed. Ongoing research at LGMD is helping to advance our understanding of wind turbine aerodynamics in this challenging environment and facilitate innovative solutions for wind energy production.

Recent studies have investigated the factors that influence wind turbine aerodynamics in the Sahara desert, including sand particles and ambient air temperature. These studies have provided valuable insights into the performance of wind turbines in this region, which is essential for optimizing wind energy production. The potential of the Sahara desert as a source of wind energy is significant, but there are unique challenges and limitations associated with wind energy production in desert climates. Therefore, it is crucial to continue research efforts to fully understand and optimize wind turbine aerodynamics in desert environments.

Recent, modest studies by LGMD have made significant contributions to the understanding of wind turbine aerodynamics, including the prediction of performance, near-wake studies, farm optimization, and more. Studies investigating the effects of sand on wind turbine aerodynamics have characterized sand particle size and distribution, and have developed accurate numerical methods for predicting the effect of sand on the aerodynamic performance of wind turbines. Similarly, studies examining the impact of temperature on wind turbine aerodynamics have explored methods for predicting this effect, such as the use of cooling systems and the optimization of blade shape and surface area.

In conclusion, the presence of sand particles in the air has a significant impact on the performance of horizontal-axis wind turbines. The power loss increases with increasing particle diameter and concentration. Particle size and concentration should be carefully considered in wind turbine design to minimize power loss. Further research is needed to develop effective mitigation measures to reduce the impact of sand particles on wind turbine performance in sandy environments.

This chapter also highlights the significant impact of roughness heights on the performance of wind turbines, especially in separated flow conditions. Therefore, it is crucial to consider the effects of roughness when designing wind turbine blades to improve their performance and efficiency.

This chapter has five specific objectives. First, it aims to provide insights into recent studies conducted by LGMD for the aerodynamics of wind turbines, including prediction of performances, near-wake studies, farm optimization, and more. Second, the chapter highlights the potential of the Sahara desert in Algeria as a source of wind energy and the unique challenges and limitations associated with wind energy production in desert climates. Third, recent studies that have investigated the effects of sand on wind turbine aerodynamics, including the characterization of sand particle size and distribution, and the development of accurate numerical methods for predicting the effect of sand on the aerodynamic performance of wind turbines, are summarized. Fourth, the chapter examines the impact of temperature on wind turbine aerodynamics and explores methods for predicting this effect, such as the use of cooling systems and the optimization of blade shape and surface area. Finally, the chapter highlights the need for further studies to fully understand and optimize wind turbine aerodynamics in desert climates and identifies advanced methods, such as computational fluid dynamics (CFD) simulations and field experiments, that can be used to study the performance of wind turbines in desert climates.

REFERENCES

1. A. Bouhelal, *Contribution to the aerodynamic study of the air-sand flow around a wind turbine blade* [PhD thesis]. Ecole Nationale Polytechnique, 2018.

2. C. Masson, A. Smaïli, C. Leclerc, "Aerodynamic analysis of HAWTs operating in unsteady conditions", *Wind Energy: An International Journal for Progress and Applications in Wind Power Conversion Technology*, vol. 4, no. 1, pp. 1–22, 2001.

3. A. Bouhelal, A. Smaili, O. Guerri, C. Masson, "Comparison of BEM and full navier-stokes CFD methods for prediction of aerodynamics performance of HAWT rotors", *Conference Comparison of BEM and full Navier-Stokes CFD methods for prediction of aerodynamics performance of HAWT rotors*. IEEE, pp. 1–6.

4. M. N. Hamlaoui, A. Smaili, H. Fellouah, "A review of stall delay models and their application on hybrid methods", *ENP Engineering Science Journal*, vol. 2, no. 2, pp. 1–6, 2022.

5. A. Bouhelal, A. Ladjal, A. Smaili, "Blade element momentum theory coupled with machine learning to predict wind turbine aerodynamic performances", *Conference Blade Element Momentum Theory Coupled with Machine Learning to Predict Wind Turbine Aerodynamic Performances*. p. 1153.

6. A. Bouhelal, A. Smaili, O. Guerri, C. Masson, "Numerical investigation of turbulent flow around a recent horizontal axis wind Turbine using low and high reynolds models",. *Journal of Applied Fluid Mechanics*, vol. 11, no. 1, pp. 151–164, 2018.

7. M. Tata, F. Massouh, I. Dobrev, A. Smaili, "Experimental and numerical analysis of a scaled wind turbine near wake", *Conference Experimental and numerical analysis of a scaled wind turbine near wake*. IEEE, pp. 1–6.

8. M. Hamlaoui, A. Smaili, I. Dobrev, M. Pereira, H. Fellouah, S. Khelladi, "Numerical and experimental investigations of HAWT near wake predictions using particle image velocimetry and actuator disk method", *Energy*, vol. 238, p. 121660, 2022.

9. A. Bouhelal, A. Smaili, O. Guerri, C. Masson, "Numerical investigations on the fluid behavior in the near wake of an experimental wind turbine model in the presence of the nacelle", *Journal of Applied Fluid Mechanics*, vol. 16, no. 1, pp. 21–33, 2022.

10. S. Zergane, A. Smaili, C. Masson, "Optimization of wind turbine placement in a wind farm using a new pseudo-random number generation method",. *Renewable Energy*, vol. 125, pp. 166–171, 2018.

11. A. Bouhelal, O. Guerri, A. Smaili, C. Masson, "Contribution to the aerodynamic study of the air-sand flow around a wind turbine blade installed in desert environment of algeria",. *Conference Contribution to the aerodynamic study of the air-sand flow around a wind turbine blade installed in desert environment of algeria*. IEEE, pp. 1–6.

12. A. Bouhelal, A. Smaïli, C. Masson, O. Guerri, "Effects of surface roughness on aerodynamic performance of horizontal axis wind turbines", *Conference Effects of surface roughness on aerodynamic performance of horizontal axis wind turbines*. pp. 18–21.

13. A. Smaili, A. Tahi, C. Massfon, "Thermal analysis of wind turbine nacelle operating in Algerian Saharan climate", *Energy Procedia*, vol. 18, pp. 187–196, 2012.

14. M. Mahdi, A. Smaili, Y. Saad, "Numerical investigations of turbulent natural convection heat transfer within a wind turbine nacelle operating in hot climate", *International Journal of Thermal Sciences*, vol. 147, p. 106143, 2020.

15. Candy Goyal, Jagpal Singh Ubhi, B. Raj, "Low leakage zero ground noise nanoscale full adder using source biasing technique", *Journal of Nanoelectronics and Optoelectronics*, American Scientific Publishers, vol. 14, pp. 360–370, Mar 2019.

16. Gurmohan Singh, R. K. Sarin, B. Raj, "A novel robust exclusive-OR function implementation in QCA nanotechnology with energy dissipation analysis", *Journal of Computational Electronics*, Springer, vol. 15, no. 2, pp. 455–465, Jun 2016.

17. Anjana Bhardwaj, Pradeep Kumar, B. Raj, Sunny Anand, "Design and performance optimization of doping-less vertical nanowire TFET using gate stack technique", *Journal of Electronic Materials (JEMS)*, Springer, vol. 41, no. 7, pp. 4005–4013, 2022.

18. Jeetendra Singh, B. Raj, "Tunnel current model of asymmetric MIM structure levying various image forces to analyze the characteristics of filamentary memristor", *Applied Physics A*, Springer, vol. 125, no. 3, p. 203.1, Feb 2019.

19. Abhishek Singh Tomar, Vijay Kumar Magraiya, B. Raj, "Scaling of access and data transistor for high performance DRAM cell design", *Quantum Matter*, vol. 2, pp. 412–416, Oct 2013.

20. Neeraj Jain, B. Raj, "Parasitic capacitance and resistance model development and optimization of raised source/drain SOI FinFET structure for analog circuit applications", *Journal of Nanoelectronics and Optoelectronins*, ASP, USA, vol. 13, pp. 531–539, Ap 2018.

21. Shradhya Singh, S. K. Vishvakarma, B. Raj, "Analytical modeling of split-gate junction-less transistor for a biosensor application", *Sensing and Bio-Sensing*, Elsevier, vol. 18, pp. 31–36, Apr 2018.

22. Maisagalla Gopal, Balwinder Raj, "Low power 8T SRAM cell design for high stability video applications", *ITSI Transaction on Electrical and Electronics Engineering*, vol. 1, no. 5, pp. 91–97, 2013.

23. Balwinder Raj, Jatin Mitra, Deepak Kumar Bihani, V. Rangharajan, A. K. Saxena, S. Dasgupta, "Analysis of noise margin, power and process variation for 32 nm FinFET based 6T SRAM cell", *Journal of Computer (JCP)*, Academy Publisher, Finland, vol. 5, no. 6, pp. 1–8, 2010.

24. Divya Sharma, Rajesh Mehra, B. Raj, "Comparative analysis of photovoltaic technologies for high efficiency solar cell design", *Superlattices and Microstructures*, Elsevier, vol. 153, p. 106861, May 2021.

25. Tanu Wadhera, Deepti Kakkar, Girish Wadhwa, B. Raj, "Recent advances and progress in development of the field effect transistor biosensor: A review", *Journal of Electronic Materials*, Springer, vol. 48, no. 12, pp. 7635–7646, Dec 2019.

26. Girish Wadhwa, Priyanka Kamboj, Balwinder Raj, "Design optimisation of junctionless TFET biosensor for high sensitivity", *Advances in Natural Sciences: Nanoscience and Nanotechnology*, vol. 10, p. 045001, 2019.

27. Priya Bansal, B. Raj, "Memristor modeling and analysis for linear dopant drift kinetics," *Journal of Nanoengineering and Nanomanufacturing*, American Scientific Publishers, vol. 6, pp. 1–7, 2016.

28. Amandeep Singh, Mamta Khosla, B. Raj, "Circuit compatible model for electrostatic doped schottky barrier CNTFET, "*Journal of Electronic Materials*, Springer, vol. 45, no. 12, pp. 4825–4835, 2016.

29. D. Vaithiyanathan, B. Raj, "Performance analysis of charge plasma induced graded channel si nanotube", *Journal of Engineering Research (JER)*, EMSME Special Issue, pp. 146–154, Aug 2021.

30. Manjit Kaur, Neena Gupta, Sanjeev Kumar, B. Raj, Arun Kumar Singh, "Comparative RF and crosstalk analysis of carbon based nano interconnects", *IET Circuits, Devices & Systems*, vol. 15, no. 6, pp. 493–503, Feb 2021.

31. Nehru Kandasamy, Firdous Ahmad, D. Ajitha, B. Raj, Nagarjuna Telagam, "Quantum dot cellular automata based scan flip flop and boundary scan register", *IETE Journal of Research*, vol. 66, pp. 535–548, 2020.

32. S. K. Sharma, B. Raj, M. Khosla, "Enhanced photosensivity of highly spectrum selective cylindrical gate In1-xGaxAs nanowire MOSFET photodetector", *Modern Physics Letter-B*, vol. 33, no. 12, p. 1950144, 2019.

33. Jeetendra Singh, B. Raj, "Design and investigation of 7T2M NVSARM with enhanced stability and temperature impact on store/restore energy", *IEEE Transactions on Very Large Scale Integration Systems*, vol. 27, no. 6, pp. 1322–1328, Jun 2019.

34. Anil Kumar Bhardwaj, Sumeet Gupta, B. Raj, Amandeep Singh, "Impact of double gate geometry on the performance of carbon nanotube field effect transistor structures for low power digital design", *Computational and Theoretical Nanoscience*, ASP, vol. 16, pp. 1813–1820, 2019.

35. Neeraj Jain, B. Raj, "Thermal stability analysis and performance exploration of asymmetrical dual-k underlap spacer (ADKUS) SOI FinFET for security and privacy applications", *Indian Journal of Pure & Applied Physics (IJPAP)*, vol. 57, pp. 352–360, May 2019.

36. Amandeep Singh, Mamta Khosla, B. Raj, "Design and analysis of dynamically configurable electrostatic doped carbon nanotube tunnel FET", *Microelectronics Journal*, Elesvier, vol. 85, pp. 17–24, Mar 2019.

37. Neeraj Jain, Balwinder Raj, "Dual-k spacer region variation at the drain side of asymmetric SOI FinFET structure: Performance analysis towards the analog/rf design applications", *Journal of Nanoelectronics and Optoelectronics*, American Scientific Publishers, vol. 14, pp. 349–359, Mar 2019.

38. Jeetendra Singh, Sanjeev Sharma, B. Raj, Mamta Khosla, "Analysis of barrier layer thickness on performance of In1-xGaxAs based gate stack cylindrical gate nanowire MOSFET," *JNO*, ASP, vol. 13, pp. 1473–1477, Oct 2018.

39. Neeraj Jain, B. Raj, "Analysis and performance exploration of high-k SOI FinFETs over the conventional low-k SOI FinFET toward analog/RF design", *Journal of Semiconductors (JoS)*, IOP Science, vol. 39, no. 12, p. 124002-1-7, Dec 2018.

40. Candy Goyal, Jagpal Singh Ubhi, B. Raj, "A reliable leakage reduction technique for approximate full adder with reduced ground bounce noise', *Journal of Mathematical Problems in Engineering*, Hindawi, vol. 2018, Article ID 3501041, p. 16, 15 Oct 2018.

41. Jeetendra Singh, B. Raj, Mamta Khosla, "Design and performance analysis of nano-scale memristor-based nonvolatile SRAM", *Journal of Sensor Letter*", American Scientific Publishers, vol. 16, pp. 798–805, Oct 2018.

42. Girish Wadhwa, B. Raj, "Parametric variation analysis of charge-plasma-based dielectric modulated JLTFET for biosensor application", *IEEE Sensor Journal*, vol. 18, no. 15, pp. 6070–6077, 2018.

43. Jeetendra Singh, B. Raj, "Comparative analysis of memristor models for memories design", *JoS*, IoP, vol. 39, no. 7, p. 074006-1-12, Jul 2018.

44. Pawandeep Kaur, Avtar Singh Buttar, B. Raj, "A comprehensive analysis of nanoscale transistor based biosensor: A review", *Indian Journal of Pure and Applied Physics*, vol. 59, pp. 304–318, Apr 2021.

45. Divya Yadav, Balwant Raj, B. Raj, "Design and simulation of low power microcontroller for IoT applications", *Journal of Sensor Letters*, ASP, vol. 18, pp. 401–409, May 2020.

46. Shailendra Singh, B. Raj,"A 2-D analytical surface potential and drain current modeling of double-gate vertical t-shaped tunnel FET", *Journal of Computational Electronics*, Springer, vol. 19, pp. 1154–1163, Apr 2020.

47. Jeetendra Singh, B. Raj, "An accurate and generic window function for nonlinear memristor model", *Journal of Computational Electronics*, Springer, vol. 18, no. 2, pp. 640–647, Jun 2019.

48. Divya Yadav, Shailesh Singh Chouhan, Santosh Kumar Vishvakarma, B. Raj, "Application specific microcontroller design for IoT based WSN", *Sensor Letter*, ASP, vol. 16, pp. 374–385, May 2018.

49. Gurmohan Singh, R. K. Sarin, B. Raj, "Fault-tolerant design and analysis of quantum-dot cellular automata based circuits", *IEEE/IET Circuits, Devices & Systems*, vol. 12, pp. 638–664, 2018.

50. Jeetendra Singh, B. Raj, "Modeling of mean barrier height levying various image forces of metal insulator metal structure to enhance the performance of conductive filament based memristor model", *IEEE Nanotechnology*, vol. 17, no. 2, pp. 268–267, Mar 2018.

51. Aakash Jain, Sanjeev Sharma, B. Raj, "Analysis of triple metal surrounding gate (TM-SG) III-V nanowire MOSFET for photosensing application", *Optoelectronics Journal*, Elsevier, vol. 26, no. 2, pp. 141–148, May 2018.

52. Aakash Jain, Sanjeev Sharma, B. Raj, "Design and analysis of high sensitivity photosensor using cylindrical surrounding gate MOSFET for low power sensor applications", *Engineering Science and Technology, an International Journal*, Elsevier, vol. 19, no. 4, pp. 1864–1870, Dec 2016.

53. Amandeep Singh, Mamta Khosla, B. Raj, "Analysis of electrostatic doped schottky barrier carbon nanotube FET for low power applications," *Journal of Materials Science: Materials in Electronics*, Springer, vol. 28, pp. 1762–1768, 2017.

54. G. Saiphani Kumar, Amandeep Singh, B. Raj, "Design and analysis of gate all around CNTFET based SRAM cell design", *Journal of Computational Electronics*, Springer, vol. 17, no. 1, pp. 138–145, Mar 2018.

55. Gurinder pal Singh, B. S. Sohi, Balwinder Raj, "Material properties analysis of graphene base transistor (GBT) for VLSI analog circuits", *Indian Journal of Pure & Applied Physics (IJPAP)*, vol. 55, pp. 896–902, Dec 2017.

56. Amandeep Singh, Mamta Khosla, B. Raj, "Comparative analysis of carbon nanotube field effect transistor and nanowire transistor for low power circuit design," *Journal of Nanoelectronics and Optoelectronics*, American Scientific Publishers, USA, vol. 11, pp. 388–393, Jun 2016.

57. Sunil Kumar, B. Raj, "Estimation of stability and performance metric for inward access transistor based 6T SRAM cell design using n-type/p-type DMDG-GDOV TFET", *IEEE VLSI Circuits and Systems Letter*, vol. 3, no. 2, pp. 25–39, Jun 2017.

58. Shashikant Sharma, Anjan Kumar, Manisha Pattanaik, B. Raj, "Forward body biased multimode multi-threshold CMOS technique for ground bounce noise reduction in static CMOS adders", *International Journal of Information and Electronics Engineering*, pp. 567–572, vol. 3, no. 3, 2013.

59. Hamendra Singh, Pankaj Kumar, Balwinder Raj, "Performance Analysis of Majority Gate SET Based 1-bit Full Adder", *International Journal of Computer and Communication Engineering* (IJCCE), IACSIT Press Singapore, ISSN: 2010-3743, Vol. 2, no. 4, 2013.

60. Anil Kumar Bhardwaj, Sumeet Gupta, B. Raj, "Investigation of parameters for schottky barrier (SB) height for schottky barrier based carbon nanotube field effect transistor device", *Journal of Nanoelectronics and Optoelectronics*, ASP, vol. 15, pp. 783–791, Jul 2020.

61. Priya Bansal, B. Raj, "Memristor: A versatile nonlinear model for dopant drift and boundary issues," *JCTN*, American Scientific Publishers, vol. 14, no. 5, pp. 2319–2325, May 2017.

62. Neeraj Jain, B. Raj, "An analog and digital design perspective comprehensive approach on Fin-FET (Fin-Field Effect transistor) technology: A review", *Reviews in Advanced Sciences and Engineering (RASE)*, ASP, vol. 5, pp. 1–14, 2016.

63. Sanjeev Sharma, B. Raj, Mamta Khosla, "Subthreshold performance of In1-xGaxAs based dual metal with gate stack cylindrical/surrounding gate nanowire MOSFET for low power analog applications", *Journal of Nanoelectronics and Optoelectronics*, American Scientific Publishers, USA, vol. 12, pp. 171–176, 2017.

64. Balwinder Raj, A. K. Saxena, S. Dasgupta, "Analytical modeling for the estimation of leakage current and subthreshold swing factor of nanoscale double gate FinFET device", *Microelectronics International, UK*, vol. 26, pp. 53–63, 2009.

65. Shailendra Singh, Girish Wadhwa, Balwinder Raj, "An analytical modeling for dual source vertical tunnel field effect transistor", *International Journal of Recent Technology and Engineering (IJRTE)*, vol. 8, no. 2, pp. 603–608, Jul 2019.

66. Shailendra Singh, B. Raj, "Design and analysis of hetrojunction vertical T-shaped tunnel field effect transistor", *Journal of Electronics Material*, Springer, vol. 48, no. 10, pp. 6253–6260, Oct 2019.

67. Candy Goyal, Jagpal Singh Ubhi, B. Raj, "A low leakage CNTFET based inexact full adder for low power image processing applications", *International Journal of Circuit Theory and Applications*, Wiley, vol. 47, no. 9, pp. 1446–1458, Sept 2019.

68. B. Raj, A. K. Saxena, S. Dasgupta, "A compact drain current and threshold voltage quantum mechanical analytical modeling for FinFETs", *Journal of Nanoelectronics and Optoelectronics (JNO)*, USA, vol. 3, no. 2, pp. 163–170, 2008.

69. Sonal Singh, Mamta Khosla, Girish Wadhwa, B. Raj, "Design and analysis of double-gate junctionless vertical TFET for gas sensing applications", *Applied Physics A*, Springer, vol. 127, no. 16, 2 Jan 2021.

70. Inderjit Singh, B. Raj, Mamta Khosla, B. Rajesh Kumar Kaushik, "Potential MRAM technologies for low power SoCs", *SPIN World Scientific Publisher*, SCIE; vol. 10, no. 4, p. 2050027, Dec 2020.

71. Shailendra Singh, B. Raj, "Parametric variation analysis on hetero-junction Vertical t-shape TFET for supressing ambipolar conduction", *Indian Journal of Pure and Applied Physics*, vol. 58, pp. 478–485, Jun 2020.

72. Shailendra Singh, Girish Wadhwa, Balwinder Raj, "Design and analysis of dual source vertical tunnel field effect transistor for high performance", *Transactions on Electrical and Electronics Materials*, Springer, vol. 21, pp. 74–82, Oct 2019.

73. Manjit Kaur, Neena Gupta, Sanjeev Kumar, B. Raj, Arun Kumar Singh, "RF performance analysis of intercalated graphene nanoribbon based global level interconnects", *Journal of Computational Electronics*, Springer, vol. 19, pp. 1002–1013, Jun 2020.

74. Girish Wadhwa, B. Raj, "Design and performance analysis of junctionless TFET biosensor for high sensitivity", *IEEE Nanotechnology*, vol. 18, pp. 567–574, 2019.

75. Jeetendra Singh, B. Raj, "Enhanced nonlinear memristor model encapsulating stochastic dopant drift", *JNO, ASP*, vol. 14, pp. 958–963, 2019.

76. Girish Wadhwa, B. Raj, "An analytical modeling of charge plasma based tunnel field effect transistor with impacts of gate underlap region", *Superlattices and Microstructures*, Elsevier, vol. 142, p. 106512, Jun 2020.

77. Shailendra Singh, B. Raj, "Modeling and simulation analysis of SiGe hetrojunction double GateVertical t-shaped tunnel FET", *Superlattices and Microstructures*, Elsevier vol. 142, p. 106496, Jun 2020.

78. Amandeep Singh, Dinesh Kumar Saini, Dinesh Agarwal, Sajal Aggarwal, Mamta Khosla, B. Raj, "Modeling and simulation of carbon nanotube field effect transistor and its circuit application," *Journal of Semiconductors (JoS)*, IOP Science, vol. 37, p. 074001-6, Jul 2016.

79. Neeraj Jain, Balwinder Raj, "Device and circuit co-design perspective comprehensive approach on FinFET technology: A review", *Journal of Electron Devices*, vol. 23, no. 1, pp. 1890–1901, 2016.

80. Sunil Kumar, B. Raj, "Analysis of I_{ON} and ambipolar current for dual-material gate-drain overlapped DG-TFET," *Journal of Nanoelectronics and Optoelectronics*, American Scientific Publishers, USA, vol. 11, pp. 323–333, Jun 2016.

81. Naveed Anjum, Tarun Bali, B. Raj, "Design and simulation of handwritten multiscript character recognition", *International Journal of Advanced Research in Computer and Communication Engineering*, vol. 2, no. 7, pp. 2544–2549, Jul 2013.

82. Sanjeev Sharma, B. Raj, Mamta Khosla, "A gaussian approach for analytical subthreshold current model of cylindrical nanowire FET with quantum mechanical effects", *Microelectronics Journal*, Elsevier, vol. 53, pp. 65–72, Apr 2016.

83. Karmjit Singh, B. Raj, "Performance and analysis of temperature dependent multi-walled carbon nanotubes as global interconnects at different technology nodes," *Journal of Computational Electronics*, Springer, vol. 14, no. 2, pp. 469–476, Jun 2015.

84. Sunil Kumar, B. Raj, "Compact channel potential analytical modeling of DG-TFET based on evanescent–mode approach," *Journal of Computational Electronics*, Springer, vol. 14, no. 2, pp. 820–827, Jul 2015.

85. Karmjit Singh, B. Raj, "Temperature dependent modeling and performance evaluation of multi-walled CNT and single-walled CNT as global

interconnects," *Journal of Electronic Materials*, Springer, vol. 44, no. 12, pp. 4825–4835, Dec 2015.

86. V. K. Sharma, M. Pattanaik, B.Raj, "INDEP approach for leakage reduction in nanoscale CMOS circuits", *International Journal of Electronics*, Taylor & Francis, vol. 102, no. 2, pp. 200–215, 2014.

87. Karmjit Singh, B. Raj, "Influence of temperature on MWCNT bundle, SWCNT bundle and copper interconnects for nanoscaled technology nodes," *Journal of Materials Science: Materials in Electronics*, Springer, vol. 26, no. 8, pp. 6134–6142, 2015.

88. Naveed Anjum, Tarun Bali, B. Raj, "Design and simulation of handwritten gurumukhi and devanagri numerical recognition", *International Journal of Computer Applications*, Foundation of Computer Science, New York, USA, vol. 73, no. 12, pp. 16–21, 2013.

89. S. Khandelwal, V. Gupta, B. Raj, R. D. "Gupta, process variability aware low leakage reliable nano scale DG-FinFET SRAM cell design technique", *Journal of Nanoelectronics and Optoelectronics*, vol. 10, no. 6, pp. 810–817, Dec 2015.

90. V. K. Sharma, M. Pattanaik, B. Raj, "ONOFIC approach: Low power high speed nanoscale VLSI circuits design", *International Journal of Electronics*, Taylor & Francis, vol. 101, no. 1, pp. 61–73, 2014.

91. S. Khandelwal, Balwinder Raj, R. D. Gupta, "FinFET based 6T SRAM cell design: Analysis of performance metric, process variation and temperature effect", *Journal of Computational and Theoretical Nanoscience*, ASP, USA, vol. 12, pp. 2500–2506, 2015.

92. Sumit Singh, Shekhar Yadav, Jagdeep Rahul, Anurag Srivastava; B. Raj, "Impact of HfO$_2$ in graded channel dual insulator double gate MOSFET", *Journal of Computational and Theoretical Nanoscience*, American Scientific Publishers, vol. 12, no. 6, pp. 950–953, Apr 2015.

93. Vijay Kumar Sharma, Manisha Pattanaik, B. Raj, "PVT variations aware low leakage INDEP approach for nanoscale CMOS circuits", *Microelectronics Reliability*, Elsevier, vol. 54, pp. 90–99, 2014.

94. B. Raj, A. K. Saxena, S. Dasgupta, "Quantum mechanical analytical modeling of nanoscale DG FinFET: Evaluation of potential, threshold voltage and source/drain resistance", *Elsevier's Journal of Material Science in Semiconductor Processing*, Elsevier, vol. 16, no. 4, pp. 1131–1137, 2013.

95. Maisagalla Gopal, Siva Sankar D Prasad, Balwinder Raj, "8T SRAM cell design for dynamic and leakage power reduction", *International Journal of Computer Applications*, Foundation of Computer Science, New York, USA, vol. 71, no. 9, pp. 43–48, Jun 2013.

96. Manisha Pattanaik, B. Raj, Shashikant Sharma, Anjan Kumar, "Diode based trimode multi-threshold CMOS technique for ground bounce noise reduction in static CMOS adders", *Advanced Materials Research*, Trans Tech Publications, Switzerland, vol. 548, pp. 885–889, 2012.

97. Balwinder Raj, A. K. Saxena, S. Dasgupta, "Nanoscale FinFET based SRAM cell design: Analysis of performance metric, process variation, underlapped FinFET and temperature effect", *IEEE Circuits and System Magazine*, vol. 11, no. 2, pp. 38–50, 2011.

98. V. K. Sharma, M. Pattanaik, Balwinder Raj, "Leakage current ONOFIC approach for deep submicron VLSI circuit design", *International Journal of*

Electrical, Computer, Electronics and Communication Engineering, World Academy of Sciences, Engineering and Technology, vol. 7, no. 4, pp. 239–244, 2013.

99. Tulika Chawla, Mamta Khosla, B. Raj, "Design and simulation of triple metal double-gate germanium on insulator vertical tunnel field effect transistor", *Microelectronics Journal*, Elsevier, vol. 114, p. 105125, Aug 2021.

100. Parminder Kaur, Sandeep Singh Gill, B. Raj, "Comparative analysis of OFETs materials and devices for sensor applications", *Journal of Silicon*, Springer, vol. 14, pp. 4463–4471, 2022.

101. Sanjeev Kumar Sharma, Parveen Kumar, Balwant Raj, B. Raj, "In$_{1-x}$Ga$_x$As double metal gate-stacking cylindrical nanowire MOSFET for highly sensitive photo detector", *Journal of Silicon*, Springer, vol. 14, pp. 3535–3541, 2022.

102. B. Raj, A. K. Saxena, S. Dasgupta, "Analytical modeling of quasi planar nanoscale double gate FinFET with source/drain resistance and field dependent carrier mobility: A quantum mechanical study", *Journal of Computer (JCP)*, Academy Publisher, Finland, vol. 4, no. 9, pp. 1–8, 2009.

103. S. Bhushan, S. Khandelwal, B. Raj, "Analyzing different mode FinFET based memory cell at different power supply for leakage reduction", *Seventh International Conference on Bio-Inspired Computing: Theories and Application, (BIC-TA 2012) Advances in Intelligent Systems and Computing*, vol. 202, pp. 89–100, 2013.

104. Jeetendra Singh, B. Raj, "Temperature dependent analytical modeling and simulations of nanoscale memristor", *Journal: Engineering Science and Technology, an International Journal*, Elsevier, vol. 21, pp. 862–868, Oct 2018.

105. Shradhya Singh, Shashi Bala, Balwant Raj, B. Raj, "Improved sensitivity of dielectric modulated junctionless transistor for nanoscale biosensor design", *Sensor Letter*, ASP, vol. 18, pp. 328–333, Apr 2020.

106. Vivek Kumar, Santosh Kumar Vishvakarma, B. Raj, "Design and performance analysis of ASIC for IoT applications", *Sensor Letter*, ASP, vol. 18, pp. 31–38, Jan 2020.

107. Akanksha Jaiswal, R. K. Sarin, B. Raj, Shikha Sukhija, "A novel circular slotted microstrip-fed patch antenna with three triangle shape defected ground structure for multiband applications", *Advanced Electromagnetic (AEM)*, vol. 7, no. 3, pp. 56–63, Aug 2018.

108. Girish Wadhwa, B. Raj, "Label free detection of biomolecules using charge-plasma-based gate underlap dielectric modulated junctionless TFET", *Journal of Electronic Materials (JEMS)*, Springer, vol. 47, no. 8, pp. 4683–4693, Aug 2018.

109. Gurmohan Singh, R. K. Sarin, B. Raj, "**Design and performance analysis of a new efficient coplanar quantum-dot cellular automata adder**", *Indian Journal of Pure & Applied Physics (IJPAP)*, vol. 55, pp. 97–103, Feb 2017.

110. Amandeep Singh, Mamta Khosla, Balwinder Raj, "Design and analysis of electrostatic doped schottky barrier CNTFET based low power SRAM," *International Journal of Electronics and Communications, (AEÜ)*, Elsevier, vol. 80, pp. 67–72, 2017.

111. Neeraj Jain, B. Raj, "Impact of underlap spacer region variation on electrostatic and analog/RF performance of symmetrical high-k SOI FinFET at 20

nm channel length", *Journal of Semiconductors (JoS)*, IOP Science, vol. 38, no. 12, p. 122002, Dec 2017.

112. Shailendra Singh, B. Raj, "Analytical modeling and simulation analysis of T-shaped III-V heterojunction Vertical T-FET", *Superlattices and Microstructures*, Elsevier, vol. 147, p. 106717, Nov 2020.

113. Gurmohan Singh, R. K. Sarin, B. Raj, "Design and analysis of area efficient QCA based reversible logic gates", *Journal of Microprocessors and Microsystems*, Elsevier, vol. 52, pp. 59–68, May 2017.

114. Amandeep Singh, Mamta Khosla, B. Raj, "Compact model for ballistic single wall CNTFET under quantum capacitance limit," *Journal of Semiconductors (JoS)*, IOP Science, vol. 37, p. 104001-8, Oct 2016.

115. Parminder Kaur, Vikas Pandey, B. Raj, "Comparative study of efficient design, control and monitoring of solar power using IoT", *Sensor Letter*, ASP vol. 18, pp. 419–426, May 2020.

116. Anil Kumar Bhardwaj, Sumeet Gupta, B. Raj, "Development & analysis of compact model for double gate schottky barrier CNTFET", *Journal of Nanoelectronics and Optoelectronics*, ASP, vol. 15, pp. 1199–1208, Aug 2020.

117. Girish Wadhwa, Priyanka Kamboj, Jeetendra Singh, B. Raj, "Design and investigation of junctionless DGTFET for biological molecule recognition", *Transactions on Electrical and Electronic Materials*, Springer, vol. 22, pp. 282–289, 2021.

118. Tulika Chawla, Mamta Khosla, B. Raj, "Optimization of double-gate dual material GeOI-vertical TFET for VLSI circuit design", *IEEE VLSI Circuits and Systems Letter*, vol. 6, no. 2, pp. 13–25, Aug 2020.

119. Sachin Kumar Verma, Shailendra Singh, Girish Wadhwa, B. Raj, "Detection of biomolecules using charge-plasma based gate underlap dielectric modulated dopingless TFET", *Transactions on Electrical and Electronic Materials (TEEM)*, Springer, vol. 21, pp. 528–535, Jun 2020.

Chapter 8

Fuel cells

Raj Kumar Saini, Amita Verma, and Piush Verma

8.1 INTRODUCTION

A device that transforms the chemical energy of a fuel (often hydrogen) and an oxidant (often oxygen) into electrical energy through two redox reactions is called a fuel cell. Fuel cells differ from most batteries because they require a continuous supply of fuel and oxygen (often from the air) to maintain the chemical reaction. Batteries, on the other hand, usually have chemical energy stored in the materials inside them. Fuel cells can generate electricity as long as fuel and oxygen are present. The first fuel cell was created by Sir William Grove in 1838. Francis Thomas Bacon developed the hydrogen-oxygen fuel cell in 1932, which was the first fuel cell to be commercially used. This fuel cell is sometimes named after its inventor, a "Bacon fuel cell". Since the mid-1960s, NASA has used alkaline fuel cells to power satellites and spacecraft. Fuel cells have also been applied for various purposes, such as primary and backup power sources for commercial, industrial, and residential buildings and remote or harsh locations. Fuel-cell vehicles, such as forklifts, cars, buses, trains, boats, motorcycles, and submarines, also employ them [1].

8.2 TYPES OF FUEL CELLS

The electrolyte that a fuel cell uses is the basis of its classification. This determines the kind of electrochemical reactions that happen in the cell, the catalysts that are needed, the operating temperature range of the cell, the fuel required, and other factors. These characteristics affect the suitability of these cells for different applications. Various types of fuel cells are being developed, each with their own advantages, limitations, and potential uses. Different fuel-cell types are explained in more detail in the following sections [2].

DOI: 10.1201/9781003487692-8

8.3 ALKALINE FUEL CELLS

The alkaline fuel cell (AFC) was one of the first types of fuel cells to be studied in depth. NASA used it for the Gemini, Apollo, and Space Shuttle missions. Francis Thomas Bacon, who lived from 1904 to 1992, made the first fuel cell with alkali electrolyte in 1939. He used electrodes that let gas pass through and potassium hydroxide as the electrolyte. This was different from the earlier fuel-cell models that used solid electrodes and acid electrolytes. Bacon also used gases under pressure, like modern fuel cells do, to stop the electrolyte from filling up the electrodes. He thought that the fuel cell with alkali electrolyte would be a good power source for the Royal Navy submarines in World War II instead of the dangerous batteries they used then [3–9].

8.3.1 Chemical reaction

Alkaline fuel cells can produce power with an efficiency of up to 70% when they work in a temperature range from normal to about 250 °C. Figure 8.1 shows how an alkaline fuel cell works. The chemical reactions that happen in an AFC are shown in Figure 8.1.

$$Anode: 2H_2(g) + 4(OH) - (ag) \rightarrow 4H_2O(l) + 4e-$$

$$Cathode: O_2(g) + 2H_2O(l) + 4e- \rightarrow 4(OH) - (ag)$$

$$Overall: 2H_2(g) + O_2(g) \rightarrow 2H_2O(l)$$

The type of electrolyte determines how oxygen reacts at the cathode to form either a hydroxide (OH-) or a carbonate ion (CO32-). The ion travels through the electrolyte to the anode, where it combines with hydrogen. AFCs use less expensive materials than other types of fuel cells [10–16].

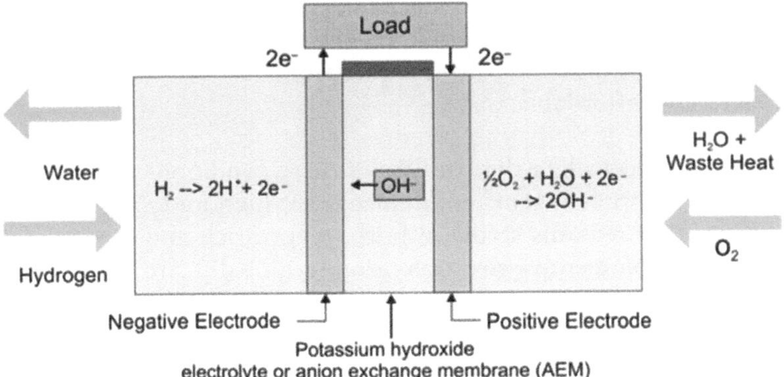

Figure 8.1 Chemical reactions in an alkaline fuel cell [3].

Previously, liquid KOH or a matrix with KOH was used as the electrolyte, while the catalyst layer could use platinum or other metals like nickel. A challenge with AFCs is that they require pure oxygen and hydrogen to be supplied to the fuel cell because they are sensitive to small amounts of carbon dioxide in the air [17–23].

8.3.2 Electrolyte layer

The electrolyte layer plays an essential role in the functioning of a fuel cell. As the fuel traverses from the fuel cell to the catalyst layer in low-temperature fuel cells, it disintegrates into protons (H+) and electrons. The electrons journey through the external circuit to energize the load, while the hydrogen protons (ions) traverse through the electrolyte until they reach the cathode, where they combine with oxygen to form water. In AFCs, similar reactions occur but in reverse order. Here, oxygen forms hydroxide first, which then combines with hydrogen at the anode [24–31]. Typically, AFCs employ an electrolyte composed of alkaline potassium hydroxide mixed with water and held in a matrix, with molarities ranging between 6 and 9. It's crucial to maintain the purity of this solution to prevent contamination of the catalyst. The electrolyte can be either mobile or stationary. In the case of a mobile electrolyte, it is circulated through the cells, facilitating waste heat and water removal from the fuel cell. The flow channels in these cells are typically large, approximately 2 to 3 mm wide, to enable rapid flow. The increased thickness of the electrolyte leads to ohmic polarization, a significant factor to consider when designing AFCs [28–32].

The electrolyte must comply with the following specifications, just like with other types of fuel cells:

- Ionic conductivity should be high.
- It should be stable in both mechanical and chemical terms.
- Electronic conductivity should be low.
- Simplicity of manufacture and accessibility.
- Preferably affordable.

Identifying an electrolyte that fulfills all these conditions can be challenging. The two most stringent requirements are high ionic conductivity and a substance that remains stable in both oxygen-rich and oxygen-depleted environments. Numerous research groups globally are exploring ways to address these constraints. One potential solution is the use of anion exchange membranes (AEMs) as solid electrolyte layers, as an alternative to liquid or matrix electrolytes. This approach is gaining traction due to its potential to overcome some of the limitations associated with traditional electrolytes [39–47].

8.3.3 AFC electrodes

Electrodes in AFCs can be either hydrophilic or hydrophobic. Hydrophilic electrodes are typically constructed from metallic materials such as nickel and nickel-based alloys. On the other hand, hydrophobic electrodes are composed of carbon combined with polytetrafluoroethylene (PTFE). These electrodes consist of multiple layers with varying porosities, which house the liquid electrolyte, fuel, and oxidant [48–55]. AFCs can utilize both precious and non-precious metal catalysts. The most frequently used non-precious metal catalysts include silver-based powders for the cathodes, with a loading of 1.5 to 2 mg Ag/cm^2, and Raney nickel for the anodes, with a loading of 120 mg Ni/cm^2.

8.3.4 Sintered nickel powder

The design of the alkaline fuel cell was intended to utilize costly metal catalysts and inexpensive components. The electrodes were fabricated from porous nickel powder that was thermally processed into a rigid form. This nickel comprised two layers of nickel powders of varying sizes to ensure optimal contact between the gas, liquid, and solid phases of the reaction [56–63].

8.3.5 Raney metals

Raney metals are highly active and porous materials, produced by combining an active metal (such as nickel) with an inactive metal (like aluminum). The combination process is executed in such a manner that the two metals retain their individual areas and properties, rather than forming an alloy. Subsequently, the mixture undergoes treatment with a potent base to eliminate the aluminum. This results in a porous substance with a large surface area. The pore size can be manipulated by adjusting the degree of mixing between the two metals [3].

8.4 PHOSPHORIC ACID FUEL CELLS (PAFCS)

Phosphoric acid fuel cells (PAFCs) operate at temperatures ranging from 150 to 200 °C (approximately 300 to 400 °F). As the name implies, PAFCs utilize phosphoric acid as an electrolyte. Hydrogen ions (protons) travel from the anode to the cathode through this electrolyte [64–71].

Electrons generated at the anode traverse an external circuit, producing electricity along the way, before returning to the cathode. Here, electrons, hydrogen ions, and oxygen combine to form water, which is subsequently expelled from the cell. The reactions are expedited by a platinum catalyst present at the electrodes. This type of fuel cell is commonly used for powering commercial buildings and is the most commercially developed. It employs concentrated phosphoric acid (H3PO4) as an electrolyte and operates at a medium temperature. PAFCs consist of two porous conducting electrodes, typically made of nickel, for charge collection. An electrode requires a catalyst to accelerate the electrochemical reaction, which is inherently slow. The electrodes are coated with a thin layer of nickel, platinum, and silver on the exterior, which act as catalysts. Nickel is often used due to its cost-effectiveness [4, 5].

Figure 8.2 depicts a schematic representation of a phosphoric acid fuel cell. The fuel is either pure hydrogen or rich hydrogen gas, and the oxidant is either oxygen or air. The following equation depicts the electrochemical reactions inside a PAFC.

$$Anode: H_2 \rightarrow 2H^+ + 2e^-$$

$$Cathode: \frac{1}{2}O_2 + 2H^+ + 2e^- \rightarrow H_2O$$

8.4.1 Advantages and disadvantages

PAFCs have a higher electrical efficiency of about 80% when they use the energy from waste heat. PAFCs are used for commercial electricity production. The PAFC can handle a carbon monoxide (CO) level of about 1.5%, which is higher than other types of fuel cells. The phosphoric acid electrolyte in the PAFC can also work above the boiling point of water, which is a good thing. The PAFC has the problem of making less power than other fuel cells with the same size and weight. The PAFC is also expensive

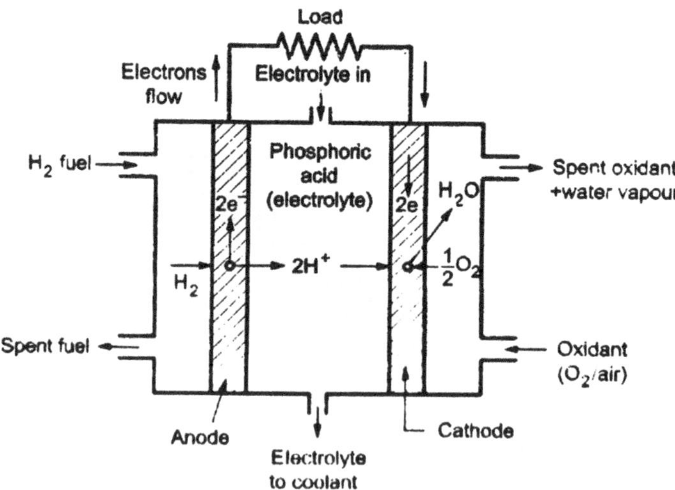

Figure 8.2 Phosphoric acid fuel cells [5].

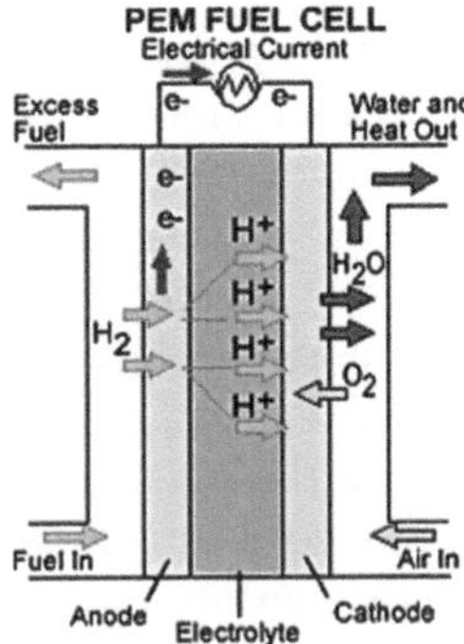

Figure 8.3 Polymer electrolyte membrane fuel cells [8].

because it needs a platinum catalyst and materials that can resist corrosion (because of the acid).

8.4.2 Applications

- Large vehicles like buses are also using PAFC, which has been used for stationary power generators with output in the 100 kW to 400 kW range.
- The Defence Research and Development Organization (DRDO) of India has created air-independent propulsion based on PAFC for use in their Kalvari-class submarines [6].

8.5 POLYMER ELECTROLYTE MEMBRANE FUEL CELLS (PEM) OR PROTON EXCHANGE MEMBRANE FUEL CELLS (PEMFCS)

Proton exchange membrane (PEM) fuel cells, also known as polymer electrolyte membrane fuel cells, are recognized for their high-power density and compact, lightweight design compared to other types of fuel cells [72–78]. They employ porous carbon electrodes infused with platinum or platinum alloy catalysts and a solid polymer functioning as an electrolyte. Their operation requires only water, atmospheric oxygen, and hydrogen. Typically, they utilize pure hydrogen sourced from reformers or storage tanks [79–84].

PEM fuel cells operate at relatively low temperatures, typically around 80 °C (176 °F). This low-temperature operation allows for a quick start-up (reducing warm-up time) and results in less wear on system components, thereby enhancing their lifespan. However, this also necessitates the use of a noble-metal catalyst (commonly platinum) to facilitate the separation of electrons and protons in hydrogen, which increases the system's cost. Furthermore, if the hydrogen is derived from a hydrocarbon fuel, an additional reactor is required to eliminate carbon monoxide from the fuel gas due to the platinum catalyst's high sensitivity to carbon monoxide poisoning. This additional reactor also contributes to the overall cost [7].

PEMFCs are devices that use hydrogen and oxygen to produce electricity and heat. They have two parts: an anode and a cathode, separated by a thin layer of plastic. Hydrogen gas goes into the anode, where it is split into positive hydrogen atoms (H+) and negative electrons (e). The positive hydrogen atoms can pass through the plastic layer, but the negative electrons cannot. The electrons have to go around the plastic layer through a wire, creating an electric current. Oxygen gas goes into the cathode, where it meets the electrons and the positive hydrogen atoms. They combine to form water (H_2O), which is the only waste product of this process [8].

Chemical reaction

$$2H2 \rightarrow 4H+ + 4e-$$

$$O2 + 4e- + 4H+ \rightarrow 2H2O$$

8.5.1 Advantages

Some of the advantages of PEMFC are that it can start working quickly, uses different kinds of hydrogen-based fuels, has a small and light design, costs less, and has a stable plastic layer. It can work with pure hydrogen, methanol, or formic acid as fuel [9].

8.5.2 Disadvantages

One of the problems with using hydrogen as a fuel is that it has to be stored in high-pressure tanks as a gas. This makes it hard to fit enough hydrogen in a car to travel long distances. Compared to cars that run on petrol or diesel, hydrogen cars can only go up to 300 miles before they need to refuel [9].

8.5.3 Applications

PEM fuel cells are devices that use hydrogen and oxygen to make electricity and heat. They can be used in different places and for different purposes. PEM fuel cells are very good for cars, trucks, and buses because they are clean, efficient, and quiet [7].

8.6 DIRECT METHANOL FUEL CELLS (DMFCS)

A direct methanol fuel cell (DMFC) is a device that turns the chemical energy of liquid methanol into electricity without any extra steps or moving parts, making it a good source of power. The DMFC fuel cells are small but strong and last for a long time [85–91]. They are the best in the market for fuel-cell systems and work without being connected to the grid or for mobile uses like important communication, IT, optronics, sensors, extra power, and more. These fuel cells use DMFC liquid fuel technology, which changes methanol into electric current directly. Using fuel cartridges, liquid methanol can be easily carried anywhere. Details of fuel cartridges and how they work are given in Figure 8.4 and Figure 8.5 [10].

8.6.1 Advantages

- They are light, powerful, and can run for a long time.
- They save a lot of fuel and need less support.
- Methanol is an easy fuel to work with, keep, and move.
- They work well even at high altitudes.
- They can handle very cold or very hot weather.
- They can be watched and controlled from a distance [10].

8.6.2 Disadvantages

Some of the problems with direct methanol fuel cells are that the plastic layer can leak, carbon monoxide can damage the cathode, methanol is hard to break down at the anode, and the system is complex to design [10].

8.6.3 Applications

Some of the devices that can use direct methanol fuel cells are battery chargers, power for testing and training tools, and power for small tactical equipment [92–97]. There are devices that can make from 25 watts to 5

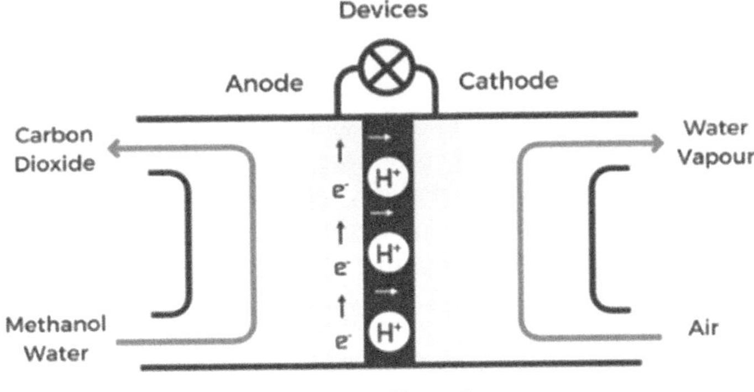

Figure 8.4 Direct methanol fuel cells [10].

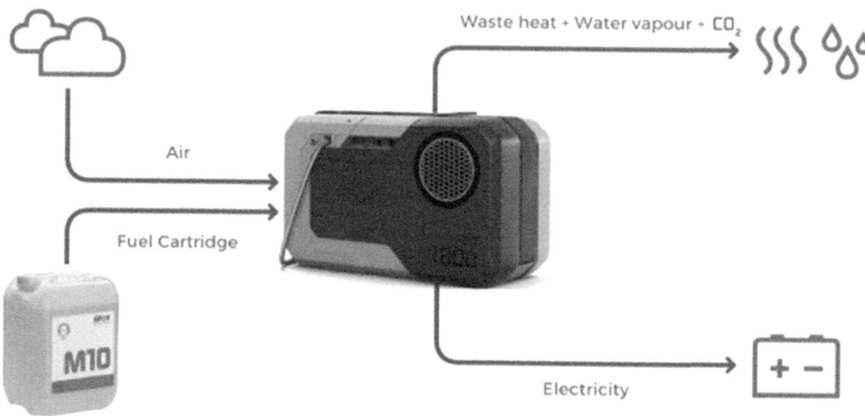

Figure 8.5 How methanol fuel cells work [10].

kilowatts of power and can run for up to 100 hours. The direct methanol fuel cell is very good at making up to 0.3 kilowatts of power. The indirect methanol fuel cell is better and cheaper at making more than 0.3 kilowatts of power. For the direct methanol fuel cell, the plastic layer can have issues if the liquid methanol-water mix in the device freezes at very low temperatures (unlike the indirect methanol fuel cell) [11].

8.7 MOLTEN-CARBONATE FUEL CELLS (MCFCS)

Molten-carbonate fuel cells (MCFCs) are fuel cells that work at very high temperatures, more than 600 °C. They can change other fuels into hydrogen by themselves, unlike many other types of fuel cells [98–104]. This is because they are hot enough to not need an extra device to make hydrogen. The material used in MCFCs is a mix of melted carbonate salts that is held in a solid ceramic layer and does not react with anything [12].

MCFCs work by moving carbonate ions from one part to another, as shown in Figure 8.6. The carbonate ions also act like an acid in the material. The gas that comes out of the anode has a lot of CO2 in it, and it goes to the cathode, where it changes into carbonate ions when it meets oxygen. The higher temperature of MCFCs makes them more efficient and able to use different kinds of fuels. But the higher temperature also makes them wear out faster and corrode more easily. In this fuel cell, the material is usually a mix of carbonates (Na, K, and Li) and is kept in a ceramic LiAlO2 layer [13].

8.7.1 Advantages

The DMFC is the most advanced fuel cell, and many people are interested in it because it has some special advantages [95–102]. It can use a lot of

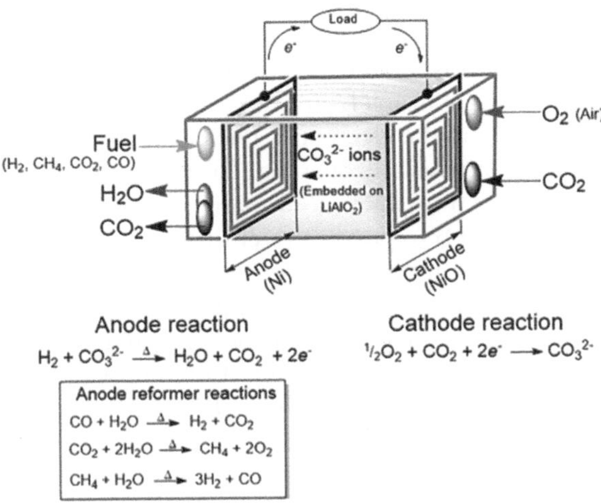

Figure 8.6 Molten-carbonate fuel cells and chemical reactions [15].

energy from the fuel, store the fuel easily as a liquid, work at a low temperature, have a simple system structure, and produce less pollution.

8.7.2 Disadvantages

Some of the problems with direct methanol fuel cells are that the methanol can leak from the anode to the cathode through the plastic layer, the carbon monoxide can harm the cathode, the methanol is difficult to split at the anode, and the system is hard to design [14].

8.7.3 Applications

MCFCs are fuel cells that work at very high temperatures. They are not cheap to make, and they need to be big to work well. This means that they are not good for small homes or businesses. But some places that need both electricity and heat, like hospitals, schools, and some big businesses, can use them. In South Korea, they have built some big MCFC power plants, like one that makes 59 MW of power and heat for a city called Hwaseong. In Europe, they are making MCFC units for boats. But the technology is still too expensive, and it needs to be made more cheaply in order to sell it to more people [12].

8.8 SOLID OXIDE FUEL CELLS (SOFCS)

Solid oxide fuel cells (SOFCs) are devices that use oxygen and fuel to make electricity and heat. They have three parts: an anode, a cathode, and a solid electrolyte in between. The anode gets fuel, and the cathode gets air. The parts are made of materials that have many holes in them, so the fuel and air can reach the electrolyte, and the waste products can leave. The electrolyte moves the oxygen atoms from the cathode to the anode, where they react with the fuel and make electrons. The electrons go through a wire to the cathode, creating an electric current. To make more power, many cells are connected together in a stack, and many stacks are put in a container to make a module. A SOFC power system has one or more modules that work with other parts to get fuel and air, use the heat, and change the electricity from DC to AC [16]. You can see how solid oxide fuel cells work in Figure 8.7. SOFCs are the best devices for changing chemical energy into electrical energy so far. They work at very high temperatures, so they need ceramic materials that can handle the heat [103–107]. The anode has nickel in it to make it better at conducting electrons and making reactions. The temperature can be different depending on the type of the fuel cell, from 600 to 1000 °C (first, second, and third, with lower temperatures for newer types). However, the high temperature can also break the ceramic parts if they change too fast. The fuels can be hydrogen and carbon monoxide, or other fuels such as diesel, natural gas, gasoline, and alcohol.

SOFCs are devices that use oxygen and fuel to make electricity and heat. They work like this:

The oxygen atoms get electrons from a wire on the cathode, which is a part that has many holes in it. The oxygen atoms become oxide atoms and go through the electrolyte, which is a solid part in between. The oxide atoms reach the anode, which is another part that has many holes in it and has fuel (hydrogen) in it. The oxide atoms react with the fuel, and make water and electrons. The electrons go back to the wire and make an electric current. The reaction also makes a lot of heat, which can be used by a

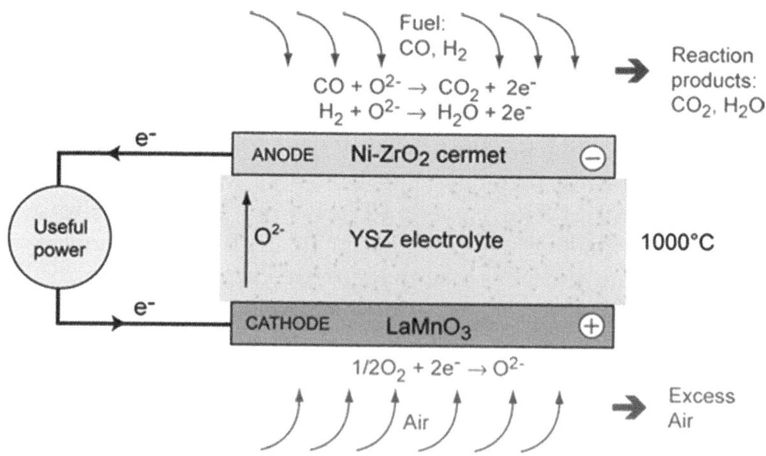

Figure 8.7 Process diagram of solid oxide fuel cell [17].

system that manages the heat. SOFCs need a long time to get hot enough to work, so they are best for places that need both heat and electricity: power plants that don't move and extra power supply. For things that move, they could use super-capacitor batteries that can start faster [17].

8.8.1 Advantages

This type of fuel cell has some good things like they can make a lot of heat and power, they can work for a long time, they can use different kinds of fuels, they make less pollution, and they are not too expensive. The main problem is that they need a lot of heat to work, so they take a long time to start, and they can have trouble with the parts fitting together or reacting with each other.

8.8.2 Disadvantages

SOFCs are fuel cells that work at very high temperatures. They have some other problems too. The parts that they use in SOFCs have to handle the heat. Some big problems that need to be fixed are their expense and the difficulty of making them.

8.8.3 Applications

Small-scale power generation, homogeneous charge compression ignition (IICCI), and gas turbine (GT) are some major examples of SOFC.

8.9 REVERSIBLE OR REGENERATIVE FUEL CELLS

A regenerative system is a system that can make and use hydrogen and oxygen with a fuel cell and an electrolyzer. A fuel cell is a device that makes electricity and water from hydrogen and oxygen. An electrolyzer is a device that makes hydrogen and oxygen from water and electricity. When there is a renewable energy source, like the sun or the wind, it gives electricity to the electrolyzer. The electrolyzer makes hydrogen and oxygen from water and stores them in the system. When there is no renewable energy source, the system uses the stored hydrogen and oxygen in the fuel cell. The fuel cell makes electricity and water from hydrogen and oxygen. This way, the system can always make electricity when it needs it. This system can help the normal grid, places that are far away from the grid, and planes that fly very high in the sky [18].

A regenerative hydrogen fuel cell is a device that uses water and electricity to make hydrogen and oxygen. A hydrogen-fueled proton exchange membrane fuel cell is a device that uses hydrogen and oxygen to make electricity and water. They are different but they can also work in reverse. When a fuel cell works in reverse, it changes the roles of the parts. The part that makes electricity becomes the part that makes hydrogen. The part that makes hydrogen becomes the part that makes electricity. The solid part in between moves the hydrogen atoms from one side to the other. When there is electricity from outside, the water on one side breaks into hydrogen and oxygen. The hydrogen atoms go to the other side and make hydrogen gas. The cell works differently in reverse than in normal. The chemical process in the reverse mode is shown by these reactions. You can see how regenerative fuel cells work in Figure 8.8.

Chemical reaction

$$At\, cathode: H_2O + 2e^- \rightarrow H_2 + O^{2-}$$
$$At\, anode: O^{2-} \rightarrow 1/2O_2 + 2e^-$$
$$Overall: H_2O \rightarrow 1/2O_2 + H_2$$

8.9.1 Advantages

Regenerative fuel cells are devices that can make and use hydrogen and oxygen to make electricity and water. They have some good things that make them better than batteries, for example, they can make more power, they can store more energy, they are light, they are cheap, they work well, they last long, and they don't harm the environment

8.9.2 Disadvantages

- They are expensive to make because they need a lot of catalysts (platinum) that are costly.
- There aren't enough places to get hydrogen.

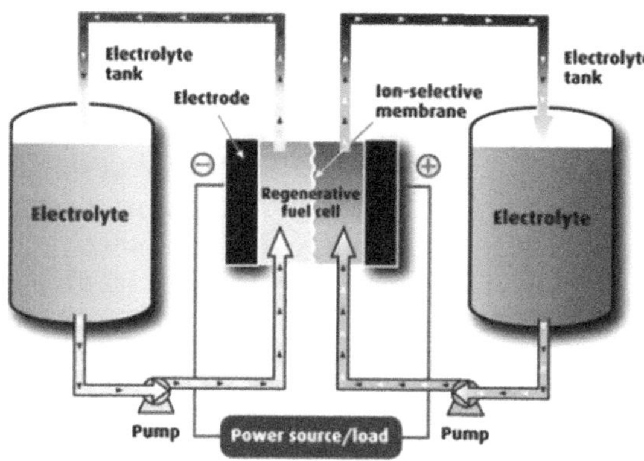

Figure 8.8 The process of regenerative fuel cells [19].

- Many of the fuel-cell technologies that are available now are still new and need to be tested more.
- Hydrogen is hard and costly to make and there is not much of it.

8.9.3 Applications

These cells are devices that can make a lot of power from hydrogen and oxygen. They can be used for many things, like making power that can move around, stay in one place, or spread out, as well as for buses, cars, planes, and electricity that can spread out [20].

8.10 CONCLUSION

Fuel cells are devices that can make power from hydrogen and oxygen. They can be used for many things, like moving things, and buildings, and storing power for the grid. They are better than burning things, which is what many power plants and cars do now. Fuel cells can make more power from the fuel and change the chemical energy in the fuel into electricity better than burning things. Fuel cells also make less pollution or no pollution at all. Hydrogen fuel cells produce only water as a byproduct, with no carbon dioxide emissions, making them environmentally friendly. They also do not release harmful pollutants that could affect human health. Additionally, fuel cells operate very quietly since they lack moving parts.

REFERENCES

1. R. R. Contreras, J. Almarza, L. Rincón, "Molten carbonate fuel cells: A technological perspective and review", *Energy Sources, Part A: Recovery, Utilization, and Environmental Effects*, pp. 1–15, 2021.
2. Z. Pu, G. Zhang, A. Hassanpour, D. Zheng, S. Wang, S. Liao, S. Sun, "Regenerative fuel cells: Recent progress, challenges, perspectives and their applications for space energy system". *Applied Energy*, vol. 283, p. 116376, 2021.
3. Candy Goyal, Jagpal Singh Ubhi, B. Raj, "Low leakage zero ground noise nanoscale full adder using source biasing technique", *Journal of Nanoelectronics and Optoelectronics*, American Scientific Publishers, vol. 14, pp. 360–370, Mar 2019.
4. Gurmohan Singh, R. K. Sarin, B. Raj, "A novel robust exclusive-OR function implementation in QCA nanotechnology with energy dissipation analysis", *Journal of Computational Electronics*, Springer, vol. 15, no. 2, pp. 455–465, Jun 2016.
5. Tanu Wadhera, Deepti Kakkar, Girish Wadhwa, B. Raj, "Recent advances and progress in development of the field effect transistor biosensor: A review", *Journal of Electronic Materials*, Springer, vol. 48, no. 12, pp. 7635–7646, Dec 2019.

6. Girish Wadhwa, Priyanka Kamboj, Balwinder Raj, "Design optimisation of junctionless TFET biosensor for high sensitivity", *Advances in Natural Sciences: Nanoscience and Nanotechnology*, vol. 10, p. 045001, 2019.

7. Priya Bansal, B. Raj, "Memristor modeling and analysis for linear dopant drift kinetics," *Journal of Nanoengineering and Nanomanufacturing*, American Scientific Publishers, vol. 6, pp. 1–7, 2016.

8. Anjana Bhardwaj, Pradeep Kumar, B. Raj, Sunny Anand, "Design and performance optimization of doping-less vertical nanowire TFET using gate stack technique", *Journal of Electronic Materials (JEMS)*, Springer, vol. 41, no. 7, pp. 4005–4013, 2022.

9. Jeetendra Singh, B. Raj, "Tunnel current model of asymmetric MIM structure levying various image forces to analyze the characteristics of filamentary memristor", *Applied Physics A*, Springer, vol. 125, no. 3, p. 203.1, Feb 2019.

10. Amandeep Singh, Mamta Khosla, B. Raj, "Circuit compatible model for electrostatic doped schottky barrier CNTFET, "*Journal of Electronic Materials*, Springer, vol. 45, no. 12, pp. 4825–4835, 2016.

11. D. Vaithiyanathan, B. Raj, "Performance analysis of charge plasma induced graded channel si nanotube", *Journal of Engineering Research (JER)*, EMSME Special Issue, pp. 146–154, Aug 2021.

12. Abhishek Singh Tomar, Vijay Kumar Magraiya, B. Raj, "Scaling of access and data transistor for high performance DRAM cell design", *Quantum Matter*, vol. 2, pp. 412–416, Oct 2013.

13. Neeraj Jain, B. Raj, "Parasitic capacitance and resistance model development and optimization of raised source/drain SOI FinFET structure for analog circuit applications", *Journal of Nanoelectronics and Optoelectronins*, ASP, USA, vol. 13, pp. 531–539, Ap 2018.

14. Shradhya Singh, S. K. Vishvakarma, B. Raj, "Analytical modeling of split-gate junction-less transistor for a biosensor application", *Sensing and Bio-Sensing*, Elsevier, vol. 18, pp. 31–36, Apr 2018.

15. Maisagalla Gopal, Balwinder Raj, "Low power 8T SRAM cell design for high stability video applications", *ITSI Transaction on Electrical and Electronics Engineering*, vol. 1, no. 5, pp. 91–97, 2013.

16. Balwinder Raj, Jatin Mitra, Deepak Kumar Bihani, V. Rangharajan, A. K. Saxena, S. Dasgupta, "Analysis of noise margin, power and process variation for 32 nm FinFET based 6T SRAM cell", *Journal of Computer (JCP)*, Academy Publisher, Finland, vol. 5, no. 6, pp. 1–8, 2010.

17. Divya Sharma, Rajesh Mehra, B. Raj, "Comparative analysis of photovoltaic technologies for high efficiency solar cell design", *Superlattices and Microstructures*, Elsevier, vol. 153, p. 106861, May 2021.

18. Pawandeep Kaur, Avtar Singh Buttar, B. Raj, "A comprehensive analysis of nanoscale transistor based biosensor: A review", *Indian Journal of Pure and Applied Physics*, vol. 59, pp. 304–318, Apr 2021.

19. Divya Yadav, Balwant Raj, B. Raj, "Design and simulation of low power microcontroller for IoT applications", *Journal of Sensor Letters*, ASP, vol. 18, pp. 401–409, May 2020.

20. Shailendra Singh, B. Raj,"A 2-D analytical surface potential and drain current modeling of double-gate vertical t-shaped tunnel FET", *Journal of Computational Electronics*, Springer, vol. 19, pp. 1154–1163, Apr 2020.

21. Jeetendra Singh, B. Raj, "An accurate and generic window function for non-linear memristor model", *Journal of Computational Electronics*, Springer, vol. 18, no. 2, pp. 640–647, Jun 2019.

22. Manjit Kaur, Neena Gupta, Sanjeev Kumar, B. Raj, Arun Kumar Singh, "Comparative RF and crosstalk analysis of carbon based nano interconnects", *IET Circuits, Devices & Systems*, vol. 15, no. 6, pp. 493–503, Feb 2021.

23. Nehru Kandasamy, Firdous Ahmad, D. Ajitha, B. Raj, Nagarjuna Telagam, "Quantum dot cellular automata based scan flip flop and boundary scan register", *IETE Journal of Research*, vol. 66, pp. 535–548, 2020.

24. S. K. Sharma, B. Raj, M. Khosla, "Enhanced photosensivity of highly spectrum selective cylindrical gate In1-xGaxAs nanowire MOSFET photodetector", *Modern Physics Letter-B*, vol. 33, no. 12, p. 1950144, 2019.

25. Jeetendra Singh, B. Raj, "Design and investigation of 7T2M NVSARM with enhanced stability and temperature impact on store/restore energy", *IEEE Transactions on Very Large Scale Integration Systems*, vol. 27, no. 6, pp. 1322–1328, Jun 2019.

26. Anil Kumar Bhardwaj, Sumeet Gupta, B. Raj, Amandeep Singh, "Impact of double gate geometry on the performance of carbon nanotube field effect transistor structures for low power digital design", *Computational and Theoretical Nanoscience*, ASP, vol. 16, pp. 1813–1820, 2019.

27. Neeraj Jain, B. Raj, "Thermal stability analysis and performance exploration of asymmetrical dual-k underlap spacer (ADKUS) SOI FinFET for security and privacy applications", *Indian Journal of Pure & Applied Physics (IJPAP)*, vol. 57, pp. 352–360, May 2019.

28. Amandeep Singh, Mamta Khosla, B. Raj, "Design and analysis of dynamically configurable electrostatic doped carbon nanotube tunnel FET", *Microelectronics Journal*, Elesvier, vol. 85, pp. 17–24, Mar 2019.

29. Neeraj Jain, Balwinder Raj, "Dual-k spacer region variation at the drain side of asymmetric SOI FinFET structure: Performance analysis towards the analog/rf design applications", *Journal of Nanoelectronics and Optoelectronics*, American Scientific Publishers, vol. 14, pp. 349–359, Mar 2019.

30. Jeetendra Singh, Sanjeev Sharma, B. Raj, Mamta Khosla, "Analysis of barrier layer thickness on performance of In1-xGaxAs based gate stack cylindrical gate nanowire MOSFET," *JNO*, ASP, vol. 13, pp. 1473–1477, Oct 2018.

31. Neeraj Jain, B. Raj, "Analysis and Performance Exploration of High-k SOI FinFETs Over the Conventional Low-k SOI FinFET toward Analog/RF Design", Journal of Semiconductors (JoS), IOP Science, vol. 39, no. 12, p. 124002-1-7, Dec 2018.

32. Candy Goyal, Jagpal Singh Ubhi, B. Raj, "A reliable leakage reduction technique for approximate full adder with reduced ground bounce noise', *Journal of Mathematical Problems in Engineering*, Hindawi, vol. 2018, Article ID 3501041, p. 16, 15 Oct 2018.

33. Jeetendra Singh, B. Raj, Mamta Khosla, "Design and performance analysis of nano-scale memristor-based nonvolatile SRAM", *Journal of Sensor Letter*", American Scientific Publishers, vol. 16, pp. 798–805, Oct 2018.

34. Girish Wadhwa, B. Raj, "Parametric variation analysis of charge-plasma-based dielectric modulated JLTFET for biosensor application", *IEEE Sensor Journal*, vol. 18, no. 15, pp. 6070–6077, 2018.

35. Jeetendra Singh, B. Raj, "Comparative analysis of memristor models for memories design", *JoS*, IoP, vol. 39, no. 7, p. 074006-1-12, Jul 2018.

36. Divya Yadav, Shailesh Singh Chouhan, Santosh Kumar Vishvakarma, B. Raj, "Application specific microcontroller design for IoT based WSN", *Sensor Letter*, ASP, vol. 16, pp. 374–385, May 2018.

37. Gurmohan Singh, R. K. Sarin, B. Raj, "Fault-tolerant design and analysis of quantum-dot cellular automata based circuits", *IEEE/IET Circuits, Devices & Systems*, vol. 12, pp. 638–664, 2018.

38. Jeetendra Singh, B. Raj, "Modeling of mean barrier height levying various image forces of metal insulator metal structure to enhance the performance of conductive filament based memristor model", *IEEE Nanotechnology*, vol. 17, no. 2, pp. 268–267, Mar 2018.

39. Aakash Jain, Sanjeev Sharma, B. Raj, "Analysis of triple metal surrounding gate (TM-SG) III-V nanowire MOSFET for photosensing application", *Opto-electronics Journal*, Elsevier, vol. 26, no. 2, pp. 141–148, May 2018.

40. Aakash Jain, Sanjeev Sharma, B. Raj, "Design and analysis of high sensitivity photosensor using cylindrical surrounding gate MOSFET for low power sensor applications", *Engineering Science and Technology, an International Journal*, Elsevier, vol. 19, no. 4, pp. 1864–1870, Dec 2016.

41. Amandeep Singh, Mamta Khosla, B. Raj, "Analysis of electrostatic doped schottky barrier carbon nanotube FET for low power applications," *Journal of Materials Science: Materials in Electronics*, Springer, vol. 28, pp. 1762–1768, 2017.

42. G. Saiphani Kumar, Amandeep Singh, B. Raj, "Design and analysis of gate all around CNTFET based SRAM cell design", *Journal of Computational Electronics*, Springer, vol. 17, no. 1, pp. 138–145, Mar 2018.

43. Gurinder pal Singh, B. S. Sohi, Balwinder Raj, "Material properties analysis of graphene base transistor (GBT) for VLSI analog circuits", *Indian Journal of Pure & Applied Physics (IJPAP)*, vol. 55, pp. 896–902, Dec 2017.

44. Amandeep Singh, Mamta Khosla, B. Raj, "Comparative analysis of carbon nanotube field effect transistor and nanowire transistor for low power circuit design," *Journal of Nanoelectronics and Optoelectronics*, American Scientific Publishers, USA, vol. 11, pp. 388–393, Jun 2016.

45. Sunil Kumar, B. Raj, "Estimation of stability and performance metric for inward access transistor based 6T SRAM cell design using n-type/p-type DMDG-GDOV TFET", *IEEE VLSI Circuits and Systems Letter*, vol. 3, no. 2, pp. 25–39, Jun 2017.

46. Shashikant Sharma, Anjan Kumar, Manisha Pattanaik, B. Raj, "Forward body biased multimode multi-threshold CMOS technique for ground bounce noise reduction in static CMOS adders", *International Journal of Information and Electronics Engineering*, pp. 567–572, vol. 3, no. 3, 2013.

47. Hamendra Singh, Pankaj Kumar, Balwinder Raj, "Performance Analysis of Majority Gate SET Based 1-bit Full Adder", *International Journal of Computer and Communication Engineering* (IJCCE), IACSIT Press Singapore, ISSN: 2010-3743, Vol. 2, no. 4, 2013.

48. Anil Kumar Bhardwaj, Sumeet Gupta, B. Raj, "Investigation of parameters for schottky barrier (SB) height for schottky barrier based carbon nanotube field effect transistor device", *Journal of Nanoelectronics and Optoelectronics*, ASP, vol. 15, pp. 783–791, Jul 2020.

49. Priya Bansal, B. Raj, "Memristor: A versatile nonlinear model for dopant drift and boundary issues," *JCTN*, American Scientific Publishers, vol. 14, no. 5, pp. 2319–2325, May 2017.

50. Neeraj Jain, B. Raj, "An analog and digital design perspective comprehensive approach on Fin-FET (Fin-Field Effect transistor) technology: A review", *Reviews in Advanced Sciences and Engineering (RASE)*, ASP, vol. 5, pp. 1–14, 2016.

51. Sanjeev Sharma, B. Raj, Mamta Khosla, "Subthreshold performance of In1-xGaxAs based dual metal with gate stack cylindrical/surrounding gate nanowire MOSFET for low power analog applications", *Journal of Nanoelectronics and Optoelectronics*, American Scientific Publishers, USA, vol. 12, pp. 171–176, 2017.

52. Balwinder Raj, A. K. Saxena, S. Dasgupta, "Analytical modeling for the estimation of leakage current and subthreshold swing factor of nanoscale double gate FinFET device", *Microelectronics International, UK*, vol. 26, pp. 53–63, 2009.

53. Shailendra Singh, Girish Wadhwa, Balwinder Raj, "An analytical modeling for dual source vertical tunnel field effect transistor", *International Journal of Recent Technology and Engineering (IJRTE)*, vol. 8, no. 2, pp. 603–608, Jul 2019.

54. Shailendra Singh, B. Raj, "Design and analysis of hetrojunction vertical T-shaped tunnel field effect transistor", *Journal of Electronics Material*, Springer, vol. 48, no. 10, pp. 6253–6260, Oct 2019.

55. Candy Goyal, Jagpal Singh Ubhi, B. Raj, "A low leakage CNTFET based inexact full adder for low power image processing applications", *International Journal of Circuit Theory and Applications*, Wiley, vol. 47, no. 9, pp. 1446–1458, Sept 2019.

56. B. Raj, A. K. Saxena, S. Dasgupta, "A compact drain current and threshold voltage quantum mechanical analytical modeling for FinFETs", *Journal of Nanoelectronics and Optoelectronics (JNO)*, USA, vol. 3, no. 2, pp. 163–170, 2008.

57. Girish Wadhwa, B. Raj, "An analytical modeling of charge plasma based tunnel field effect transistor with impacts of gate underlap region", *Superlattices and Microstructures*, Elsevier, vol. 142, p. 106512, Jun 2020.

58. Shailendra Singh, B. Raj, "Modeling and simulation analysis of SiGe hetrojunction double GateVertical t-shaped tunnel FET", *Superlattices and Microstructures*, Elsevier vol. 142, p. 106496, Jun 2020.

59. Amandeep Singh, Dinesh Kumar Saini, Dinesh Agarwal, Sajal Aggarwal, Mamta Khosla, B. Raj, "Modeling and simulation of carbon nanotube field effect transistor and its circuit application," *Journal of Semiconductors (JoS)*, IOP Science, vol. 37, p. 074001-6, Jul 2016.

60. Neeraj Jain, Balwinder Raj, "Device and circuit co-design perspective comprehensive approach on FinFET technology: A review", *Journal of Electron Devices*, vol. 23, no. 1, pp. 1890–1901, 2016.

61. Sunil Kumar, B. Raj, "Analysis of I_{ON} and ambipolar current for dual-material gate-drain overlapped DG-TFET," *Journal of Nanoelectronics and Optoelectronics*, American Scientific Publishers, USA, vol. 11, pp. 323–333, Jun 2016.

62. Naveed Anjum, Tarun Bali, B. Raj, "Design and simulation of handwritten multiscript character recognition", *International Journal of Advanced Research in Computer and Communication Engineering*, vol. 2, no. 7, pp. 2544–2549, Jul 2013.

63. Sanjeev Sharma, B. Raj, Mamta Khosla, "A gaussian approach for analytical subthreshold current model of cylindrical nanowire FET with quantum mechanical effects", *Microelectronics Journal*, Elsevier, vol. 53, pp. 65–72, Apr 2016.

64. Karmjit Singh, B. Raj, "Performance and analysis of temperature dependent multi-walled carbon nanotubes as global interconnects at different technology nodes," *Journal of Computational Electronics*, Springer, vol. 14, no. 2, pp. 469–476, Jun 2015.

65. Sunil Kumar, B. Raj, "Compact channel potential analytical modeling of DG-TFET based on evanescent–mode approach," *Journal of Computational Electronics*, Springer, vol. 14, no. 2, pp. 820–827, Jul 2015.

66. Karmjit Singh, B. Raj, "Temperature dependent modeling and performance evaluation of multi-walled CNT and single-walled CNT as global interconnects," *Journal of Electronic Materials*, Springer, vol. 44, no. 12, pp. 4825–4835, Dec 2015.

67. V. K. Sharma, M. Pattanaik, B.Raj, "INDEP approach for leakage reduction in nanoscale CMOS circuits", *International Journal of Electronics*, Taylor & Francis, vol. 102, no. 2, pp. 200–215, 2014.

68. Karmjit Singh, B. Raj, "Influence of temperature on MWCNT bundle, SWCNT bundle and copper interconnects for nanoscaled technology nodes," *Journal of Materials Science: Materials in Electronics*, Springer, vol. 26, no. 8, pp. 6134–6142, 2015.

69. Naveed Anjum, Tarun Bali, B. Raj, "Design and simulation of handwritten gurumukhi and devanagri numerical recognition", *International Journal of Computer Applications*, Foundation of Computer Science, New York, USA, vol. 73, no. 12, pp. 16–21, 2013.

70. S. Khandelwal, V. Gupta, B. Raj, R. D. "Gupta, process variability aware low leakage reliable nano scale DG-FinFET SRAM cell design technique", *Journal of Nanoelectronics and Optoelectronics*, vol. 10, no. 6, pp. 810–817, Dec 2015.

71. V. K. Sharma, M. Pattanaik, B. Raj, "ONOFIC approach: Low power high speed nanoscale VLSI circuits design", *International Journal of Electronics*, Taylor & Francis, vol. 101, no. 1, pp. 61–73, 2014.

72. S. Khandelwal, Balwinder Raj, R. D. Gupta, "FinFET based 6T SRAM cell design: Analysis of performance metric, process variation and temperature effect", *Journal of Computational and Theoretical Nanoscience*, ASP, USA, vol. 12, pp. 2500–2506, 2015.

73. Sumit Singh, Shekhar Yadav, Jagdeep Rahul, Anurag Srivastava; B. Raj, "Impact of HfO$_2$ in graded channel dual insulator double gate MOSFET", *Journal of Computational and Theoretical Nanoscience*, American Scientific Publishers, vol. 12, no. 6, pp. 950–953, Apr 2015.

74. Vijay Kumar Sharma, Manisha Pattanaik, B. Raj, "PVT variations aware low leakage INDEP approach for nanoscale CMOS circuits", *Microelectronics Reliability*, Elsevier, vol. 54, pp. 90–99, 2014.

75. B. Raj, A. K. Saxena, S. Dasgupta, "Quantum mechanical analytical modeling of nanoscale DG FinFET: Evaluation of potential, threshold voltage and source/drain resistance", *Elsevier's Journal of Material Science in Semiconductor Processing*, Elsevier, vol. 16, no. 4, pp. 1131–1137, 2013.

76. Maisagalla Gopal, Siva Sankar D Prasad, Balwinder Raj, "8T SRAM cell design for dynamic and leakage power reduction", *International Journal of Computer Applications*, Foundation of Computer Science, New York, USA, vol. 71, no. 9, pp. 43–48, Jun 2013.

77. Manisha Pattanaik, B. Raj, Shashikant Sharma, Anjan Kumar, "Diode based trimode multi-threshold CMOS technique for ground bounce noise reduction in static CMOS adders", *Advanced Materials Research*, Trans Tech Publications, Switzerland, vol. 548, pp. 885–889, 2012.

78. Balwinder Raj, A. K. Saxena, S. Dasgupta, "Nanoscale FinFET based SRAM cell design: Analysis of performance metric, process variation, underlapped FinFET and temperature effect", *IEEE Circuits and System Magazine*, vol. 11, no. 2, pp. 38–50, 2011.

79. V. K. Sharma, M. Pattanaik, Balwinder Raj, "Leakage current ONOFIC approach for deep submicron VLSI circuit design", *International Journal of Electrical, Computer, Electronics and Communication Engineering, World Academy of Sciences, Engineering and Technology*, vol. 7, no. 4, pp. 239–244, 2013.

80. Tulika Chawla, Mamta Khosla, B. Raj, "Design and simulation of triple metal double-gate germanium on insulator vertical tunnel field effect transistor", *Microelectronics Journal*, Elsevier, vol. 114, p. 105125, Aug 2021.

81. Parminder Kaur, Sandeep Singh Gill, B. Raj, "Comparative analysis of OFETs materials and devices for sensor applications", *Journal of Silicon*, Springer, vol. 14, pp. 4463–4471, 2022.

82. Sanjeev Kumar Sharma, Parveen Kumar, Balwant Raj, B. Raj, "In$_{1-x}$Ga$_x$As double metal gate-stacking cylindrical nanowire MOSFET for highly sensitive photo detector", *Journal of Silicon*, Springer, vol. 14, pp. 3535–3541, 2022.

83. B. Raj, A. K. Saxena, S. Dasgupta, "Analytical modeling of quasi planar nanoscale double gate FinFET with source/drain resistance and field dependent carrier mobility: A quantum mechanical study", *Journal of Computer (JCP)*, Academy Publisher, Finland, vol. 4, no. 9, pp. 1–8, 2009.

84. S. Bhushan, S. Khandelwal, B. Raj, "Analyzing different mode FinFET based memory cell at different power supply for leakage reduction", *Seventh International Conference on Bio-Inspired Computing: Theories and Application, (BIC-TA 2012) Advances in Intelligent Systems and Computing*, vol. 202, pp. 89–100, 2013.

85. Jeetendra Singh, B. Raj, "Temperature dependent analytical modeling and simulations of nanoscale memristor", *Journal: Engineering Science and Technology, an International Journal*, Elsevier, vol. 21, pp. 862–868, Oct 2018.

86. Shradhya Singh, Shashi Bala, Balwant Raj, B. Raj, "Improved sensitivity of dielectric modulated junctionless transistor for nanoscale biosensor design", *Sensor Letter*, ASP, vol. 18, pp. 328–333, Apr 2020.

87. Vivek Kumar, Santosh Kumar Vishvakarma, B. Raj, "Design and performance analysis of ASIC for IoT applications", *Sensor Letter*, ASP, vol. 18, pp. 31–38, Jan 2020.

88. Akanksha Jaiswal, R. K. Sarin, B. Raj, Shikha Sukhija, "A novel circular slotted microstrip-fed patch antenna with three triangle shape defected ground structure for multiband applications", *Advanced Electromagnetic (AEM)*, vol. 7, no. 3, pp. 56–63, Aug 2018.

89. Girish Wadhwa, B. Raj, "Label free detection of biomolecules using charge-plasma-based gate underlap dielectric modulated junctionless TFET", *Journal of Electronic Materials (JEMS)*, Springer, vol. 47, no. 8, pp. 4683–4693, Aug 2018.

90. Gurmohan Singh, R. K. Sarin, B. Raj, **"Design and performance analysis of a new efficient coplanar quantum-dot cellular automata adder"**, *Indian Journal of Pure & Applied Physics (IJPAP)*, vol. 55, pp. 97–103, Feb 2017.

91. Amandeep Singh, Mamta Khosla, Balwinder Raj, "Design and analysis of electrostatic doped schottky barrier CNTFET based low power SRAM," *International Journal of Electronics and Communications, (AEÜ)*, Elsevier, vol. 80, pp. 67–72, 2017.

92. Parminder Kaur, Vikas Pandey, B. Raj, "Comparative study of efficient design, control and monitoring of solar power using IoT", *Sensor Letter*, ASP vol. 18, pp. 419–426, May 2020.

93. Anil Kumar Bhardwaj, Sumeet Gupta, B. Raj, "Development & analysis of compact model for double gate schottky barrier CNTFET", *Journal of Nanoelectronics and Optoelectronics*, ASP, vol. 15, pp. 1199–1208, Aug 2020.

94. Girish Wadhwa, Priyanka Kamboj, Jeetendra Singh, B. Raj, "Design and investigation of junctionless DGTFET for biological molecule recognition", *Transactions on Electrical and Electronic Materials*, Springer, vol. 22, pp. 282–289, 2021.

95. Tulika Chawla, Mamta Khosla, B. Raj, "Optimization of double-gate dual material GeOI-vertical TFET for VLSI circuit design", *IEEE VLSI Circuits and Systems Letter*, vol. 6, no. 2, pp. 13–25, Aug 2020.

96. Sachin Kumar Verma, Shailendra Singh, Girish Wadhwa, B. Raj, "Detection of biomolecules using charge-plasma based gate underlap dielectric modulated dopingless TFET", *Transactions on Electrical and Electronic Materials (TEEM)*, Springer, vol. 21, pp. 528–535, Jun 2020.

97. Neeraj Jain, B. Raj, "Impact of underlap spacer region variation on Electrostatic and analog/RF performance of symmetrical high-k SOI FinFET at 20 nm channel length", Journal of Semiconductors (JoS), IOP Science, vol. 38, no. 12, p. 122002, Dec 2017.

98. Shailendra Singh, B. Raj, "Analytical modeling and simulation analysis of T-shaped III-V heterojunction Vertical T-FET", *Superlattices and Microstructures*, Elsevier, vol. 147, p. 106717, Nov 2020.

99. Gurmohan Singh, R. K. Sarin, B. Raj, "Design and analysis of area efficient QCA based reversible logic gates", *Journal of Microprocessors and Microsystems*, Elsevier, vol. 52, pp. 59–68, May 2017.

100. Amandeep Singh, Mamta Khosla, B. Raj, "Compact model for ballistic single wall CNTFET under quantum capacitance limit," *Journal of Semiconductors (JoS)*, IOP Science, vol. 37, p. 104001-8, Oct 2016.

101. Sonal Singh, Mamta Khosla, Girish Wadhwa, B. Raj, "Design and analysis of double-gate junctionless vertical TFET for gas sensing applications", *Applied Physics A*, Springer, vol. 127, no. 16, 2 Jan 2021.

102. Inderjit Singh, B. Raj, Mamta Khosla, B. Rajesh Kumar Kaushik, Potential MRAM Technologies for Low Power SoCs, SPIN World Scientific Publisher, SCIE; vol. 10, no. 04, pp. 2050027, Dec 2020.

103. Girish Wadhwa, B. Raj, "Design and performance analysis of junctionless TFET biosensor for high sensitivity", *IEEE Nanotechnology*, vol. 18, pp. 567–574, 2019.

104. Jeetendra Singh, B. Raj, "Enhanced nonlinear memristor model encapsulating stochastic dopant drift", *JNO*, ASP, vol. 14, pp. 958–963, 2019.

105. Shailendra Singh, B. Raj, "Parametric variation analysis on hetero-junction Vertical t-shape TFET for supressing ambipolar conduction", *Indian Journal of Pure and Applied Physics*, vol. 58, pp. 478–485, Jun 2020.

106. Shailendra Singh, Girish Wadhwa, Balwinder Raj, "Design and analysis of dual source vertical tunnel field effect transistor for high performance", *Transactions on Electrical and Electronics Materials*, Springer, vol. 21, pp. 74–82, Oct 2019.

107. Manjit Kaur, Neena Gupta, Sanjeev Kumar, B. Raj, Arun Kumar Singh, "RF performance analysis of intercalated graphene nanoribbon based global level interconnects", *Journal of Computational Electronics*, Springer, vol. 19, pp. 1002–1013, Jun 2020.

Chapter 9

Energy-efficient SRAM cell design

Pawanram and Balwinder Raj

9.1 6T SRAM CELL USED IN DESIGN

A 6T SRAM circuit is a type of static random access memory (SRAM) circuit that uses six transistors (hence the name "6T") to store each memory bit, as shown in Figure 9.1. The six transistors are arranged in a cross-coupled inverter configuration with two access transistors for reading and writing the data [1–5].

Here's a brief description of the function of each transistor in a 6T SRAM cell:

Two n-channel access transistors (M5 and M6) connect the memory cell to the bitlines for read and write operations [6–13].

Two n-channel pass transistors (M3 and M4) provide a path for the data to flow from the storage nodes to the bitlines during a read operation.

Two p-channel load transistors (M1 and M2) act as pull-up resistors and provide a stable voltage to the cross-coupled inverters.

During a write operation, the bitlines are precharged to a known voltage level, and then the access transistors are activated to allow the data to be written to the storage nodes [14–19]. During a read operation, the bitlines are again precharged, and the access transistors are activated to allow the data to flow through the pass transistors to the bitlines, which can be sensed to determine the stored data. The SRAM sense amplifier circuit is given in Figure 9.2.

The design of a 6T SRAM circuit involves various trade-offs between performance, power consumption, and area. It is a complex process that requires careful consideration of factors such as transistor sizing, voltage levels, and layout optimization [20–25]. Advanced process technologies, such as FinFETs and nanowire FETs, have enabled the design of high-performance and low-power 6T SRAM circuits for modern processors and memory devices.

DOI: 10.1201/9781003487692-9

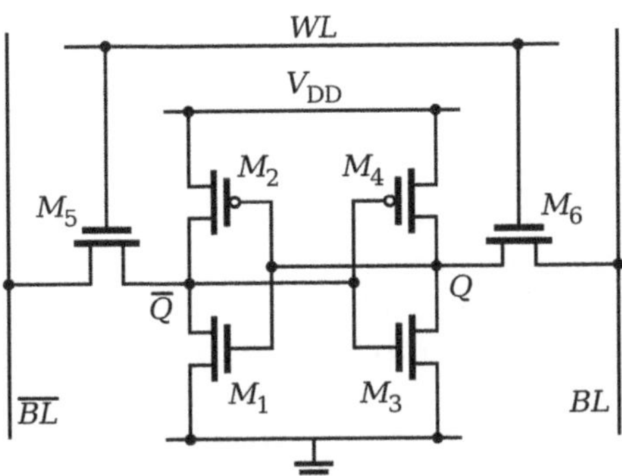

Figure 9.1 6T SRAM cell.

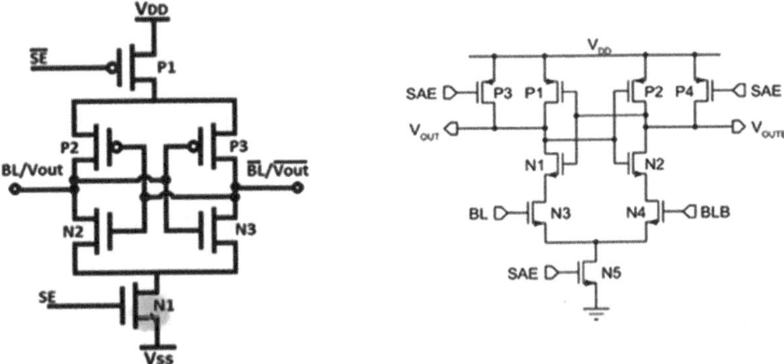

Figure 9.2 SRAM sense amplifier.

9.2 CARRIER SENSE AMPLIFIERS

Carrier sense amplifiers (CSAs) are an important component in 6T SRAM circuits, as they are used to amplify and sense the small voltage differences between the bitlines that store the binary data [26–32].

In an SRAM circuit, the stored data is represented by the voltage difference between the two bitlines that are connected to each SRAM cell. When reading the stored data, the voltage difference between the two bitlines is small, typically a few millivolts, which makes it difficult to detect and amplify. The CSA is used to sense and amplify this small voltage difference, and to provide a reliable output signal that can be interpreted by the control logic [33–37].

A typical CSA circuit consists of a pair of differential amplifiers, one for each bitline, which amplifies the voltage difference between the two bitlines. The differential amplifiers are connected to a latch circuit, which holds the output voltage steady until the next read operation. The output of the CSA circuit is then fed to the control logic, which interprets the amplified voltage difference as logic high or low, depending on the threshold voltage.

The CSA is a critical component in the read operation of an SRAM circuit, as it provides a reliable and accurate way of reading the stored data. The CSA must be carefully designed to provide high gain and low noise while also being able to operate at high speeds to meet the timing requirements of modern electronic devices.

In a 6T SRAM circuit, the read operation is initiated by applying a voltage to the wordline of the cell to be read, which activates the access transistors and connects the storage nodes to the bitlines. The stored data in the SRAM cell is represented by the voltage difference between the two bitlines, which is typically a few millivolts [38–44].

To read the stored data, the voltage difference between the two bitlines needs to be sensed and amplified. This is where the CSAs come into play. The CSAs are connected to the two bitlines and amplify the voltage difference between them. The amplified output from the CSAs is then fed to the sense amplifiers, which further amplify the voltage difference and convert it into a logic-level signal that can be interpreted by the control logic.

Here is a step-by-step explanation of the read operation in a 6T SRAM circuit using CSAs:

i. Wordline activation: A voltage is applied to the wordline of the SRAM cell to be read, which activates the access transistors and connects the storage nodes to the bitlines.

ii. Precharge: The bitlines are precharged to a known voltage level, typically halfway between the high and low voltage levels.

iii. CSA amplification: The voltage difference between the two bitlines is sensed and amplified by the CSAs, which are connected to the two bitlines. The CSAs amplify the voltage difference between the two bitlines and provide an amplified output.

iv. Sense amplification: The amplified output from the CSAs is fed to the sense amplifiers, which further amplify the voltage difference and convert it into a logic-level signal that can be interpreted by the control logic.

v. Output: The output from the sense amplifiers is then fed to the control logic, which interprets the amplified voltage difference as a logic high or low, depending on the threshold voltage. The output signal is then used for further processing or storage.

Overall, the read operation in a 6T SRAM circuit using CSAs is a critical component in the operation of the circuit, as it provides a reliable and accurate way of reading the stored data from the SRAM cells [45–51].

CSAs are an important component in 6T SRAM circuits, as they are used to amplify and sense the small voltage differences between the bitlines that store the binary data.

In an SRAM circuit, the stored data is represented by the voltage difference between the two bitlines that are connected to each SRAM cell. When reading the stored data, the voltage difference between the two bitlines is small, typically a few millivolts, which makes it difficult to detect and amplify. The CSA is used to sense and amplify this small voltage difference and to provide a reliable output signal that can be interpreted by the control logic.

A typical CSA circuit consists of a pair of differential amplifiers, one for each bitline, which amplifies the voltage difference between the two bitlines. The differential amplifiers are connected to a latch circuit, which holds the output voltage steady until the next read operation. The output of the CSA circuit is then fed to the control logic, which interprets the amplified voltage difference as logic high or low, depending on the threshold voltage [52–57].

The CSA is a critical component in the read operation of an SRAM circuit, as it provides a reliable and accurate way of reading the stored data. The CSA must be carefully designed to provide high gain and low noise, while also being able to operate at high speeds to meet the timing requirement of modern electronic devices. CSAs are an important component in 6T SRAM circuits, as they are used to amplify and sense the small voltage differences between the bitlines that store the binary data [58–63].

It can be used to verify the functionality of the circuit and optimize its performance before fabrication.

9.3 DESIGN CONSIDERATIONS

When designing a 6T SRAM circuit in LTspice, some of the key parameters that need to be considered include:

i. MOSFET models: The MOSFET (metal-oxide-semiconductor field-effect transistor) models used in the design should be chosen carefully to ensure that they accurately reflect the behavior of the physical devices used in the circuit.

ii. Wordline voltage: The voltage applied to the wordline should be high enough to activate the access transistors and connect the storage nodes to the bitlines, but not so high as to cause excessive power dissipation [64–72].

iii. Bitline precharge voltage: The bitlines should be precharged to a suitable voltage level to ensure that the stored data can be accurately read.

iv. Sense amplifier threshold voltage: The threshold voltage of the sense amplifier should be set correctly to ensure that the amplified voltage difference between the two bitlines can be accurately detected.

v. CSA design parameters: The design parameters for the CSAs should be optimized to ensure that the voltage difference between the two bitlines is accurately sensed and amplified.

   vi. Capacitance values: The capacitance values of the storage nodes and bitlines should be chosen to ensure that the circuit operates correctly and has the desired speed and power consumption characteristics.

   vii. Resistance values: The resistance values of the load resistors and other resistors used in the circuit should be chosen carefully to ensure that the circuit operates correctly and has the desired speed and power consumption characteristics.

Overall, the design parameters for a 6T SRAM circuit in LTspice will depend on the specific requirements of the circuit, such as the desired operating speed, power consumption, and storage capacity, as well as the physical characteristics of the MOSFET devices used in the circuit [73–79]. A thorough analysis of the circuit's performance and optimization of the design parameters using LTspice simulations can help ensure that the circuit meets the desired specifications.

9.3.1 Design parameters for 6T SRAM in LTspice

When designing a 6T SRAM circuit in LTspice, there are several design parameters that need to be considered. Here are some of the key parameters to keep in mind:

   i. MOSFET models: The MOSFET models used in the 6T SRAM circuit should be chosen carefully to ensure accurate simulation results. LTspice comes with a built-in library of MOSFET models that can be used, or custom models can be created as needed.

   ii. Transistor sizing: The size of the transistors used in the SRAM circuit is an important design parameter that affects the speed and power consumption of the circuit. The sizing should be optimized based on the specific requirements of the circuit [80–84].

   iii. Power supply voltage: The power supply voltage used in the circuit should be chosen based on the desired performance and power consumption of the circuit.

   iv. Wordline voltage: The voltage applied to the wordline during read and write operations is a critical parameter that affects the stability and reliability of the SRAM circuit. The voltage should be chosen based on the specific requirements of the circuit.

   v. Bitline capacitance: The capacitance of the bitlines is an important design parameter that affects the read and write times of the circuit. The capacitance should be optimized based on the specific requirements of the circuit.

   vi. Sense amplifier design: The design of the sense amplifier used in the SRAM circuit is another important parameter that affects the read speed and reliability of the circuit. The sense amplifier should be optimized based on the specific requirements of the circuit.

vii. Noise immunity: The SRAM circuit should be designed to be immune to noise and other sources of interference that can affect the stability and reliability of the circuit. The noise immunity of the circuit should be tested and optimized as needed.

By carefully considering these design parameters and optimizing the SRAM circuit in LTspice, designers can create a highly reliable and efficient circuit that meets the specific requirements of the target application [85–89].

The 6T SRAM cell is one of the most commonly used memory cells in modern digital circuits. Here are some key design considerations for the 6T SRAM cell:

i. Read and write stability: The 6T SRAM cell must be designed to ensure that the stored data can be read accurately and that new data can be written without causing unintentional changes to the existing data.

ii. Power consumption: Power consumption is a critical consideration for any digital circuit, and the 6T SRAM cell is no exception. Designers should aim to minimize power consumption while maintaining adequate read and write stability.

iii. Area efficiency: The size of the SRAM cell affects the overall size of the memory array, so designers should strive to minimize the cell size while maintaining the desired performance characteristics.

iv. Speed: The speed of the SRAM cell affects the overall speed of the memory array. Designers should aim to optimize the cell design for high-speed operation.

v. Noise immunity: The SRAM cell must be designed to minimize the effects of external noise sources, such as electromagnetic interference (EMI) and crosstalk from neighboring cells [90–96].

vi. Process variations: The performance of the SRAM cell can be affected by process variations during manufacturing. Designers should account for these variations and aim to create a design that is robust against process variations.

vii. Yield: Finally, designers must consider yield, which refers to the percentage of working cells in a memory array. Yield is affected by factors such as process variations and design robustness. Designers must strive to maximize yield to reduce manufacturing costs.

9.4 SYSTEM MODELING AND SIMULATIONS

9.4.1 Circuit implementation

The circuit in Figure 9.3 shows the design developed for the proposed neural network using LTspice software. The values of voltages and currents

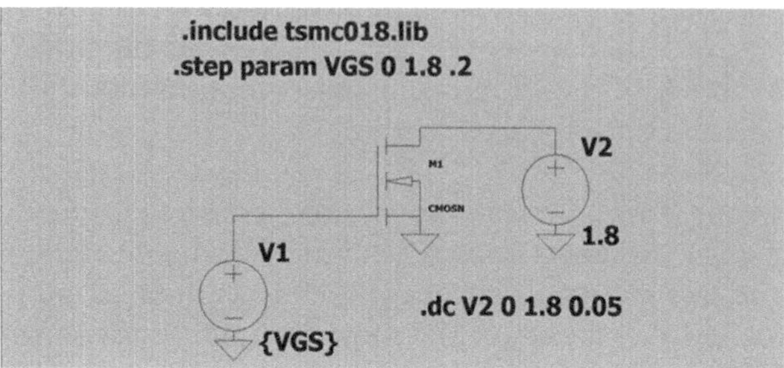

Figure 9.3 NMOS transistor circuit.

decide the design parameters length (*L*) and width (*W*) of the transistors. The channel length and width are calculated using the simulation of the circuit using LTspice and predicted using the proposed neural network [97–102].

9.4.2 **NMOS transistor circuit**

An NMOS (N-channel metal-oxide-semiconductor) transistor is a type of field-effect transistor (FET) used in integrated circuit technology. NMOS transistors are commonly used in digital logic circuits, as shown in Figure 9.3. Here's a brief overview of the NMOS transistor circuit:

  i. **Structure**: An NMOS transistor consists of three terminals: source, drain, and gate. It is made up of a silicon substrate with a layer of insulating material (usually silicon dioxide) and a gate electrode made of metal. The source and drain are typically doped regions in the silicon substrate.
  ii. **Operation**: NMOS transistors work on the principle of controlling the flow of current between the source and drain terminals using a voltage applied to the gate terminal. When a positive voltage is applied to the gate relative to the source, it creates an electric field in the silicon substrate, which forms a conductive channel between the source and drain. This allows current to flow from the source to the drain [103–108].
  iii. **Voltage levels**: NMOS transistors are typically used in digital circuits. They are turned on when the gate voltage (V_G) is sufficiently higher than the source voltage (V_S). When V_G is greater than a threshold voltage (V_{th}), the transistor is in the "on" state, and current can flow between the source and drain. When V_G is lower than V_{th}, the transistor is in the "off" state, and there is negligible current flow.
  iv. **Logic gates**: NMOS transistors are often used to create logic gates, such as NAND, NOR, and NOT gates. In these gates, the presence or absence of voltage on the gate terminal controls the logic operation.

v. **Inverters**: An NMOS transistor can be used as an inverter, where the input signal is applied to the gate, and the output is taken from the drain terminal. When the input is high, the output is low, and vice versa.

vi. **Advantages and disadvantages**: NMOS transistors are efficient for switching applications and are typically used in complementary metal-oxide-semiconductor (CMOS) technology. They consume less power when off compared to PMOS (P-channel metal-oxide-semiconductor) transistors but can be slower. The main disadvantage is that they require a positive gate-source voltage to operate.

NMOS transistors, along with PMOS transistors, are the building blocks of most digital integrated circuits, providing the foundation for complex digital systems. When combined with CMOS technology, they offer a balance of performance, power efficiency, and noise immunity, making them essential components in modern electronics.

9.4.3 NMOS transistor characteristics

NMOS transistors exhibit various characteristics that define their behavior in electronic circuits, as shown in Figure 9.4. Here's a brief overview of some key NMOS transistor characteristics:

i. **Threshold voltage (Vth)**: The threshold voltage is the minimum gate-source voltage (Vgs) required to turn on the NMOS transistor. Below this threshold voltage, the transistor remains in the "off" state, and no significant current flows from the source to the drain [109–111].

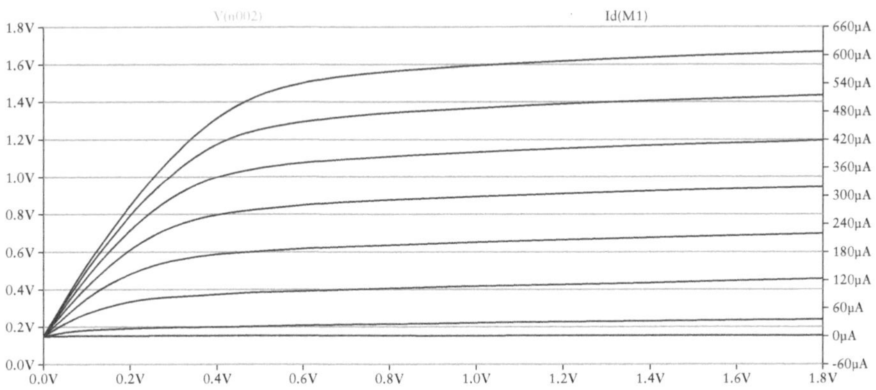

Figure 9.4 NMOS transistor characteristics.

ii. **On-resistance (Rds(on))**: When an NMOS transistor is in the "on" state, it acts as a low-resistance channel between the source and drain terminals. The on-resistance, denoted as Rds(on), represents the resistance of the conducting channel when the transistor is fully on. Lower Rds(on) values indicate better conduction and reduced power dissipation.

iii. **Transconductance (gm)**: Transconductance is a measure of how the NMOS transistor's drain current (Id) changes in response to changes in the gate-source voltage (Vgs). It is a measure of the transistor's ability to amplify small input voltage variations. Higher transconductance implies better amplification characteristics.

iv. **Drain current (Id)**: The drain current is the current that flows from the source to the drain when the NMOS transistor is in the "on" state. The drain current is a function of the gate-source voltage and the transistor's characteristics.

v. **Cut-off and saturation regions**: NMOS transistors have different operating regions. In the cut-off region, the transistor is off, and no significant current flows. In the saturation region, the transistor is fully on, and it behaves like a low-resistance switch, allowing a relatively high current to flow.

vi. **Voltage dependence**: NMOS transistors are voltage-dependent devices. The gate-source voltage determines their behavior. When Vgs is greater than the threshold voltage (Vth), the transistor turns on. Conversely, when Vgs is less than Vth, the transistor is off.

vii. **Inversion**: NMOS transistors operate by creating an inversion layer or channel in the silicon substrate beneath the gate when Vgs is sufficiently higher than Vth. This inversion layer provides a conductive path for current flow between the source and drain.

viii. **Voltage and current relationships**: NMOS transistors follow certain mathematical relationships, such as the drain current being proportional to $(Vgs - Vth)^2$ when the transistor is in the saturation region. These relationships are essential for circuit analysis and design.

ix. **High input impedance**: NMOS transistors typically have a high input impedance, making them suitable for use in digital logic gates and amplification stages.

x. **Low power consumption in the off state**: When NMOS transistors are off, they have minimal current leakage, which helps conserve power in electronic circuits.

Understanding these characteristics is crucial for designing and analyzing circuits that incorporate NMOS transistors, whether they are used in digital logic gates, amplifiers, or other applications. NMOS transistors, in combination with PMOS transistors, form the basis of CMOS technology, which is widely used in modern integrated circuits [112–115].

9.4.4 PMOS transistor circuit

A PMOS transistor is another type of field-effect transistor used in integrated circuit technology, as shown in Figure 9.5, and it complements the operation of NMOS transistors. The following paragraphs provide a brief overview of a PMOS transistor circuit.

9.4.4.1 Structure

Source, drain, and gate: Like NMOS transistors, PMOS transistors have three terminals: source, drain, and gate. However, they operate differently. The source and drain are typically doped regions in a silicon substrate, and the gate is made of metal and separated from the substrate by an insulating layer, usually silicon dioxide.

9.4.4.2 Operation

Voltage polarities: PMOS transistors operate using opposite voltage polarities to NMOS transistors. In a PMOS transistor, the gate voltage (V_G) is lower than the source voltage (V_S) to turn the transistor on. When V_G is lower than a certain threshold voltage (V_th), the transistor is in the "on" state, allowing current to flow between the source and drain.

Inversion layer: When the gate voltage is below the threshold, the PMOS transistor forms a conductive inversion layer between the source and drain, allowing current to flow when the transistor is on.

9.4.4.3 Logic levels and CMOS

PMOS transistors are often used in conjunction with NMOS transistors to create complementary CMOS technology. In CMOS, NMOS and PMOS transistors are combined to create digital logic gates, such as NAND, NOR, and NOT gates.

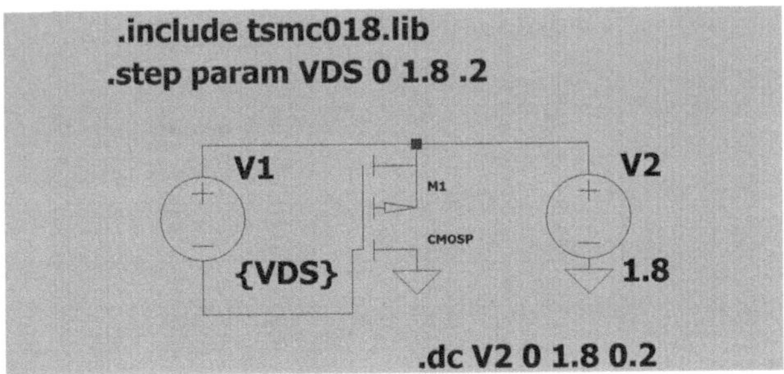

Figure 9.5 PMOS transistor circuit.

The PMOS transistor is used in CMOS to implement the complement of a logic function (e.g., an NMOS inverter has a PMOS complement). When the NMOS transistor is off (output high), the PMOS transistor is on (output low), and vice versa.

9.4.4.4 Advantages and disadvantages

i. PMOS transistors are efficient for certain applications and are often used in low power circuits. They consume less power when off compared to NMOS transistors but can be slower.
ii. One disadvantage of PMOS transistors is that they require a negative gate-source voltage to operate, which can complicate circuit design in comparison to NMOS transistors.

9.4.4.5 Complementary operation

i. PMOS transistors and NMOS transistors are complementary in their operation. When an NMOS transistor is on, a PMOS transistor is off, and vice versa. This complementary operation makes them suitable for implementing energy-efficient digital logic circuits.
ii. In CMOS technology, both PMOS and NMOS transistors are used in pairs to create logic gates with very low static power consumption, making them ideal for modern integrated circuits.

PMOS transistors have specific characteristics, as shown in Figure 9.6, that define their behavior in electronic circuits. The following paragraphs provide a brief overview of some key PMOS transistor characteristics.

i. **Threshold voltage (Vth):** The threshold voltage (Vth) is the minimum gate-source voltage (Vgs) required to turn on a PMOS transistor. Unlike NMOS transistors, PMOS transistors require a negative

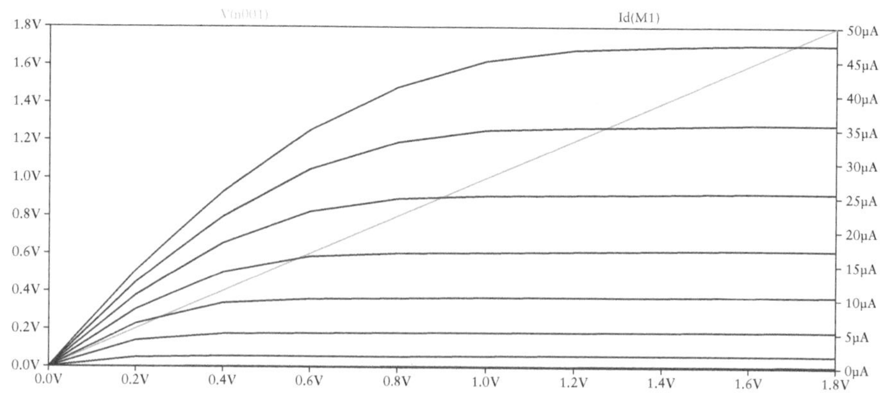

Figure 9.6 PMOS transistor characteristics.

gate-source voltage to operate. When the gate voltage is more negative (i.e., less positive) than the source voltage by at least Vth, the transistor turns on.

ii. **On-resistance (Rds(on))**: In the "on" state, a PMOS transistor forms a low-resistance channel between the source and drain terminals. The on-resistance, represented as Rds(on), indicates the resistance of this conducting channel. Lower Rds(on) values signify better conduction and reduced power dissipation.

iii. **Transconductance (gm)**: Transconductance measures the sensitivity of the drain current (Id) to changes in the gate-source voltage (Vgs). Higher transconductance implies better amplification characteristics.

iv. **Drain current (Id)**: The drain current is the current flowing from the source to the drain when the PMOS transistor is in the "on" state. The drain current depends on the gate-source voltage and the transistor's characteristics.

v. **Cut-off and saturation regions**: Similar to NMOS transistors, PMOS transistors have different operating regions. In the cut-off region, the transistor is off, and negligible current flows. In the saturation region, the transistor is fully on, and it acts as a low-resistance path for current flow.

vi. **Voltage dependence**: PMOS transistors are voltage-dependent devices. The gate-source voltage determines their operation. When the gate-source voltage is more negative than the source voltage by at least Vth, the transistor turns on.

vii. **Inversion**: PMOS transistors operate by creating an inversion layer or channel in the silicon substrate beneath the gate when the gate-source voltage is sufficiently more negative than Vth. This inversion layer provides a conductive path for current flow between the source and drain.

viii. **Voltage and current relationships**: PMOS transistors follow mathematical relationships similar to NMOS transistors. For instance, the drain current is proportional to $(Vgs - Vth)^2$ when the transistor is in the saturation region. These relationships are essential for circuit analysis and design.

ix. **High input impedance**: PMOS transistors typically exhibit a high input impedance, making them suitable for use in digital logic gates and amplification stages.

x. **Low power consumption in the off state**: When PMOS transistors are off, they have minimal current leakage, helping conserve power in electronic circuits.

Understanding these characteristics is crucial for designing and analyzing circuits that incorporate PMOS transistors, whether they are used in digital logic gates, amplifiers, or other applications. When combined with NMOS transistors, PMOS transistors are integral to CMOS technology, which is widely used in modern integrated circuits for its energy efficiency and versatility in digital circuit design.

9.4.5 CMOS inverter circuits

A CMOS inverter, as seen in Figure 9.7, is one of the fundamental building blocks in digital integrated circuit design. It consists of both PMOS and NMOS transistors to achieve high-performance digital logic operations. The following paragraphs provide an overview of CMOS inverter circuits.

9.4.5.1 Components

   i. **PMOS transistor:** The PMOS transistor is connected between the power supply voltage (V_DD) and the output node.

   ii. **NMOS Transistor:** The NMOS transistor is connected between the output node and ground (0V).

   iii. **Input (IN):** The input signal is applied to the gates of both the PMOS and NMOS transistors.

   iv. **Output (OUT):** The output of the CMOS inverter is taken from the junction between the drain of the NMOS transistor and the source of the PMOS transistor.

9.4.5.2 Operation

   i. **When input is low (0):** When the input is at a logic low (0), the NMOS transistor turns off because its gate-source voltage is below its threshold voltage (V_th). Simultaneously, the PMOS transistor turns on because the gate-source voltage for the PMOS transistor is high enough to exceed its threshold voltage. As a result, the connection

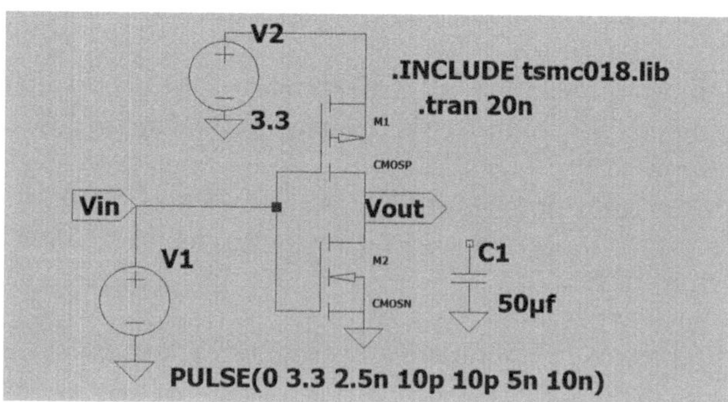

Figure 9.7 CMOS inverter circuits.

between the output node and V_DD is established through the PMOS transistor. This drives the output to a logic high (1).

ii. **When input is high (1)**: When the input is at a logic high (1), the NMOS transistor turns on because its gate-source voltage exceeds the threshold voltage. The PMOS transistor turns off because its gate-source voltage is too low. Now, the connection between the output node and V_DD is interrupted, and the connection between the output node and ground is established through the NMOS transistor. This drives the output to a logic low (0).

9.4.5.3 CMOS inverter input and output waveforms

9.4.5.3.1 Characteristics

i. **High input impedance**: The input of the CMOS inverter has a high input impedance due to the NMOS transistor, which makes it compatible with various driving circuits, as shown in Figure 9.8.

ii. **Low output impedance**: The output has a low output impedance because the NMOS and PMOS transistors are in series when one is on and the other is off. This low output impedance ensures that the CMOS inverter can drive other circuits effectively.

iii. **Low power consumption**: CMOS inverters are known for their low power consumption because they consume power only during the brief transition period when switching from one logic state to another. When they are in a stable state (either high or low), they draw minimal power.

iv. **Complementary operation**: The PMOS and NMOS transistors operate in a complementary manner, which minimizes static power consumption. When one transistor is on (conducting), the other is off, leading to little current flow between the power supply and ground.

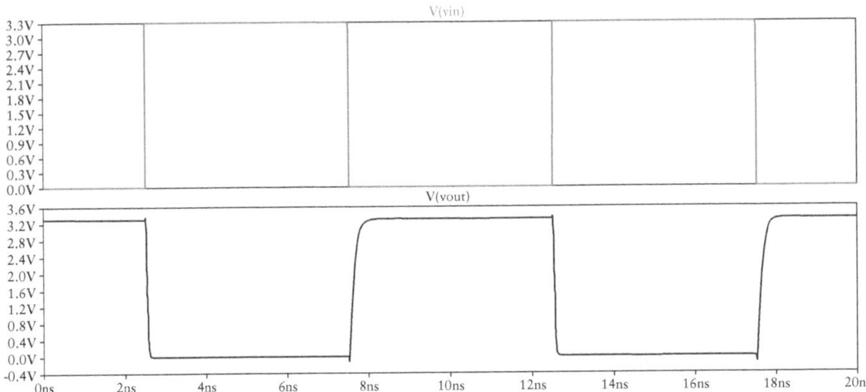

Figure 9.8 CMOS inverter input and output waveforms.

CMOS inverters are the basis for many other logic gates and complex digital circuits. They are widely used in integrated circuits (ICs) and microprocessors due to their excellent noise immunity, low power consumption, and high switching speed. Additionally, they are crucial in constructing digital logic gates, such as NAND, NOR, XOR, and others, by interconnecting multiple CMOS inverters.

9.4.5 Circuit diagram of 6T SRAM cell

The 6T SRAM cell presented in Figure 9.9 is a fundamental unit of memory storage in modern ICs. It is comprised of six transistors, which are used to store a single bit of data.

9.4.5.1 Components

i. Bitline: The bitline represents the data storage and data retrieval line. It is used to write data into the SRAM cell and read data from it.
ii. **Wordline**: The wordline is used to control access to the SRAM cell. When the wordline is activated, it allows data to be read from or written into the cell.
iii. **M1 and M2 (access transistors)**: These are typically NMOS transistors. M1 connects the bitline to the internal storage nodes when the

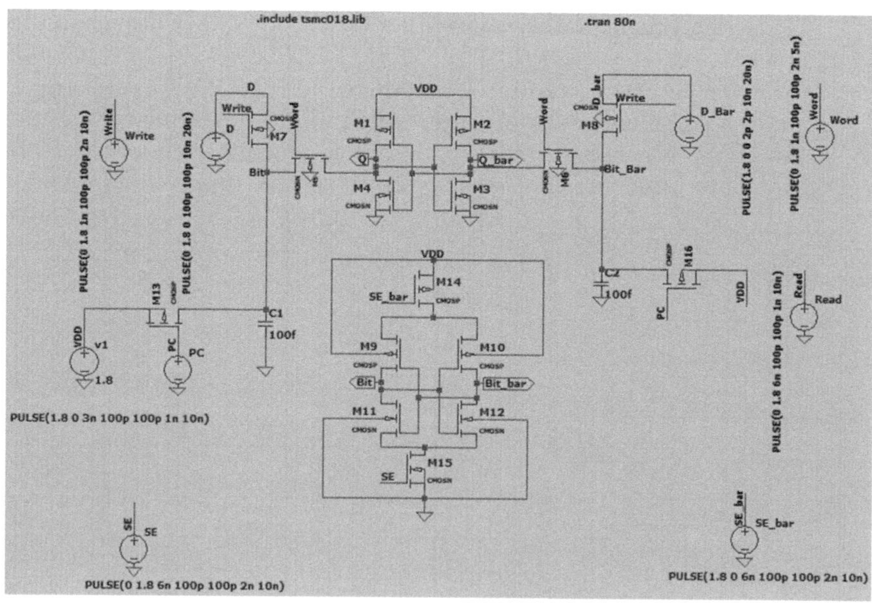

Figure 9.9 Circuit diagram for a 6T SRAM cell using carrier sense amplifier for read operation.

wordline is active, allowing data to be read from or written to the SRAM cell. M2 is used for access to the complementary data bit.

iv. **M3 and M4 (pull-down transistors):** These are also NMOS transistors. M3 and M4 help discharge the stored data when the SRAM cell is being accessed or overwritten. They ensure that the SRAM cell is stable and doesn't enter an ambiguous state.

v. **M5 and M6 (cross-coupled inverters):** M5 and M6 are typically PMOS transistors. They form a latch that stores the data in the SRAM cell. The gates of M5 and M6 are connected to each other's drains, creating positive feedback, which maintains the stored data.

9.4.5.2 Operation

- When the wordline is activated, it controls the access transistors M1 and M2. M1 connects the bitline to the internal storage nodes, and M2 connects the complementary data bit.
- The cross-coupled inverters M5 and M6 store the data. The state of the cell is determined by the electrical state of the cross-coupled inverters. If M5 is on and M6 is off, the SRAM cell stores a logic high (1); if M6 is on and M5 is off, it stores a logic low (0).
- Reading data from the SRAM cell involves evaluating the electrical state of the cross-coupled inverters by sensing the data bit and its complement on the bitline and complementary bitline.
- Writing data to the SRAM cell involves precharging the bitline and bitline-complement, and then activating the wordline to allow the access transistors M1 and M2 to change the state of the cross-coupled inverters.

The 6T SRAM cell is essential in the construction of cache memory, register files, and other high-speed memory elements in microprocessors and other digital systems due to its speed and stability.

9.5 WRITE AND READ OPERATION WAVEFORMS

A 6T SRAM circuit is a type of memory cell that is used to store binary information in electronic devices. The 6T SRAM cell consists of six transistors that are connected in a cross-coupled inverter configuration, as well as two access transistors that are used for reading and writing data. SRAM write and read operation waveforms are given in Figures 9.10 and 9.11.

Here is a step-by-step explanation of how a 6T SRAM circuit works:

- Initialization: The SRAM circuit is initialized by setting the two storage nodes to a known voltage level (usually high and low) using the two access transistors (M5 and M6) that are connected to the bitlines.

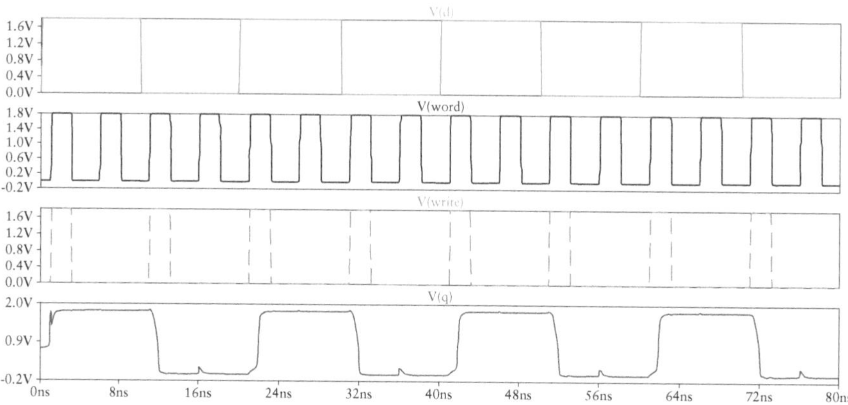

Figure 9.10 Write operation waveform

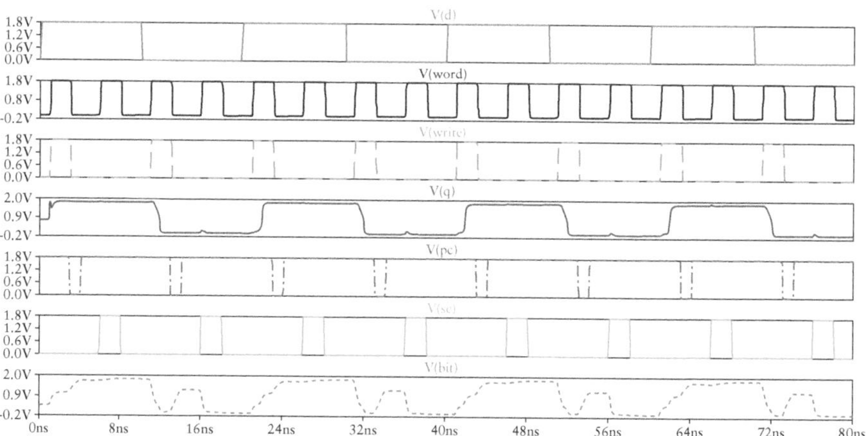

Figure 9.11 Read operation waveform.

- Writing: To write data to the SRAM circuit, a voltage is applied to one of the bitlines, while the other bitline is grounded. This activates the corresponding access transistor (M5 or M6) and allows the data to flow through the transistor to the storage node. The cross-coupled inverters then latch the data, and the voltage levels on the storage nodes are maintained as long as the power is on.

- Reading: To read data from the SRAM circuit, the bitlines are precharged to a known voltage level. Then, the access transistors are activated by applying a voltage to their gates. This allows the data to flow through the pass transistors (M3 and M4) to the bitlines, which can be sensed to determine the stored data. The stored data is the voltage level difference between the two bitlines, which can be amplified and sensed using sense amplifiers.

Overall, the 6T SRAM circuit is a high-speed and low-power memory cell that is widely used in electronic devices such as microprocessors, memory modules, and digital cameras. Its simple and robust design allows it to store data reliably and efficiently, while its fast access time makes it suitable for real-time processing and high-speed data transfers.

9.6 PROPOSED NEURAL NETWORK MODEL

A neural network generally refers to a system that consists of many neurons. It produces the best possible output though there is a change in the input. Generally, it consists of source nodes which include an input layer, one or more hidden layers, and an output layer. In general, in neural network architecture, the set of sensory nodes known as input neurons constitute the input layer, one more computational hidden layer, and a summed output layer. The output of the neural network is decided by the activation function and threshold. It has been observed from the literature that the sigmoidal function can perform better than the binary and step functions. The basic neural network with three inputs and one out is shown in Figure 9.12. In Figure 9.12, x_1, x_2, and x_3 are the inputs to the neurons, and w_1, w_2, and w_3 are the weights to the respective inputs. The activation function is used to calculate the output of the neural network using the threshold value. X_1

In this chapter, the NMOS1 MOSFET model is used for design and analysis. The input parameters such as gate-to-source (V_{GS}), drain-to-source (V_{DS}), and output drain current (I_D) are considered as inputs to the neural network by which the design parameters channel length (L) and width (W) are estimated. The proposed neural network architecture is shown in Figure 9.13.

Training and testing data are obtained by various simulations of MOSFET (n-channel and p-channel) in an LTspice software environment. The activation of the hidden neuron is the function of weighted inputs and bias. It can be calculated as $x_i = w_1 V_{GS} + w_2 V_{ds} + w_3 I_d + \theta_j$.

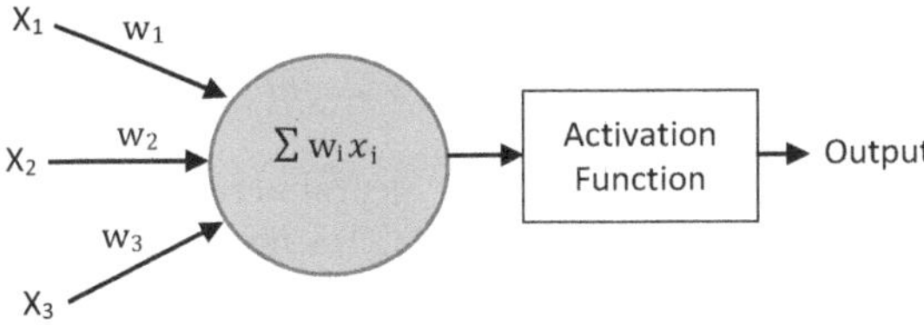

Figure 9.12 Basic block diagram neuron model.

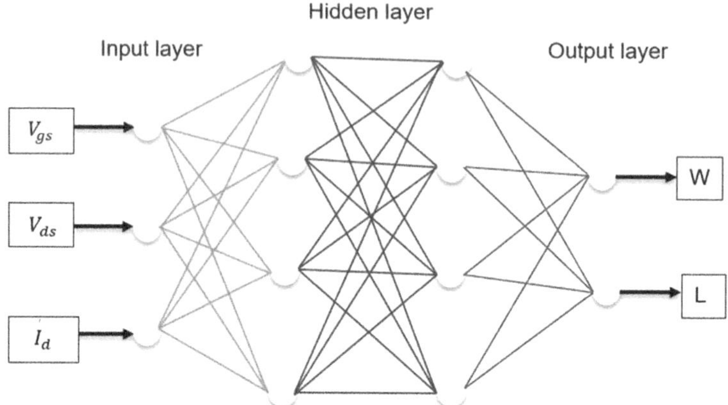

Figure 9.13 Proposed neural network model for circuit design and analysis.

9.7 CONCLUSION

This chapter presents a comprehensive design approach for an SRAM cell. The basic building blocks of the SRAM are designed and simulated using the LTspice software. The aspect ratio of each MOSFET is calculated using the simulation. Moreover, the channel length and width are designed using the artificial neural network. The various experimentations are carried out using the MOSFET, inverter, and SRAM circuits. It is observed from the results that the presented neural network model can predict the design parameters such as channel length and width accurately, and it is close to simulated results. Thus, the proposed approach can be used to design analog and digital VLSI circuits. This proposed design will be beneficial for the scientific community working in this research area.

REFERENCES

1. G. Kothapalli, "Artificial neural networks as aids in circuit design," *Microelectronics Journal*, vol. 26, no. 6, pp. 569–578, 1995.
2. J. Langeheine, S. Fölling, K. Meier, J. Schemmel, "Towards a silicon primordial soup: A fast approach to hardware evolution with a VLSI transistor array," *International Conference on Evolvable Systems*, 2000, pp. 123–132, Springer, Berlin, Heidelberg.
3. M. Avcı, M. Y. Babaç, T. Yıldırım, "Neural network based transistor modeling and aspect ratio estimation for yital 1.5 micron process," *In Third IEEE International Conference on Electronic Engineering*, ELECO, 2003, pp. 54–57.
4. M. Avci, T. Yildirim, "Neural network based MOS transistor geometry decision for TSMC 0.18 μ process technology," *In The International Conference on Computational Science*, 2006, pp. 615–622, Springer, Berlin, Heidelberg.

5. T. H. Borgstrom, M. Ismail, S. B. Bibyk, "Programmable current- mode neural network for implementation in analogue MOS VLSI," *IEE Proceedings G (Circuits, Devices System)*, vol. 137, no. 2, pp. 175–184, 1990.

6. M. Hayati, A. Rezaei, M. Seifi, "CNT-MOSFET modeling based on artificial neural network: Application to simulation of nanoscale circuits," *Solid-State Electron*, vol. 54, no. 1, pp. 52–57, 2010.

7. M. Avci, M. Y. Babac, T. Yildirim, "Neural network based MOSFET channel length and width decision method for analogue integrated circuits," *International Journal of Electronics*, vol. 92, no. 5, pp. 281–293, 2005.

8. N. Kumar, V. H. Gaidhane, R. K. Mittal, "Cloud-based electricity consumption analysis using neural network," *International Journal of Computers and Applications*, vol. 62, no. 1, pp. 45–56, 2020.

9. V. H. Gaidhane, N. Kumar, R. K. Mittal, J. Rajevenceltha, "An efficient approach for cement strength prediction," *International Journal of Computers and Applications*, pp. 1–11, 2019.

10. V. H. Gaidhane, Y. V. Hote, Vijander Singh, "Emotion recognition using eigenvalues and Levenberg–Marquardt algorithm-based classifier," *Sādhanā*, vol. 41, no. 4, pp. 415–423, 2016.

11. Anjana Bhardwaj, Pradeep Kumar, B. Raj, Sunny Anand, "Design and performance optimization of doping-less vertical nanowire TFET using gate stack technique", *Journal of Electronic Materials (JEMS)*, Springer, vol. 41, no. 7, pp. 4005–4013, 2022.

12. Jeetendra Singh, B. Raj, "Tunnel current model of asymmetric MIM structure levying various image forces to analyze the characteristics of filamentary memristor", *Applied Physics A*, Springer, vol. 125, no. 3, p. 203.1, Feb 2019.

13. Candy Goyal, Jagpal Singh Ubhi, B. Raj, "Low leakage zero ground noise nanoscale full adder using source biasing technique", *Journal of Nanoelectronics and Optoelectronics*, American Scientific Publishers, vol. 14, pp. 360–370, Mar 2019.

14. Gurmohan Singh, R. K. Sarin, B. Raj, "A novel robust exclusive-OR function implementation in QCA nanotechnology with energy dissipation analysis", *Journal of Computational Electronics*, Springer, vol. 15, no. 2, pp. 455–465, Jun 2016.

15. Tanu Wadhera, Deepti Kakkar, Girish Wadhwa, B. Raj, "Recent advances and progress in development of the field effect transistor biosensor: A review", *Journal of Electronic Materials*, Springer, vol. 48, no. 12, pp. 7635–7646, Dec 2019.

16. Girish Wadhwa, Priyanka Kamboj, Balwinder Raj, "Design optimisation of junctionless TFET biosensor for high sensitivity", *Advances in Natural Sciences: Nanoscience and Nanotechnology*, vol. 10, p. 045001, 2019.

17. Priya Bansal, B. Raj, "Memristor modeling and analysis for linear dopant drift kinetics," *Journal of Nanoengineering and Nanomanufacturing*, American Scientific Publishers, vol. 6, pp. 1–7, 2016.

18. Amandeep Singh, Mamta Khosla, B. Raj, "Circuit compatible model for electrostatic doped schottky barrier CNTFET, "*Journal of Electronic Materials*, Springer, vol. 45, no. 12, pp. 4825–4835, 2016.

19. D. Vaithiyanathan, B. Raj, "Performance analysis of charge plasma induced graded channel si nanotube", **Journal of Engineering Research** *(JER)*, EMSME Special Issue, pp. 146–154, Aug 2021.

20. Abhishek Singh Tomar, Vijay Kumar Magraiya, B. Raj, "Scaling of access and data transistor for high performance DRAM cell design", *Quantum Matter*, vol. 2, pp. 412–416, Oct 2013.
21. Neeraj Jain, B. Raj, "Parasitic capacitance and resistance model development and optimization of raised source/drain SOI FinFET structure for analog circuit applications", *Journal of Nanoelectronics and Optoelectronins*, ASP, USA, vol. 13, pp. 531–539, Ap 2018.
22. Shradhya Singh, S. K. Vishvakarma, B. Raj, "Analytical modeling of split-gate junction-less transistor for a biosensor application", *Sensing and Bio-Sensing*, Elsevier, vol. 18, pp. 31–36, Apr 2018.
23. Maisagalla Gopal, Balwinder Raj, "Low power 8T SRAM cell design for high stability video applications", *ITSI Transaction on Electrical and Electronics Engineering*, vol. 1, no. 5, pp. 91–97, 2013.
24. Balwinder Raj, Jatin Mitra, Deepak Kumar Bihani, V. Rangharajan, A. K. Saxena, S. Dasgupta, "Analysis of noise margin, power and process variation for 32 nm FinFET based 6T SRAM cell", *Journal of Computer (JCP)*, Academy Publisher, Finland, vol. 5, no. 6, pp. 1–8, 2010.
25. Divya Sharma, Rajesh Mehra, B. Raj, "Comparative analysis of photovoltaic technologies for high efficiency solar cell design", *Superlattices and Microstructures*, Elsevier, vol. 153, p. 106861, May 2021.
26. Pawandeep Kaur, Avtar Singh Buttar, B. Raj, "A comprehensive analysis of nanoscale transistor based biosensor: A review", *Indian Journal of Pure and Applied Physics*, vol. 59, pp. 304–318, Apr 2021.
27. Divya Yadav, Balwant Raj, B. Raj, "Design and simulation of low power microcontroller for IoT applications", *Journal of Sensor Letters*, ASP, vol. 18, pp. 401–409, May 2020.
28. Shailendra Singh, B. Raj,"A 2-D analytical surface potential and drain current modeling of double-gate vertical t-shaped tunnel FET", *Journal of Computational Electronics*, Springer, vol. 19, pp. 1154–1163, Apr 2020.
29. Jeetendra Singh, B. Raj, "An accurate and generic window function for non-linear memristor model", *Journal of Computational Electronics*, Springer, vol. 18, no. 2, pp. 640–647, Jun 2019.
30. Manjit Kaur, Neena Gupta, Sanjeev Kumar, B. Raj, Arun Kumar Singh, "Comparative RF and crosstalk analysis of carbon based nano interconnects", *IET Circuits, Devices & Systems*, vol. 15, no. 6, pp. 493–503, Feb 2021.
31. Nehru Kandasamy, Firdous Ahmad, D. Ajitha, B. Raj, Nagarjuna Telagam, "Quantum dot cellular automata based scan flip flop and boundary scan register", *IETE Journal of Research*, vol. 66, pp. 535–548, 2020.
32. S. K. Sharma, B. Raj, M. Khosla, "Enhanced photosensivity of highly spectrum selective cylindrical gate In1-xGaxAs nanowire MOSFET photodetector", *Modern Physics Letter-B*, vol. 33, no. 12, p. 1950144, 2019.
33. Jeetendra Singh, B. Raj, "Design and investigation of 7T2M NVSARM with enhanced stability and temperature impact on store/restore energy", *IEEE Transactions on Very Large Scale Integration Systems*, vol. 27, no. 6, pp. 1322–1328, Jun 2019.
34. Anil Kumar Bhardwaj, Sumeet Gupta, B. Raj, Amandeep Singh, "Impact of double gate geometry on the performance of carbon nanotube field effect transistor structures for low power digital design", *Computational and Theoretical Nanoscience*, ASP, vol. 16, pp. 1813–1820, 2019.

35. Neeraj Jain, B. Raj, "Thermal stability analysis and performance exploration of asymmetrical dual-k underlap spacer (ADKUS) SOI FinFET for security and privacy applications", *Indian Journal of Pure & Applied Physics (IJPAP)*, vol. 57, pp. 352–360, May 2019.
36. Amandeep Singh, Mamta Khosla, B. Raj, "Design and analysis of dynamically configurable electrostatic doped carbon nanotube tunnel FET", *Microelectronics Journal*, Elesvier, vol. 85, pp. 17–24, Mar 2019.
37. Neeraj Jain, Balwinder Raj, "Dual-k spacer region variation at the drain side of asymmetric SOI FinFET structure: Performance analysis towards the analog/rf design applications", *Journal of Nanoelectronics and Optoelectronics*, American Scientific Publishers, vol. 14, pp. 349–359, Mar 2019.
38. Jeetendra Singh, Sanjeev Sharma, B. Raj, Mamta Khosla, "Analysis of barrier layer thickness on performance of In1-xGaxAs based gate stack cylindrical gate nanowire MOSFET," *JNO*, ASP, vol. 13, pp. 1473–1477, Oct 2018.
39. Neeraj Jain, B. Raj, "Analysis and performance exploration of high-k SOI FinFETs over the conventional low-k SOI FinFET toward analog/RF design", *Journal of Semiconductors (JoS)*, IOP Science, vol. 39, no. 12, p. 124002-1-7, Dec 2018.
40. Candy Goyal, Jagpal Singh Ubhi, B. Raj, "A reliable leakage reduction technique for approximate full adder with reduced ground bounce noise', *Journal of Mathematical Problems in Engineering*, Hindawi, vol. 2018, Article ID 3501041, p. 16, 15 Oct 2018.
41. Jeetendra Singh, B. Raj, Mamta Khosla, "Design and performance analysis of nano-scale memristor-based nonvolatile SRAM", *Journal of Sensor Letter*", American Scientific Publishers, vol. 16, pp. 798–805, Oct 2018.
42. Girish Wadhwa, B. Raj, "Parametric variation analysis of charge-plasma-based dielectric modulated JLTFET for biosensor application", *IEEE Sensor Journal*, vol. 18, no. 15, pp. 6070–6077, 2018.
43. Jeetendra Singh, B. Raj, "Comparative analysis of memristor models for memories design", *JoS*, IoP, vol. 39, no. 7, p. 074006-1-12, Jul 2018.
44. Divya Yadav, Shailesh Singh Chouhan, Santosh Kumar Vishvakarma, B. Raj, "Application specific microcontroller design for IoT based WSN", *Sensor Letter*, ASP, vol. 16, pp. 374–385, May 2018.
45. Gurmohan Singh, R. K. Sarin, B. Raj, "Fault-tolerant design and analysis of quantum-dot cellular automata based circuits", *IEEE/IET Circuits, Devices & Systems*, vol. 12, pp. 638–664, 2018.
46. Jeetendra Singh, B. Raj, "Modeling of mean barrier height levying various image forces of metal insulator metal structure to enhance the performance of conductive filament based memristor model", *IEEE Nanotechnology*, vol. 17, no. 2, pp. 268–267, Mar 2018.
47. Aakash Jain, Sanjeev Sharma, B. Raj, "Analysis of triple metal surrounding gate (TM-SG) III-V nanowire MOSFET for photosensing application", *Optoelectronics Journal*, Elsevier, vol. 26, no. 2, pp. 141–148, May 2018.
48. Aakash Jain, Sanjeev Sharma, B. Raj, "Design and analysis of high sensitivity photosensor using cylindrical surrounding gate MOSFET for low power sensor applications", *Engineering Science and Technology, an International Journal*, Elsevier, vol. 19, no. 4, pp. 1864–1870, Dec 2016.
49. Amandeep Singh, Mamta Khosla, B. Raj, "Analysis of electrostatic doped schottky barrier carbon nanotube FET for low power applications," *Journal*

of Materials Science: Materials in Electronics, Springer, vol. 28, pp. 1762–1768, 2017.

50. G. Saiphani Kumar, Amandeep Singh, B. Raj, "Design and analysis of gate all around CNTFET based SRAM cell design", *Journal of Computational Electronics*, Springer, vol. 17, no. 1, pp. 138–145, Mar 2018.

51. Gurinder pal Singh, B. S. Sohi, Balwinder Raj, "Material properties analysis of graphene base transistor (GBT) for VLSI analog circuits", *Indian Journal of Pure & Applied Physics (IJPAP)*, vol. 55, pp. 896–902, Dec 2017.

52. Amandeep Singh, Mamta Khosla, B. Raj, "Comparative analysis of carbon nanotube field effect transistor and nanowire transistor for low power circuit design," *Journal of Nanoelectronics and Optoelectronics*, American Scientific Publishers, USA, vol. 11, pp. 388–393, Jun 2016.

53. Sunil Kumar, B. Raj, "Estimation of stability and performance metric for inward access transistor based 6T SRAM cell design using n-type/p-type DMDG-GDOV TFET", *IEEE VLSI Circuits and Systems Letter*, vol. 3, no. 2, pp. 25–39, Jun 2017.

54. Shashikant Sharma, Anjan Kumar, Manisha Pattanaik, B. Raj, "Forward body biased multimode multi-threshold CMOS technique for ground bounce noise reduction in static CMOS adders", *International Journal of Information and Electronics Engineering*, pp. 567–572, vol. 3, no. 3, 2013.

55. Hamendra Singh, Pankaj Kumar, Balwinder Raj, "Performance Analysis of Majority Gate SET Based 1-bit Full Adder", *International Journal of Computer and Communication Engineering* (IJCCE), IACSIT Press Singapore, ISSN: 2010-3743, Vol. 2, no. 4, 2013.

56. Anil Kumar Bhardwaj, Sumeet Gupta, B. Raj, "Investigation of parameters for schottky barrier (SB) height for schottky barrier based carbon nanotube field effect transistor device", *Journal of Nanoelectronics and Optoelectronics*, ASP, vol. 15, pp. 783–791, Jul 2020.

57. Priya Bansal, B. Raj, "Memristor: A versatile nonlinear model for dopant drift and boundary issues," *JCTN*, American Scientific Publishers, vol. 14, no. 5, pp. 2319–2325, May 2017.

58. Neeraj Jain, B. Raj, "An analog and digital design perspective comprehensive approach on Fin-FET (Fin-Field Effect transistor) technology: A review", *Reviews in Advanced Sciences and Engineering (RASE)*, ASP, vol. 5, pp. 1–14, 2016.

59. Sanjeev Sharma, B. Raj, Mamta Khosla, "Subthreshold performance of In1-xGaxAs based dual metal with gate stack cylindrical/surrounding gate nanowire MOSFET for low power analog applications", *Journal of Nanoelectronics and Optoelectronics*, American Scientific Publishers, USA, vol. 12, pp. 171–176, 2017.

60. Balwinder Raj, A. K. Saxena, S. Dasgupta, "Analytical modeling for the estimation of leakage current and subthreshold swing factor of nanoscale double gate FinFET device", *Microelectronics International, UK*, vol. 26, pp. 53–63, 2009.

61. Shailendra Singh, Girish Wadhwa, Balwinder Raj, "An analytical modeling for dual source vertical tunnel field effect transistor", *International Journal of Recent Technology and Engineering (IJRTE)*, vol. 8, no. 2, pp. 603–608, Jul 2019.

62. Shailendra Singh, B. Raj, "Design and analysis of hetrojunction vertical T-shaped tunnel field effect transistor", *Journal of Electronics Material*, Springer, vol. 48, no. 10, pp. 6253–6260, Oct 2019.

63. Candy Goyal, Jagpal Singh Ubhi, B. Raj, "A low leakage CNTFET based inexact full adder for low power image processing applications", *International Journal of Circuit Theory and Applications*, Wiley, vol. 47, no. 9, pp. 1446–1458, Sept 2019.

64. B. Raj, A. K. Saxena, S. Dasgupta, "A compact drain current and threshold voltage quantum mechanical analytical modeling for FinFETs", *Journal of Nanoelectronics and Optoelectronics (JNO)*, USA, vol. 3, no. 2, pp. 163–170, 2008.

65. Girish Wadhwa, B. Raj, "An analytical modeling of charge plasma based tunnel field effect transistor with impacts of gate underlap region", *Superlattices and Microstructures*, Elsevier, vol. 142, p. 106512, Jun 2020.

66. Shailendra Singh, B. Raj, "Modeling and simulation analysis of SiGe hetrojunction double GateVertical t-shaped tunnel FET", *Superlattices and Microstructures*, Elsevier vol. 142, p. 106496, Jun 2020.

67. Amandeep Singh, Dinesh Kumar Saini, Dinesh Agarwal, Sajal Aggarwal, Mamta Khosla, B. Raj, "Modeling and simulation of carbon nanotube field effect transistor and its circuit application," *Journal of Semiconductors (JoS)*, IOP Science, vol. 37, p. 074001-6, Jul 2016.

68. Neeraj Jain, Balwinder Raj, "Device and circuit co-design perspective comprehensive approach on FinFET technology: A review", *Journal of Electron Devices*, vol. 23, no. 1, pp. 1890–1901, 2016.

69. Sunil Kumar, B. Raj, "Analysis of I_{ON} and ambipolar current for dual-material gate-drain overlapped DG-TFET," *Journal of Nanoelectronics and Optoelectronics*, American Scientific Publishers, USA, vol. 11, pp. 323–333, Jun 2016.

70. Naveed Anjum, Tarun Bali, B. Raj, "Design and simulation of handwritten multiscript character recognition", *International Journal of Advanced Research in Computer and Communication Engineering*, vol. 2, no. 7, pp. 2544–2549, Jul 2013.

71. Sanjeev Sharma, B. Raj, Mamta Khosla, "A gaussian approach for analytical subthreshold current model of cylindrical nanowire FET with quantum mechanical effects", *Microelectronics Journal*, Elsevier, vol. 53, pp. 65–72, Apr 2016.

72. Karmjit Singh, B. Raj, "Performance and analysis of temperature dependent multi-walled carbon nanotubes as global interconnects at different technology nodes," *Journal of Computational Electronics*, Springer, vol. 14, no. 2, pp. 469–476, Jun 2015.

73. Sunil Kumar, B. Raj, "Compact channel potential analytical modeling of DG-TFET based on evanescent–mode approach," *Journal of Computational Electronics*, Springer, vol. 14, no. 2, pp. 820–827, Jul 2015.

74. Karmjit Singh, B. Raj, "Temperature dependent modeling and performance evaluation of multi-walled CNT and single-walled CNT as global interconnects," *Journal of Electronic Materials*, Springer, vol. 44, no. 12, pp. 4825–4835, Dec 2015.

75. V. K. Sharma, M. Pattanaik, B.Raj, "INDEP approach for leakage reduction in nanoscale CMOS circuits", *International Journal of Electronics*, Taylor & Francis, vol. 102, no. 2, pp. 200–215, 2014.

76. Karmjit Singh, B. Raj, "Influence of temperature on MWCNT bundle, SWCNT bundle and copper interconnects for nanoscaled technology nodes," *Journal of Materials Science: Materials in Electronics*, Springer, vol. 26, no. 8, pp. 6134–6142, 2015.

77. Naveed Anjum, Tarun Bali, B. Raj, "Design and simulation of handwritten gurumukhi and devanagri numerical recognition", *International Journal of Computer Applications*, Foundation of Computer Science, New York, USA, vol. 73, no. 12, pp. 16–21, 2013.

78. S. Khandelwal, V. Gupta, B. Raj, R. D. "Gupta, process variability aware low leakage reliable nano scale DG-FinFET SRAM cell design technique", *Journal of Nanoelectronics and Optoelectronics*, vol. 10, no. 6, pp. 810–817, Dec 2015.

79. V. K. Sharma, M. Pattanaik, B. Raj, "ONOFIC approach: Low power high speed nanoscale VLSI circuits design", *International Journal of Electronics*, Taylor & Francis, vol. 101, no. 1, pp. 61–73, 2014.

80. S. Khandelwal, Balwinder Raj, R. D. Gupta, "FinFET based 6T SRAM cell design: Analysis of performance metric, process variation and temperature effect", *Journal of Computational and Theoretical Nanoscience*, ASP, USA, vol. 12, pp. 2500–2506, 2015.

81. Sumit Singh, Shekhar Yadav, Jagdeep Rahul, Anurag Srivastava; B. Raj, "Impact of HfO_2 in graded channel dual insulator double gate MOSFET", *Journal of Computational and Theoretical Nanoscience*, American Scientific Publishers, vol. 12, no. 6, pp. 950–953, Apr 2015.

82. Vijay Kumar Sharma, Manisha Pattanaik, B. Raj, "PVT variations aware low leakage INDEP approach for nanoscale CMOS circuits", *Microelectronics Reliability*, Elsevier, vol. 54, pp. 90–99, 2014.

83. B. Raj, A. K. Saxena, S. Dasgupta, "Quantum mechanical analytical modeling of nanoscale DG FinFET: Evaluation of potential, threshold voltage and source/drain resistance", *Elsevier's Journal of Material Science in Semiconductor Processing*, Elsevier, vol. 16, no. 4, pp. 1131–1137, 2013.

84. Maisagalla Gopal, Siva Sankar D Prasad, Balwinder Raj, "8T SRAM cell design for dynamic and leakage power reduction", *International Journal of Computer Applications*, Foundation of Computer Science, New York, USA, vol. 71, no. 9, pp. 43–48, Jun 2013.

85. Manisha Pattanaik, B. Raj, Shashikant Sharma, Anjan Kumar, "Diode based trimode multi-threshold CMOS technique for ground bounce noise reduction in static CMOS adders", *Advanced Materials Research*, Trans Tech Publications, Switzerland, vol. 548, pp. 885–889, 2012.

86. Balwinder Raj, A. K. Saxena, S. Dasgupta, "Nanoscale FinFET based SRAM cell design: Analysis of performance metric, process variation, underlapped FinFET and temperature effect", *IEEE Circuits and System Magazine*, vol. 11, no. 2, pp. 38–50, 2011.

87. V. K. Sharma, M. Pattanaik, Balwinder Raj, "Leakage current ONOFIC approach for deep submicron VLSI circuit design", *International Journal of Electrical, Computer, Electronics and Communication Engineering*, World Academy of Sciences, Engineering and Technology, vol. 7, no. 4, pp. 239–244, 2013.

88. Tulika Chawla, Mamta Khosla, B. Raj, "Design and simulation of triple metal double-gate germanium on insulator vertical tunnel field effect transistor", *Microelectronics Journal*, Elsevier, vol. 114, p. 105125, Aug 2021.

89. Parminder Kaur, Sandeep Singh Gill, B. Raj, "Comparative analysis of OFETs materials and devices for sensor applications", *Journal of Silicon*, Springer, vol. 14, pp. 4463–4471, 2022.

90. Sanjeev Kumar Sharma, Parveen Kumar, Balwant Raj, B. Raj, "In$_{1-x}$Ga$_x$As double metal gate-stacking cylindrical nanowire MOSFET for highly sensitive photo detector", *Journal of Silicon*, Springer, vol. 14, pp. 3535–3541, 2022.

91. B. Raj, A. K. Saxena, S. Dasgupta, "Analytical modeling of quasi planar nanoscale double gate FinFET with source/drain resistance and field dependent carrier mobility: A quantum mechanical study", *Journal of Computer (JCP)*, Academy Publisher, Finland, vol. 4, no. 9, pp. 1–8, 2009.

92. S. Bhushan, S. Khandelwal, B. Raj, "Analyzing different mode FinFET based memory cell at different power supply for leakage reduction", *Seventh International Conference on Bio-Inspired Computing: Theories and Application, (BIC-TA 2012) Advances in Intelligent Systems and Computing*, vol. 202, pp. 89–100, 2013.

93. Jeetendra Singh, B. Raj, "Temperature dependent analytical modeling and simulations of nanoscale memristor", *Journal: Engineering Science and Technology, an International Journal*, Elsevier, vol. 21, pp. 862–868, Oct 2018.

94. Shradhya Singh, Shashi Bala, Balwant Raj, B. Raj, "Improved sensitivity of dielectric modulated junctionless transistor for nanoscale biosensor design", *Sensor Letter*, ASP, vol. 18, pp. 328–333, Apr 2020.

95. Vivek Kumar, Santosh Kumar Vishvakarma, B. Raj, "Design and performance analysis of ASIC for IoT applications", *Sensor Letter*, ASP, vol. 18, pp. 31–38, Jan 2020.

96. Akanksha Jaiswal, R. K. Sarin, B. Raj, Shikha Sukhija, "A novel circular slotted microstrip-fed patch antenna with three triangle shape defected ground structure for multiband applications", *Advanced Electromagnetic (AEM)*, vol. 7, no. 3, pp. 56–63, Aug 2018.

97. Girish Wadhwa, B. Raj, "Label free detection of biomolecules using charge-plasma-based gate underlap dielectric modulated junctionless TFET", *Journal of Electronic Materials (JEMS)*, Springer, vol. 47, no. 8, pp. 4683–4693, Aug 2018.

98. Manjit Kaur, Neena Gupta, Sanjeev Kumar, B. Raj, Arun Kumar Singh, "RF performance analysis of intercalated graphene nanoribbon based global level interconnects", *Journal of Computational Electronics*, Springer, vol. 19, pp. 1002–1013, Jun 2020.

99. Girish Wadhwa, B. Raj, "Design and performance analysis of junctionless TFET biosensor for high sensitivity", *IEEE Nanotechnology*, vol. 18, pp. 567–574, 2019.

100. Jeetendra Singh, B. Raj, "Enhanced nonlinear memristor model encapsulating stochastic dopant drift", *JNO*, ASP, vol. 14, pp. 958–963, 2019.

101. **Gurmohan Singh, R. K. Sarin, B. Raj, "Design and performance analysis of a new efficient coplanar quantum-dot cellular automata adder",** *Indian Journal of Pure & Applied Physics (IJPAP)*, vol. 55, pp. 97–103, Feb 2017.

102. Amandeep Singh, Mamta Khosla, Balwinder Raj, "Design and analysis of electrostatic doped schottky barrier CNTFET based low power SRAM,"

International Journal of Electronics and Communications, (AEÜ), Elsevier, vol. 80, pp. 67–72, 2017.

103. Parminder Kaur, Vikas Pandey, B. Raj, "Comparative study of efficient design, control and monitoring of solar power using IoT", *Sensor Letter*, ASP vol. 18, pp. 419–426, May 2020.

104. Anil Kumar Bhardwaj, Sumeet Gupta, B. Raj, "Development & analysis of compact model for double gate schottky barrier CNTFET", *Journal of Nanoelectronics and Optoelectronics*, ASP, vol. 15, pp. 1199–1208, Aug 2020.

105. Girish Wadhwa, Priyanka Kamboj, Jeetendra Singh, B. Raj, "Design and investigation of junctionless DGTFET for biological molecule recognition", *Transactions on Electrical and Electronic Materials*, Springer, vol. 22, pp. 282–289, 2021.

106. Tulika Chawla, Mamta Khosla, B. Raj, "Optimization of double-gate dual material GeOI-vertical TFET for VLSI circuit design", *IEEE VLSI Circuits and Systems Letter*, vol. 6, no. 2, pp. 13–25, Aug 2020.

107. Sachin Kumar Verma, Shailendra Singh, Girish Wadhwa, B. Raj, "Detection of biomolecules using charge-plasma based gate underlap dielectric modulated dopingless TFET", *Transactions on Electrical and Electronic Materials (TEEM)*, Springer, vol. 21, pp. 528–535, Jun 2020.

108. Neeraj Jain, B. Raj, "Impact of underlap spacer region variation on electrostatic and analog/RF performance of symmetrical high-k SOI FinFET at 20 nm channel length", *Journal of Semiconductors (JoS)*, IOP Science, vol. 38, no. 12, p. 122002, Dec 2017.

109. Shailendra Singh, B. Raj, "Analytical modeling and simulation analysis of T-shaped III-V heterojunction Vertical T-FET", *Superlattices and Microstructures*, Elsevier, vol. 147, p. 106717, Nov 2020.

110. Gurmohan Singh, R. K. Sarin, B. Raj, "Design and analysis of area efficient QCA based reversible logic gates", *Journal of Microprocessors and Microsystems*, Elsevier, vol. 52, pp. 59–68, May 2017.

111. Amandeep Singh, Mamta Khosla, B. Raj, "Compact model for ballistic single wall CNTFET under quantum capacitance limit," *Journal of Semiconductors (JoS)*, IOP Science, vol. 37, p. 104001-8, Oct 2016.

112. Sonal Singh, Mamta Khosla, Girish Wadhwa, B. Raj, "Design and analysis of double-gate junctionless vertical TFET for gas sensing applications", *Applied Physics A*, Springer, vol. 127, no. 16, 2 Jan 2021.

113. Inderjit Singh, B. Raj, Mamta Khosla, B. Rajesh Kumar Kaushik, "Potential MRAM technologies for low power SoCs", *SPIN World Scientific Publisher*, *SCIE*; vol. 10, no. 4, p. 2050027, Dec 2020.

114. Shailendra Singh, B. Raj, "Parametric variation analysis on hetero-junction Vertical t-shape TFET for supressing ambipolar conduction", *Indian Journal of Pure and Applied Physics*, vol. 58, pp. 478–485, Jun 2020.

115. Shailendra Singh, Girish Wadhwa, Balwinder Raj, "Design and analysis of dual source vertical tunnel field effect transistor for high performance", *Transactions on Electrical and Electronics Materials*, Springer, vol. 21, pp. 74–82, Oct 2019.